Lecture Notes in Computer S

Edited by G. Goos, J. Hartmanis, and J.

Springer
Berlin
Heidelberg
New York
Barcelona
Hong Kong
London
Milan
Paris
Tokyo

Maura Cerioli   Gianna Reggio (Eds.)

Recent Trends in Algebraic Development Techniques

15th International Workshop, WADT 2001
Joint with the CoFI WG Meeting
Genova, Italy, April 1-3, 2001
Selected Papers

Springer

Series Editors

Gerhard Goos, Karlsruhe University, Germany
Juris Hartmanis, Cornell University, NY, USA
Jan van Leeuwen, Utrecht University, The Netherlands

Volume Editors

Maura Cerioli
Gianna Reggio
Università di Genova
DISI - Dipartimento e Informatica di Scienze dell'Informazione
Via Dodecaneso, 35, 16146 Genova, Italy
E-mail: {cerioli, reggio}@disi.unige.it

Cataloging-in-Publication Data applied for

Die Deutsche Bibliothek - CIP-Einheitsaufnahme

Recent trends in algebraic development techniques : 15th international
workshop ; selected papers / WADT 2001 joint with the CoFI WG Meeting,
Genova, Italy, April 1 - 3, 2001. Maura Cerioli ; Gianna Reggio (ed.). -
Berlin ; Heidelberg ; New York ; Barcelona ; Hong Kong ; London ; Milan ;
Paris ; Tokyo : Springer, 2002
   (Lecture notes in computer science ; Vol. 2267)
   ISBN 3-540-43159-4

CR Subject Classification (1998):F.3.1, F.4, I.1, D.2.1

ISSN 0302-9743
ISBN 3-540-43159-4 Springer-Verlag Berlin Heidelberg New York

Springer-Verlag Berlin Heidelberg New York
a member of BertelsmannSpringer Science+Business Media GmbH

http://www.springer.de

© Springer-Verlag Berlin Heidelberg 2002
Printed in Germany

Typesetting: Camera-ready by author, data conversion by Steingräber Satztechnik GmbH, Heidelberg
Printed on acid-free paper   SPIN 10846050   06/3142   5 4 3 2 1 0

Preface

The algebraic approach to system specification and development, born in the early 1970s as a formal method for abstract data types, encompasses today, at the beginning of the third millennium, the formal design of integrated hardware and software systems, new specification frameworks and programming paradigms, and a wide range of application areas. The workshops on Algebraic Development Techniques, initiated in 1982 as the workshop on Abstract Data Types, are a prominent forum to present and discuss current research in that area.

The 15th International Workshop on Algebraic Development Techniques (WADT 2001) took place in Genova, as a satellite of ETAPS 2001, on April 1–3, 2001, and was organized by Maura Cerioli and Gianna Reggio.

From 1991 to 1995 WADT workshops were held jointly with the General Workshop of the ESPRIT Basic Research Working Group COMPASS. Following this tradition, the WADT 2001 was held jointly with the General Workshop of the ESPRIT Working Group CoFI. The Common Framework Initiative, started in 1995 and funded since 1998 as CoFI WG, is centered around the definition of the Common Algebraic Specification Language (CASL).

The program started with a full day tutorial on the CASL, followed by 32 presentations, several of them on the CASL as well, organized in parallel sessions during the following two days. The parallel sessions were devoted to: logics and proofs, concurrent processes, institutions and categories, applications and case studies, higher-order and parameterized specifications, static analysis, software architectures, graph and transformation rules.

The main topics of the workshop were:

- algebraic specification
- other approaches to formal specification
- specification languages and methods
- term rewriting and proof systems
- specification development systems (concepts, tools, etc.)

The program committee invited submissions of full papers for possible inclusion in this volume, on the basis of the abstracts and the presentations at WADT 2001.

All the submissions were subject to careful refereeing, and the selection of papers was made following further discussion by the full program committee.

We are extremely grateful to all workshop participants, to the (other) members of the program committee, and to the external referees for their contribution to the scientific quality of the workshop and of this volume.

The WADT series is sponsored by IFIP WG1.3 on Foundations of System Specification (see `http://www.brics.dk/~pdm/IFIP-WG1.3`).

November 2001   Maura Cerioli and Gianna Reggio

Organization

Program Committee

Michel Bidoit   (Cachan, France)
Maura Cerioli   (Genova, Italy)
Hans-Jörg Kreowski   (Bremen, Germany)
Peter Mosses, *chair*   (Aarhus, Denmark)
Fernando Orejas   (Barcelona, Spain)
Francesco Parisi-Presicce   (Roma, Italy)
Gianna Reggio   (Genova, Italy)
Donald Sannella   (Edinburgh, Scotland)
Andrzej Tarlecki   (Warsaw, Poland)

External Referees

Egidio Astesiano

Marek Bednarczyk

Benjamin Blanc

Alexandre Boisseau

Tomasz Borzyszkowski

Paolo Bottoni

Paolo Cenciarelli

Hartmut Ehrig

Steven Eker

Jean Goubault-Larrecq

Piotr Hoffman

Manuel Koch

Mikolaj Konarski

Alexander Kurz

Sabine Kuske

Slawomir Lasota

Christoph Lüth

Dirk Pattinson

Wiesław Pawłowski

Jean-Claude Reynaud

Markus Roggenbach

Sponsoring Institutions

CoFI (Common Framework Initiative), ESPRIT Working Group 29432.
DISI, Università degli Studi di Genova.
IFIP WG1.3 on Foundations of System Specification.

Table of Contents

Interactive Rule-Based Specification with an Application to Visual Language Definition

Roswitha Bardohl[1], Martin Große-Rhode[1], and Marta Simeoni[2]

[1] Institut für Softwaretechnik und Theoretische Informatik, TU Berlin,
{rosi,mgr}@cs.tu-berlin.de
[2] Dipartimento di Informatica, Università Cà Foscari di Venezia,
simeoni@dsi.unive.it

Abstract. In a rule-based approach the computation steps of a system are specified by rules that completely define how the system's state may change. For open systems a more liberal approach is required, where the state changes are only partly specified, and – interactively – other components may contribute further information on how the transformation is defined completely. In this paper we introduce a formal model for interactive rule-based specifications, where states are modelled as partial algebras and transformations are given by internal algebra rewritings and arbitrary external components. As an application we discuss how visual languages can be defined in this framework. Thereby the internal (logical) representations of visual expressions are transformed by rewriting rules, whereas their layouts are obtained interactively by external components like a constraint solver or a user working with a display and a mouse.

1 Introduction

In a traditional rule-based specification of a system a set of rules is given to describe the possible state changes of the system. Thereby the rules completely describe the relation of the initial and final states of each transformation step. For open systems, however, it is more adequate to specify the effect of a transformation step only partly by a rule. The transformation is determined completely only in interaction with the environment the system is placed in. That means, the impact of the environment on the local state changes has to be taken into account.

In a software environment for a visual language, for example, there might be an editor offering rules for creating items. Such an item would be offered when the rule is applied, but not directly placed into the figure. The actual position must be chosen (via a mouse click) by the user, or might be computed by a constraint solver that chooses one of the possible positions that satisfies the given constraints. In both cases the rule applied by the rule-based part of the system just yields the existence of the item, usually together with a predefined shape or other attributes, whereas the remaining attributes – like the concrete position and size – are set externally, by the user, the constraint solver, or another component.

M. Cerioli and G. Reggio (Eds.): WADT/CoFI 2001, LNCS 2267, pp. 1–20, 2002.
© Springer-Verlag Berlin Heidelberg 2002

In this paper we introduce a formal algebraic approach for the precise specification of such *interactive rule-based behaviours*, where the effect of a rule is determined by two components. The internal part is modelled by the rewriting of algebras as formal models of states, as defined in [Gro99]. The basic idea for the rewriting is to present algebras by their sets of elements and their functions, where the latter are also considered as sets. Then simple set operations like subtraction, intersection and union can be used to rewrite the presentations, which induces the rewriting of the algebras. The interactive, external part of the transformation is embedded into this approach by designating parts of the algebras to be defined by the external component. That means, the signature is divided into an *internal* part that is updated by applying the algebra rewrite rules and an *external* part that is updated by the external component.

As an important application domain we investigate environments for visual languages in this framework. To treat visual languages properly it is important to distinguish the logical (grammatical) structure of an expression and its visual structure, given by its layout. Layout operations connect these two layers in that each logical item is mapped to a graphical one. This basic structure is used in [Bar00] to define a generic visual editor for the generation of visual languages and manipulation of their expressions. It is generic in the sense that it accepts language definitions as input and delivers concrete visual editors for these languages as output. The distinctive feature of logical and visual structure is that the semantics of a visual expression is completely determined by the logical structure. If, for example, the crossing of lines expresses semantic information it must be represented in the logical structure. If it occurs just due to lack of space or cannot be avoided due to topological properties, the intersection point is only present in the visual structure. Furthermore, the visual layout of an expression may change without affecting its information (logical structure). For instance, classes in a class diagram may be represented by rectangles in one style and rectangles with rounded corners in another style.

The paper is organized as follows. In the next section presentations as technical means for the rewriting of algebras are introduced. In Sect. 3 then the interactive transformation of algebras with internal algebra rewriting and external transformation by some other component is discussed. For our main application, visual languages, an example is given in Sect. 4 that illustrates the formal concepts. First an algebraic specification of class diagrams is given, containing their logical structure, a graphics domain, and layout operations. Then their interactive generation and manipulation with algebra rewrite rules for the logical part and a constraint solver or a user acting with a mouse to determine the layout is discussed. In Sect. 5 we give a conclusion, sketch further extensions of the framework, and discuss the relation with other approaches.

2   Presentations

In this section we recall from [Gro99] the basic concepts concerning presentations of partial algebras that are the technical means to define the rewriting of algebras. The basic idea is to (re-) present partial algebras by a family of sets (their carrier sets) and, in addition, a set (re-) presenting the functions of a partial algebra via their input/output pairs. The latter are given in the form of equations $f(a_1, \ldots, a_n) = b$, where f is a function symbol and $a_1, \ldots, a_n$ and b are elements of the corresponding carrier sets. According to the terminology of universal algebra the elements of the carrier sets in such a presentation are called *generators* and the equations are called *relations*. In general arbitrary equations are allowed in a presentation, not only *function entries* $f(a_1, \ldots, a_n) = b$.

Simple set operations like subtraction, intersection, and union can be used then to transform presentations by removing or adding generators and/or relations. As shown in Prop. 1 each presentation induces a partial algebra. Thus to rewrite an algebra it is first translated into a presentation, this presentation is transformed via the set operations, and then the transformed presentation is translated into a partial algebra again which yields the result of the rewriting. (The rewriting procedure is discussed in the following section.)

Definition 1 (Presentation). *Let* $\Sigma = (S, F)$ *be an algebraic signature. A* Σ-presentation $P = (P_S; P_E)$ *is given by an S-indexed set* $P_S = (P_s)_{s \in S}$, *the generators, and a set* $P_E \subseteq Eqns_\Sigma(P_S)$, *the relations. A presentation is functional if* P_E *is a set of function entries over* P_S, *i.e.,*

$$P_E \subseteq \{f(a) = b \mid f : w \to v \in F, \ a \in P_w, \ b \in P_v\} \subseteq Eqns_\Sigma(P_S).$$

A morphism of Σ-presentations $p : (P_S; P_E) \to (P'_S; P'_E)$ *is an S-indexed function* $p = (p_s : P_s \to P'_s)_{s \in S}$ *such that* $P_E[p] \subseteq P'_E$, *where* _[p] *denotes the substitution of generators according to p.* Σ-presentations and morphisms yield *the category* **Pres**(Σ). *Finally let* **Pres**$(\Gamma) =$ **Pres**(Σ) *for each partial equational specification* $\Gamma = (\Sigma, CE)$ *extending* Σ *by conditional equations CE w.r.t.* Σ.

Presentations can be restricted to subsets of generators by deleting all relations that contain generators not contained in the subset. This will be used to model the implicit removal of function entries containing elements that have been deleted. On the other hand, presentations can be extended to larger signatures by adding empty sets of generators for the new sorts.

Definition 2 (Restriction and Extension). *Let* $P = (P_S; P_E)$ *be a* Σ-*presentation.*

1. *Given an S-indexed subset* $Q_S \subseteq P_S$ *the restriction* $P|_{Q_S}$ *is defined as the* Σ-*presentation given by*
   $$P|_{Q_S} = (Q_S; P_E \cap Eqns_\Sigma(Q_S)).$$
2. *Given a signature extension* $\Sigma' \supseteq \Sigma$ *with* $\Sigma = (S, F)$, $\Sigma' = (S', F')$ *the extension* $P^{\Sigma'} = (P_S^{\Sigma'}; P_E^{\Sigma'})$ *of P to* Σ' *is defined as the* Σ'-*presentation given by*
   $$P_{s'}^{\Sigma'} = \begin{cases} P_{s'} & \text{if } s' \in S \\ \emptyset & \text{else} \end{cases} \quad (s' \in S')$$
   $$P_E^{\Sigma'} = P_E.$$

Consider now a partial equational specification $\Gamma = (\Sigma, CE)$ extending the signature Σ by some conditional Σ-equations CE. To each Γ-presentation $P = (P_S; P_E)$ there is a *smallest* (free generated) partial Γ-algebra A^P that contains the generators P_S and satisfies the relations (equations) P_E. If P contains only function entries $f(a) = b$ as relations and these are consistent (in the sense that $(f(a) = b) \in P_E$ and $(f(a) = b') \in P_E$ implies $b = b'$) it yields the partial Σ-algebra A^P given by

$$A_s^P = P_s \qquad\qquad (s \in S)\,,$$
$$f^{A^P}(a) = b \ \ \text{iff} \ \ (f(a) = b) \in P_E \quad (f \in F)\,.$$

If furthermore A^P already satisfies the conditional equations CE this yields the partial Γ-algebra induced by P. If the equations are not satisfied, however, or if there are inconsistent function entries, further elements may be generated and some generators or generated elements may be identified.

This property is formally stated as the existence of a free functor from the category of Γ-presentations $\mathbf{Pres}(\Gamma)$ to the category of partial Γ-algebras $\mathbf{PAlg}(\Gamma)$, i.e., a left adjoint to the *presentation functor* that maps partial Γ-algebras to Γ-presentations.

Proposition 1. *Let* $\Gamma = (\Sigma, CE)$ *be a partial equational specification. The* presentation functor $Pres_\Gamma : \mathbf{PAlg}(\Gamma) \to \mathbf{Pres}(\Gamma)$, *given by*

$$Pres_\Gamma(A_S, A_F) = (A_S, \{f(a) = b \mid f^A(a) = b\}), \ \text{and}$$
$$Pres_\Gamma(h) = h$$

has a left adjoint $PAlg_\Gamma : \mathbf{Pres}(\Gamma) \to \mathbf{PAlg}(\Gamma)$ *that satisfies* $PAlg_\Gamma \circ Pres_\Gamma \cong Id_{\mathbf{PAlg}(\Gamma)}$.

As mentioned above, if P is given by consistent function entries only and conditional equations are not considered (or already satisfied) the partial algebra induced by P carries exactly the same information as P itself. This is made precise in the following corollary.

Corollary 1. *Let* Σ *be an algebraic signature and* $P = (P_S; P_E)$ *be a* Σ-*presentation. If* P *is consistently functional, i.e.,* P *is functional and satisfies*

$$(f(a) = b) \in P_E \wedge (f(a) = b') \in P_E \Rightarrow b = b'$$
$$\text{for all } f : w \to v \in F, \ a \in P_w, \ b, b' \in P_v,$$

then $Pres_\Sigma(PAlg_\Sigma(P)) \cong P$.

3   Interactive Transformation

In this section we describe the formal algebraic approach for the specification of interactive rule-based behaviours, where the effect of a transformation is determined by the integration of an internal mechanism and an external one. More precisely, the internal part of the transformation is modelled by the rewriting of partial algebras, while the external part is embedded into the transformation by designating parts of the algebra to be defined by the external component. This allows the formalization of the complete framework independently of any particular mechanism to deal with the external part of the transformation.

A signature for interactive transformation reflects the distinction between the internal and external mechanisms by designating two different algebraic signatures: one designating the internal part (*internal signature*) and an extension of it including also the parts of the algebra to be defined externally (*complete signature*). Beyond these algebraic parts an interactive transformation signature also provides names of actions with parameter type lists. The actions can be used then to trigger the application of algebra rewrite rules.

Definition 3 (Interactive Transformation Signature). *An* interactive transformation signature $R\Sigma = (\Sigma_{in}, \Sigma, A)$ *is given by algebraic signatures* $\Sigma_{in} = (S_{in}, F_{in})$, *the* internal signature, *and* $\Sigma = (S, F)$, *the* complete signature, *with* $\Sigma_{in} \subseteq \Sigma$, *and a family* $A = (A_w)_{w \in S^*}$ *of sets of action names, called the* action signature.

The sorts and functions in $S - S_{in}$ *and* $F - F_{in}$ *respectively are called* external sorts *and* functions. *An action name* $a \in A_w$ *is also denoted by* $(a : w) \in A$.

3.1 Internal Transformation by Algebra Rewriting

Partial algebras are rewritten via rules which are essentially given by pairs of presentations. The *left hand side* of a rule specifies which elements and function entries are to be removed from the algebra representing the actual state; therefore it is required to be functional. Its *right hand side* specifies the elements and relations that are to be added. The generators that occur in the left hand side of a rule play the role of variables that are matched to the actual state. If a variable occurs in both parts of a rule the corresponding element is preserved. Such variables are used as context for the rewriting that describes how the right hand side is embedded into the remainder of the actual state after the removal of the left hand side. According to the distinction between internal and complete signature described above, only elements and function entries belonging to the internal signature can be used in the left and right hand sides of a rewrite rule. In addition to the rule body, given by its left and right hand sides as discussed just now, rules are equipped with a logical formula that specifies the positive and/or negative conditions for the rule application. Moreover, a formal action expression is added that binds the rule to the action whose behaviour is being specified and instantiates to the label of the transformation step.

Definition 4 (Rewrite Rule). *Let* $R\Sigma = (\Sigma_{in}, \Sigma, A)$ *be an interactive transformation signature. A* rewrite rule $r = (a(\bar{x}) \triangleq Cond \Rightarrow (P_l \to P_r))$ *w.r.t.* $R\Sigma$ *is given by*

- *an action name* $a : s_1 \ldots s_n \in A$,
- *a list of variables* $\bar{x} = (x_1, \ldots, x_n)$, *the formal parameters of* a,

(this formal action expression yields the action label*)*

- a functional Σ_{in}-presentation $P_l = (X_l, E_l)$,
- an arbitrary Σ_{in}-presentation $P_r = (X_r, E_r)$,

(the left and right hand side *of the rule)*

- a condition *Cond, given by a set of variables* X_F *and formula* F *with free variables in* $X_F \cup X_l \cup X_r$.

These components must satisfy the condition that

- $x_i \in (X_F \cup X_l)_{s_i}$ *for each* $i \in \{1, \ldots, n\}$,

i.e., the formal parameters of the action are contained in the sets of (free) variables of the condition and the left hand side.

Adding rewrite rules to an interactive transformation signature to specify the behaviour of its actions yields an interactive transformation specification.

Definition 5 (Interactive Transformation Specification). *An* interactive transformation specification $R\Gamma = (\Gamma_{in}, \Gamma, A, R)$ *is given by algebraic specifications* $\Gamma_{in} = (\Sigma_{in}, CE_{in})$ *and* $\Gamma = (\Sigma, CE)$ *with* $\Gamma_{in} \subseteq \Gamma$, *and a set* R *of* $R\Sigma$*-rewrite rules, where the interactive transformation signature* $R\Sigma$ *is given by* $R\Sigma = (\Sigma_{in}, \Sigma, A)$.

Consider now the application of a rewrite rule $r = (a(\bar{x}) \triangleq Cond \Rightarrow (P_l \to P_r))$ to a partial Σ-algebra A. First the free variables X_F of the condition $Cond = (X_F, F)$ and X_l of the left hand side $P_l = (X_l, E_l)$ must be instantiated in A via a mapping $m : X \to A$. Since the formal parameters x_i of the action expression $a(\bar{x}) = a(x_1, \ldots, x_n)$ are contained in $X_F \cup X_l$ this yields also a corresponding action instance $a(m(\bar{x})) = a(m(x_1), \ldots, m(x_n))$. The mapping m is a *match* of r in A if the condition and the equations of the left hand side are satisfied in A w.r.t. the instantiation m (i.e., $A, m \models Cond$ and $A, m \models E_l$). Only matches yield rewriting steps, other instantiations of the variables are not admissible.

To describe the effect of the application of r w.r.t. a match m in A the following symmetric differences and intersections are needed. They correspond to the parts that shall be deleted (X_l^0, E_l^0), retained (X_c, E_c), and added (X_r^0, E_r^0) respectively.

$$X_l^0 = X_l - X_r \qquad X_c = X_l \cap X_r \qquad X_r^0 = X_r - X_l$$
$$E_l^0 = E_l - E_r \qquad E_c = E_l \cap E_r \qquad E_r^0 = E_r - E_l$$

The internal rewriting step $r/m : A \Rightarrow_{in} iB$ rewrites then A into an intermediate state iB in three steps:

1. the subtraction (deletion) of the image of (X_l^0, E_l^0) under m in the presentation (A_S, A_E) of A,
2. the addition of (X_r^0, E_r^0), and

3. the free construction of the partial Σ-algebra from this presentation. (Since iB is only an intermediate state the conditional equations CE are not considered yet.)

The carrier sets iB_S of the intermediate result of the rewriting are thus given by $iB_S = (A_S - m(X_l^0)) \uplus X_r^0$. Its generating set of relations is given by $iB_E = (A_E - E_l^0[m]) \cup E_r^0[m]$, corresponding to the subtraction of the function entries E_l^0 and the addition of the relations E_r^0. Finally, all relations still containing variables from A_S that do not belong to iB_S must be removed in order to obtain a well defined presentation again. Thus the deletion of an element a of A has the side effect of removing all function entries containing a in any position.

Definition 6 (Internal Algebra Rewriting). *Let $R\Gamma$ be an interactive transformation specification and $r = (a(\bar{x}) \triangleq Cond \Rightarrow (P_l \rightarrow P_r))$ be a rewrite rule w.r.t. $R\Sigma$. Furthermore let A be a partial Γ-algebra with $Pres_\Sigma(A) = (A_S; A_E)$ and let $m : P_l^\Sigma \rightarrow (A_S, A_E)$ be a Σ-presentation morphism such that $A, m \models F$. (That means, m is a match for r in A.)*

Then the internal rewriting step $a(m(\bar{x})) : A \Rightarrow_{in} iB$ rewrites A into the partial Σ-algebra iB defined by

$$iB = PAlg_\Sigma(iB_S, iB_E)$$
$$iB_S = (A_S - m(X_l^0)) \uplus X_r^0$$
$$iB_E = ((A_E - E_l^0[m]) \cup E_r^0[m])|_{iB_S}$$

3.2 External Transformation by Other Components

The internal algebra rewriting step only transforms (explicitly) the internal parts of a given algebra, i.e., the ones corresponding to the internal signature. (There may be side effects, however, induced by functions from the internal to the external part.) The intermediate state iB obtained by this rewriting is then further transformed by some external component. The final output of the transformation step is required to be a partial Γ-algebra, i.e., the conditional equations must hold. The external component might use all the information given in iB, but may change only its external parts, i.e., the ones corresponding to $\Sigma - \Sigma_{in}$. ¿From the most abstract point of view, the behaviour of an external component is given thus by a mapping from $\mathbf{PAlg}(\Sigma)$ to $\mathbf{PAlg}(\Gamma)$ that preserves the Σ_{in}-reducts of partial Σ-algebras. This transformation is appended to the internal rewrite step.

Definition 7 (Interactive Transformation). *Let $R\Gamma = (\Gamma_{in}, \Gamma, A, R)$ be an interactive transformation specification and $Ext : |\mathbf{PAlg}(\Sigma)| \rightarrow |\mathbf{PAlg}(\Gamma)|$ a mapping with $V_{\Sigma_{in}}(Ext(C)) = V_{\Sigma_{in}}(C)$ for all $C \in |\mathbf{PAlg}(\Sigma)|$. For each partial Γ-algebra A, $R\Sigma$-rewrite rule $r = (a(\bar{x}) \triangleq Cond \Rightarrow (P_l \rightarrow P_r))$, and match m of r in A, the interactive transformation step $a(m(\bar{x})) : A \Rightarrow B$ transforms A into the output state $B \in |\mathbf{PAlg}(\Gamma)|$ given by $B = Ext(iB)$, where iB is the internal output state of the internal rewriting $a(m(\bar{x})) : A \Rightarrow_{in} iB$.*

Like the internal rewriting step the behaviour of the external component may be defined in terms of presentations. That means, the presentation format can be used as an interface to connect the external component with the interactive transformation, as indicated in Fig. 1.

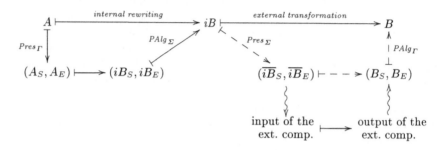

Fig. 1. Transformation step with internal algebra rewriting, external transformation by another component, and its interface to the presentation format

The decision to apply first the internal rewriting and then the external transformation mechanism seems to be non-symmetric. It reflects the idea that the transformation is guided by the rule-based component of the interactive system. On the other hand, in a sequence of transformation steps just the alternation of internal and external transformations remains, i.e. the order in which the internal and external transformations are applied does not matter. Parallel execution of internal and external steps has not been taken into the definition since it does not reflect interactivity, as desired in our applications. (Note that this does not exclude *independent* sequential internal and external transformations representing an interleavings of semantically independent or parallel steps.)

4   An Application: Specification of Visual Languages

In this section we discuss visual languages as an application domain for interactive rule based specifications. This application can be used to provide a formal basis for the construction of visual languages environments, as introduced in [Bar00].

Defining a visual expression (i.e., a sentence belonging to a visual language) means to deal with the description of two different aspects: its logical structure and its visual structure, i.e., its layout. These two aspects have to be connected by layout operations associating each logical item with a graphical one. The semantics of a visual expression is completely determined by its logical structure, the layout makes it visual.

We proceed by showing how to apply the interactive rule based specification framework to the definition of a concrete visual language, namely (a part of) the UML class diagram language (see [BRJ99]), consisting of classes and associations. We present first the algebraic specification of the language together with some rewriting rules for the underlying logic structures, then we describe the interactive generation of class diagrams where the visual layout is determined by two possible external components.

4.1   Algebraic Specification of Class Diagrams

The specification of the class diagram language comprises the two parts discussed in section 3, the *internal* and the *complete* specification, where the latter adds the externally defined parts. In addition we distinguish a *static* part of the internal signature to denote those parts of algebras that never change. In our case this is basically given by the domain of graphic elements. ¿From the formal point of view this distinction of a static part is just a comment, because in general rewrite rules might not preserve these parts.

– The *static part* provides the graphics elements, like rectangles, arrows, etc, and operations for their construction and manipulation. Rectangles for example are created by the operation *rect* with parameters for their left upper corners (of type *Point*), width, and height (of type *Real*). This graphic domain should provide all operations to define the desired layouts of the logical elements of the language and express relationships or constraints like: the point lies on the border of the rectangle etc. When using concrete graphic tools the signature should correspond of course to the interfaces of these tools.

  The sorts *Real*, *Point*, and *Graphic* are introduced in the static part, where *Graphic* here stands for a general sort for graphical elements and *Point* indicates positions on the display. Furthermore the static part provides built-in data types like *Strings* to express attributes of the logical elements like their names.

  sorts: *Real, Point, Graphic, String, . . .*

  funs: (graphic operations and constructors)
  $width : Graphic \rightarrow Real$
  $height : Graphic \rightarrow Real$
  $rect : Point,\ Real,\ Real,\ \rightarrow Graphic$
  $line : Point,\ Point \rightarrow Graphic$

  (predicates)
  $on\_border : Point,\ Graphic$

– The *internal part* of the signature includes the static part and extends it by a specification of the logical structure of the language. The sorts *Class* and *Assoc* are introduced, corresponding to the logical objects to be specified. Moreover, the operations allow for assigning names to logical elements and defining associations as structural relationships between classes.

sorts: $Class, Assoc$

funs: (logical: giving names to objects)
$c\_name : Class \rightarrow String$
$a\_name : Assoc \rightarrow String$

(logical: building structural relationships)
$a\_begin : Assoc \rightarrow Class$
$a\_end : Assoc \rightarrow Class$

- The *complete* specification of the class diagram language extends the internal part by an external part that specifies the layout functions. These have to be defined by an arbitrary external component.

  The basic idea thereby is to associate with each logical item some *attachment points* that yield enough information to obtain the complete layout by a corresponding parameterized layout operation. For instance, classes are visualized by rectangles whose width and height is determined by the size of the graphical representation of their names. Thus it suffices to specify a point (the attachment point of the class) where the rectangle shall be positioned. For an association two attachment points are specified. Its layout is then given by a line from the first to the second point.

  To formulate the constraints the predicate *on_border* is used. It allows us to specify that the association symbols are drawn in the correct way, i.e., that the lines representing them graphically start and end at the border of the corresponding class symbols.

  Finally, the external part comprises the functions connecting the logical and graphical structures: each logical element is associated with a layout, i.e., a graphic element. These functions are defined in terms of conditional equations that use the graphics constructors and the attachment points of the logical elements. (In the layout definitions the definedness predicate $\downarrow$ is used. Only if the attachment points are defined the layout can be computed.)

funs: (attachment points: to be set externally)
$c\_pos : Class \rightarrow Point$
$a\_bpos : Assoc \rightarrow Point$
$a\_epos : Assoc \rightarrow Point$

(graphical layout functions)
$c\_layout : Class \rightarrow Graphic$
$a\_layout : Assoc \rightarrow Graphic$
$s\_layout : String \rightarrow Graphic$

axioms: (constraints)
$on\_border(a\_bpos(a), c\_layout(a\_begin(a)))$
$on\_border(a\_epos(a), c\_layout(a\_end(a)))$
$width(a\_layout(a)) \geq width(s\_layout(a\_name(a)))$

(layout definition)
$c\_pos(c) \downarrow \Rightarrow$

$$c\_layout(c) = rect(c\_pos(c), width(s\_layout(c\_name(c))),$$
$$height(s\_layout(c\_name(c))))$$

$$a\_bpos(a) \downarrow \wedge a\_epos(a) \downarrow \Rightarrow$$
$$a\_layout(a) = line(a\_bpos(a), a\_epos(a))$$

This completes the specification of our class diagram language. Each visual expression of the language is completely specified by an algebra of the given specification.

Now we have to define the rewriting rules for the manipulation of the logical structures of visual expressions. We present here just some of them. (Layouts are considered in Sect. 4.2).

The first rule allows the insertion of a new class object. There are no preconditions to be fulfilled in order to add a new class into the logical structure: the *insert_class* rule just specifies in the right-hand side the new class element, and defines its associated name.

acts: $insert\_class : String$
rule: $insert\_class(cn) \triangleq (\emptyset; true) \Rightarrow$
$(\emptyset; \emptyset) \rightarrow (c : Class; c\_name(c) = cn)$

The *insert_assoc* rule is similar: the only precondition for its application is the existence of two classes. Since these also yield the context where to insert the new association (both logically and graphically) they are given both in the left and the right hand side of the rule. That means that they will be preserved by a rule application. The new association, its name and the connections with the source and target classes are specified in the right-hand side.

acts: $insert\_assoc : Class, Class, String$
rule: $insert\_assoc(c_1, c_2, an) \triangleq (\emptyset; true) \Rightarrow$
$(c_1, c_2 : Class; \emptyset) \rightarrow$
$(c_1, c_2 : Class, a : Assoc;$
$a\_name(a) = an, a\_begin(a) = c_1, a\_end(a) = c_2)$

Conversely, when removing an association all *relationships* to other elements, like its name, the beginning and the ending class, and its layout, are removed. (These elements themselves are of course not removed.) This is due to the restriction step in the construction of the internal algebra rewriting that removes all relations that contain removed elements (see Def. 6).

acts: $delete\_association : Assoc$
rule: $delete\_assoc(a) \triangleq (\emptyset; true) \Rightarrow (a : Assoc; \emptyset) \rightarrow (\emptyset; \emptyset)$

The deletion of a class is more complicated, since a class can be removed from a visual expression only if it is not related with any association. This is exactly the application condition specified in the rule *delete_class*. If the condition holds, the class element is deleted (by specifying it only in the left-hand side of the rule). As above, the connection with its name is deleted automatically, too.

acts: $delete\_class : Class$
rule: $delete\_class(c) \triangleq$
                 $(c : Class; \not\exists a \in Assoc : a\_begin(a) = c \ \lor \ a\_end(a) = c) \ \Rightarrow$
                 $(c : Class; \emptyset) \to (\emptyset; \emptyset)$

All the above rules deal with the creation or deletion of logical objects. However, also rules for moving objects have to be provided. In our framework such rules are empty, since the logical structure of the language does not change. For instance, the *move_class* rule is defined as follows:

acts: $move\_class : Class$
rule: $move\_class(c) \triangleq (\emptyset; true) \Rightarrow (\emptyset; \emptyset) \to (\emptyset; \emptyset)$

We conclude this section by showing an example of an internal algebra rewriting given by a rule application. Consider the insertion of an association as depicted in Fig. 4.1

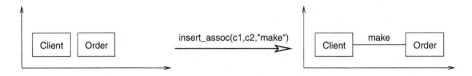

Fig. 2. Insertion of an association

The visual expression on the left side of the figure can be represented by the algebra defined as follows. We do not present the graphical part explicitly. The logical structure and the layout are given by:

$Class$	$\{x_1, x_2\}$	(there are two classes ...)
$Assoc$	$\emptyset$	(... and no association)
c_name	$c\_name(x_1) = \text{``}Client\text{''}, \ c\_name(x_2) = \text{``}Order\text{''}$	
		(each class has a name)
a_name	—	(these functions are undefined ...)
a_begin	—	(... because there are no associations)
a_end	—	
c_pos	$c\_pos(x_1) = (0.75, 2.25), \ c\_pos(x_2) = (3.75, 2.25)$	
a_bpos	—	
a_epos	—	
c_layout	$c\_layout(x_1) = rect((0.75, 2.25), 2.5, 1.5),$	
	$c\_layout(x_2) = rect((3.75, 2.25), 2.5, 1.5)$	
a_layout	—	

Applying the *insert_assoc* rule as shown in fig. 4.1 yields the following algebra as intermediate state of the transformation. A new association a is introduced, corresponding to the variable a that appears only in the right hand side of the rule. The corresponding equations connect a with its name *"make"* and the two classes x_1 and x_2 given to the *insert_assoc* action as parameters. The external parts are again defined as before. In particular, there is no attachment point and no layout for the new association yet.

Class	$\{x_1, x_2\}$	(as before)
Assoc	$\{a\}$	(the new association)
c_name	$c\_name(x_1) =$ *"Client"*, $c\_name(x_2) =$ *"Order"*	
	(as before)	
a_name	$a\_name(a) =''$ *make''*	
a_begin	$a\_begin(a) = x_1$	
a_end	$a\_begin(a) = x_2$	
	(according to the equations in the rhs of the rule)	
c_pos	$c\_pos(x_1) = (0.75, 2.25)$, $c\_pos(x_2) = (3.75, 2.25)$	
a_bpos	$-$	
a_epos	$-$	
c_layout	$c\_layout(x_1) = rect((0.75, 2.25), 2.5, 1.5)$,	
	$c\_layout(x_2) = rect((3.75, 2.25), 2.5, 1.5)$	
a_layout	$-$	

4.2   Interactive Definition of Visual Layouts

After the internal rewriting of the logical structure of a diagram its layout still has to be determined. In the specification designated attachment points have been introduced whose values completely determine the layout of the logical elements via the layout functions. Thus it suffices to deliver the values for the attachment points of all objects (classes and associations) that exist in a given state. We discuss two external components here that can be used for this purpose, a graphical constraint solver and a user interacting with the system via a display and a mouse.

As mentioned in Sect. 3.2 an external component can be connected to the interactive transformation mechanism by plugging it into the interface given by the presentation format (cf. Fig. 1). That means, we must define how a Σ-presentation yields an input for the component and how its output is translated into a Σ-presentation. In the following we define this connection for the graphical constraint solver and the clicking user.

The input of a constraint solver is given by a list of typed variables and a set of constraints. In our example (and for the constraint solver PARCON used in [Bar00] for this purpose) all variables are of type *Point* or *Real*, and the constraints are predicates or equations w.r.t. the functions provided by the graphic domain. The output of the constraint solver is given by a binding of values to the variables that respects the constraints. Usually this is given by a list of values that in comparison with the list of variables yields the association of the values to the variables.

The basic idea for the connection now is to use ground terms corresponding to the attachment points of the logical elements as names of the variables for the constraint solver. That means, the terms $c\_pos(c)$ for all $c \in iB_{Class}$ and $a\_bpos(a), a\_epos(a)$ for all $a \in iB_{Assoc}$ are used as variable names. On the other hand, the output of the constraint solver, i.e., the binding, can be translated into a set of Σ-equations $\{c\_pos(c) = t_c, \ldots\}$, where t_c is a ground term w.r.t. the graphics signature representing the value obtained by the binding for the variable named $c\_pos(c)$. (Thereby we assume that the signature is sufficiently expressive to obtain such a ground term for each value the constraint solver could possibly deliver.) Now the old function entries (equations) for the attachment points and layout operations are removed from the presentation $(\overline{iB}_S, \overline{iB}_E)$ obtained by the internal rewriting step and replaced by the equations $\{c\_pos(c) = t_c | c \in iB_{Class}\}$ and $\{a\_bpos(a) = t_a, a\_epos(a) = t'_a \mid a \in iB_{Assoc}\}$ for the attachment points obtained from the binding. In this way the missing information on the new positions of the logical elements is added. Note, however, that at this point the layout functions are completely undefined; the corresponding function entries have been removed.

The replacement of the old function entries for the attachment functions and layout operations by the new equations for the attachment points from the binding defines the behaviour of the constraint solver as a mapping on Σ-presentations, as required in Def. 7 (implicitly) and shown in Fig. 1.

The generation of the partial algebra $B = PAlg_\Gamma(B_S, B_E)$ in the interactive transformation step finally defines the layout functions according to the equations given in the specification Γ. Since the binding delivered by the constraint solver respects the constraints given in the specification, these are also satisfied by construction. Note that in this construction the layout of the whole diagram may be altered, since all attachment points are given to the constraint solver, whence all positions are determined anew. This is realistic, since the constraints are global w.r.t. the diagram. The introduction of a new association for example might require to move another class, and the renaming of a class might lead to a larger rectangle which might have side effects on its associations and the classes positioned close to it.

Consider as example again the application of the rule $insert\_assoc(x_1, x_2,$ "$make$") shown in Fig. 4.1. The constraint solver must move at least one of the two classes in order to display the association with its name properly in between them, as specified in the last relation in the constraints part of the specification. A possible output of the complete transformation step is given by the following algebra, with corresponding updates in the attachment points and layout functions.

Class	$\{x_1, x_2\}$
Assoc	$\{a\}$
c_name	$c_name(x_1) = $ "Client", $c_name(x_2) = $ "Order"

a_name	$a\_name(a) ="$ make$"$
a_begin	$a\_begin(a) = x_1$
a_end	$a\_begin(a) = x_2$
c_pos	$c\_pos(x_1) = (0.75, 2.25),\ c\_pos(x_2) = (6.75, 2.25)$
	(new position for x_2 !)
a_bpos	$a\_bpos(a) = (3.25, 1.5)$
a_epos	$a\_epos(a) = (6.75, 1.5)$
	(attachment points for a)
c_layout	$c\_layout(x_1) = rect((0.75, 2.25), 2.5, 1.5),$
	$c\_layout(x_2) = rect((6.75, 2.25), 2.5, 1.5)$
	(according to the new position of x_2)
a_layout	$a\_layout(a) = line((3.25, 1.5), (6.75, 1.5))$
	(the derived layout for a)

The embedding of a constraint solver into the interactive transformation can be defined more generally as follows. First a set of functions from the external signature is designated to be defined by the constraint solver. In our example these were the attachment points resp. the corresponding position functions, marked in the signature as *to be set externally*. The input variables of the constraint solver are then given by the set of all terms that consist of such a designated function symbol applied to an element of the actual state. Thereby the functions have to be compatible with the constraint solvers interface in the sense that their types are accepted (PARCON accepts real numbers and points for example). Then an appropriate set of equations is designated to obtain the constraints (marked as *constraints* in our example specification). Again, these have to be compatible with the input format of the constraint solver.

To let a user determine the positions of the logical elements is realized analogously to the connection of the constraint solver. In this case, however, we assume that the layout is changed only locally. The basic idea is again to obtain the new values of the attachment points for a logical element and replace the corresponding equations in the presentation $(\overline{iB}_S, \overline{iB}_E)$ obtained in the internal rewriting step. In this scenario an internal rewrite rule is always triggered by the user by pushing a button for one of the actions (*insert_class* etc.) and providing the parameters for this action. Then a preliminary (blinking or washy) layout appears somewhere on the screen and the user has to designate one or more points on the screen via her mouse to fix the layout. The number of points thereby depends on the specification, i.e., the number of attachment points for the concerned object (one for a class, two for an association etc.) Each mouse click is then interpreted as an equation, analogous to the translation of bindings to equations discussed above for the constraint solver.

5   Conclusion

In this paper we have introduced a formal model for interactive rule-based specifications of open systems. The states of a component are represented as partial algebras, their transformations are obtained in two consecutive steps. The internal transformation of a state of the considered component is specified by rewriting the internal part of algebra by the application of algebra rewrite rules, the external ones are added by an external mechanism whose behaviour is only given abstractly as a mapping on algebras. In this framework we have discussed a formal model of a generic editor for visual languages that uses a graphical constraint solver as external mechanism for the computation of graphical layouts.

5.1   Extensions

As already remarked in the discussion of the example some extensions to the pure formal framework would support its applicability. In most applications it is convenient for example to distinguish – beyond internal and complete signature – further layers of the overall specification. Firstly, *static* parts may be distinguished by a subsignature $\Sigma_{static} \subseteq \Sigma_{in}$ that designates all parts of the states that are assumed to immutable, i.e., do not change when rules are applied. For instance, built-in or pervasive data types like integers, strings, or booleans are usually considered as static. In our example we used a *graphics*-algebra modelling a real plane, points, and various figures, as well as corresponding operations. Obviously, these sets and operations should always be the same. The designation of a static subsignature (or specification) yields a proof obligation for the rules since in general it cannot be assured that the rules indeed preserve the static part. Thus their consistency has to be shown.

Secondly, some functions may be derivable from other more basic, but also mutable functions. In our example the layout functions have been derived from the attachment points. This derived part can be distinguished by a superspecification $\Gamma_{der} \supseteq \Gamma$. Usually it will have the same set of sorts as Γ but introduce further functions with conditional equations that define them w.r.t. the ones in Γ. The specification extension $\Gamma \subseteq \Gamma_{der}$ induces a free functor F_{der} that yields the semantics of the derived functions. Note that in this case the semantics are always defined, since a free functor exists for each specification extension. Nevertheless, consistency and completeness of the extension are not guaranteed in general, which yields another proof obligation.

The interactive rewriting w.r.t. a *stratified transformation specification* $(\Gamma_{static} \subseteq \Gamma_{in} \subseteq \Gamma \subseteq \Gamma_{der}, A, R)$ is then defined for partial Γ_{der}-algebras and corresponding matches. It uses the interactive rewriting w.r.t. $(\Gamma_{in} \subseteq \Gamma)$ as defined above and extends it as follows. At the beginning a partial Γ_{der}-algebra $\tilde{A}$ is restricted to its Γ-part, which yields the input for the interactive transformation step, and at the end the free construction $B \mapsto F_{der}(B) = \tilde{B}$ is added to obtain the final result. That means, all derived functions are computed anew w.r.t. the new basic mutable functions, corresponding to the definitions given in the specification Γ_{der}.

5.2   Related Work

One of the main sources for the development of the interactive transformation specification framework has been the search for a complete and adequate formal model of the generic interactive visual editor environment GENGED. The GENGED environment proposed in [Bar00] has been developed for the visual definition of visual languages and corresponding visual editors. The definition of a visual language as well as the manipulation of visual expressions is based there on algebraic graph transformation and graphical constraint solving. Similar to formal textual languages, a visual language is defined by an alphabet and a grammar. The alphabet is represented by a graph structure signature, i.e., an algebraic signature with unary operation symbols, and a constraint satisfaction problem for the admissible layouts. Accordingly, the grammar is represented by a graph structure grammar, where the constraint satisfaction problems derived from the alphabet are satisfied for each visual expression in the grammar.

The graph structure signature for the alphabet represents both the logical and the visual part. The logical part is thereby defined by an attributed graph structure signature. This is extended by an algebraic signature for the visual part. The specification of the logical part of a visual language corresponds to the *internal part* of an interactive transformation signature, whereas the specification of the visual part corresponds to the *static part* and the *external part*. The approach using attributed graph structures and corresponding grammars, however, imposed design decisions on the formal model that turned out not to be feasible. In particular, in [Bar00] the connection between the visual part and the external component could be treated semi-formally only.

In the GENGED environment, the graphical constraint solver PARCON [Gri96] is used to give the values (positions and sizes) for the visual part, i.e., for the layouts. Visual expressions are constructed by applying grammar rules according to the so-called *Single-Pushout* approach of graph transformation (see [LKW93]). The user of GENGED therefore is supposed to have some knowledge about the treatment of side effects in this approach, which are not explicitly specified. In contrast to that, everything is explicitly specified in the formal approach we propose here. The only side effect is the automatic removal of relations (function entries) that contain elements that have been deleted. (In the single pushout approach also elements with identities can be removed implicitly as a side effect sometimes.)

In the literature one can find many approaches for specifying visual languages and creating editors for them ([MMW98]). These formalisms range from early approaches like array and web grammars, positional grammars, relational grammars, constraint multiset grammars, several types of graph grammars, logic-based approaches and algebraic approaches. The existence of many formalisms on the one hand gives rise to a lot of possible classification criteria [Schi98,Bur01] and, on the other hand, makes it difficult to decide about *the best* approach. Such a decision depends on the purpose of the approaches, for example, whether a visual or textual definition of a VL is in the fore, or which kind of editing mode (freehand or syntax-directed editing) is supported in a graphical editor, or – if available – which kind of formal representation model is used.

Most tools for creating freehand editors analyze diagrams directly and avoid to create a formal representation model like a graph structure in [Bar00] or an algebra as presented in this paper. Possibly, freehand editing is desired in a graphical editor because a user can create and modify diagrams unrestrictedly; but these diagrams may contain errors that have to be recognized by a parser. In contrast, syntax-directed editing provides a set of editing commands which transforms correct diagrams into other correct diagrams; but the user is restricted to these commands.

The main aim of this paper has been the (visual) specification of visual languages which may be the basis for syntax-directed editing similar to [Bar00]. Furthermore, the approach presented here was motivated by [Bar00] in order to provide a formal approach for the logical part of a visual language and the visual part as well. Moreover, in contrast to the graph transformation formalism used in [Bar00] which has some side effects according to the formalism, in this paper all the features of a visual language are specified explicitly. A further difference between [Bar00] and our approach presented so far is given by the kind of algebras. In [Bar00] the algebras describing the logical part of a visual language are restricted to total algebras whereas we considered partial algebras.

A similar approach where expecially different kinds of graphics and graphical constraints are investigated is presented in [CLOT97,CP00]. In relation with our approach it can define an external device more concretely.

Beyond the several kinds of formalisms used for visual language specification, we have to mention the VAS (visual algebraic specification) formalism proposed in [DÜ96]. The VAS formalism is not only used for the specification of syntax but also semantics of visual languages. However, in this approach relations are not considered due to the logical part of a language and moreover, the semantics of a language is defined by evaluating terms over an algebraic specification according to conditional equations.

Concerning other formal approaches to the rule-based specification of open systems the *pull-back* approach to graph rewriting (see [EHLO98,EHLO99]) must be mentioned. Also in this approach the effect of a rule application is not completely specified. However, as opposed to the consecutive internal and external substeps in the interactive transformation specification approach only single global steps are considered, i.e., the impact of the environment is implicitly incorporated into the rule application in a parallel action. Moreover, this impact cannot be specified. Instead, each rule gives rise to (infinitely) many possible state transformations. Thus, beyond the use of graphs instead of algebras as state representations, this approach can be considered as dual w.r.t. the treatment of the other components.

The algebra rewriting formalism used for the internal rewriting steps is of course similar to the abstract state machine (ASM) approach (see [BH98]). We have chosen the former since we have been looking for a clean formal model incorporating function updates, creation and deletion of elements, and imposing constraints directly. In fact, algebra rewriting supports arbitrary transformations of algebras in classes defined by arbitrary sets of conditional equations. the categorical approach allows us to reason about properties of transformation systems abstractly and avoid encodings of the desired structure. On the other hand, ASMs are supported by tools for the specification, analysis, and simulation. Thus to apply the interactive rule-based specification framework our abstract algebra rewriting concepts should be mapped to the ASM framework to make use of this tool support.

References

[Bar00]   R. Bardohl "Visual Definition of Visual Languages based on Algebraic Graph Transformation" *Phd Thesis, Kovac Verlag* (2000)

[BH98]   E. Börger and J.K. Huggins. Abstract State Machines 1988–1998. *Bull. EATCS 64*, pages 71–87, 1998. Commented ASM Bibliography.

[BRJ99]   G. Booch, J. Rumbaugh, I. Jacobson "The Unified Modeling Language User Guide" *Addison-Wesley* (1999)

[Bur01]   M. Burnett. Visual Language Research Bibliography.
          URL: http://www.cs.orst.edu/~burnett/vpl.html.

[CLOT97] G. Costagliola, A. De Lucia, S. Orefice, and G. Tortora. A Framework of Syntactic Models for the Implementation of Visual Languages. In *Proc. IEEE Symp. on Visual Languages*, 1997.

[CP00]   G. Costagliola, and G. Polese. Extended Positional Grammars. In *Proc. IEEE Symp. on Visual Languages*, 2000.

[DÜ96]   T.B. Dinesh and S.M. Üsküdarlı. Specifying Input and Output of Visual Languages. In *Proc. of the AVI'96 Workshop Theory of Visual Languages*, Gubbio, Italy, May 1996.

[EEKR99] H. Ehrig, G. Engels, H.-J. Kreowski, and G. Rozenberg (eds.). *Handbook of Graph Grammars and Computing by Graph Transformation. Volume 2: Applications, Languages and Tools.* World Scientific, 1999.

[EHLO98] H. Ehrig, R. Heckel, M Llabres, and F. Orejas. Construction and characterization of double-pullback graph transitions. In G. Engels and G. Rozenberg, editors, *Proc. 6th Int. Workshop on Theory and Applications of Graph Transformation (TAGT'98)*, number tr-ri-98-201 in Reihe Informatik, pages 308-315. Universität-Gesamthochschule Paderborn, Fachbereich Mathematik-Informatik, 1998.

[EHLO99] H. Ehrig, R. Heckel, M Llabres, and F. Orejas. Basic properties of double-pullback graph transitions. Technical Report 99-02, Technical University of Berlin, 1999.

[ERT98]   C. Ermel, M. Rudolf, and G. Taentzer. The AGG-Approach: Language and Tool Environment. In [EEKR99], pages 551-604.

[Gri96]   P. Griebel. *ParCon – Paralleles Lösen von grafischen Constraints.* PhD thesis, Paderborn University, February 1996.

[Gro99]   M. Große-Rhode "Specification of State Based Systems by Algebra Rewrite Systems and Refinements" *Tech. Report TU-Berlin 99-04* (1999)

[LKW93]   M. Löwe, M. Korff, and A. Wagner. An Algebraic Framework for the Transformation of Attributed Graphs. In M.R. Sleep, M.J. Plasmeijer, and M.C. van Eekelen, editors, *Term Graph Rewriting: Theory and Practice*, pages 185-199. John Wiley & Sons Ltd, 1993.

20   R. Bardohl, M. Große-Rhode, and M. Simeoni

[MM98]   K. Marriott and B. Meyer (eds.). *Visual Language Theory*. Springer, 1998.
[MMW98] K. Marriott, B. Meyer, and K. Wittenburg. A Survey of Visual Language Specification and Recognition. In [MM98], pages 5-86.
[Schi98]   J. Schiffer. *Visuelle Programmierung*. Addison-Wesley, 1998.

Algebraic Abstractions[*]

Michel Bidoit and Alexandre Boisseau

Laboratoire Spécification et Vérification (LSV), CNRS & ENS de Cachan, France
{bidoit,boisseau}@lsv.ens-cachan.fr

Abstract. In this paper we study abstraction techniques for verification problems of the form $\mathcal{C} \models \phi$, where $\mathcal{C}$ is a first-order structure and ϕ is a first-order formula (both w.r.t. a given signature Σ). This study is motivated by the need of such abstractions for the automatic verification of properties of cryptographic protocols, which in our approach are modeled by first-order structures. Our so-called algebraic abstractions will be correct by construction and optimal in some certain technical sense. Moreover, we provide guidelines to design specific algebraic abstractions suited for verification problems corresponding to cryptographic protocols.

1 Introduction and Motivations

Formal verification of cryptographic protocols is nowadays an important issue due to their crucial rôle in the design of security or electronic commerce systems. A cryptographic protocol is usually described by a sequence of steps, where each step describes a message exchange between two principals.[1] Let us consider for instance the following simplified cryptographic protocol (see e.g. [4]):

$$1)\ A \to S : \langle A, B \rangle$$
$$2)\ S \to A : \{\langle K_B, B \rangle\}_{K_S^{-1}}$$

This protocol would be used in a public key cryptography system, where each participant has a private key (which should remain secret) and a public key (which can be provided to the other participants). To start a private session with B, the participant A should first obtain, from a server S, the public key of B. Hence A first sends a corresponding request to S (step 1 above). When S receives such a request, S computes his answer by encrypting the pair $\langle$public key of B, B's name$\rangle$ with his private key K_S^{-1} and sends it to A (step 2 above).

---

[*] This work is partially supported by the RNTL project EVA funded by the French Ministry of Research.

[1] A principal corresponds to a specific rôle in the cryptographic system. For instance, in the following example, there are two principals, one for the key server, one for the participants requesting a key. There may be several participants interacting according to the same rôle, they are all instances of the same principal. In informal explanations it is very frequent not to distinguish between participants and principals.

M. Cerioli and G. Reggio (Eds.): WADT/CoFI 2001, LNCS 2267, pp. 21–47, 2002.

The aim of this encryption step is not to ensure the secrecy of the message, since anybody can decrypt it using the public key of S, but to ensure authentication: S is the unique participant able to encrypt a message with K_S^{-1}.

Verification of such a cryptographic protocol amounts to prove that some properties about all its possible executions hold, for example:

– at the end of each execution, A receives the key he asked for and no other,[2]
– during each execution, K_S^{-1} remains secret.

The first property reflects in a certain sense the aim of the protocol. The second one ensures that the private key of S (which plays an important rôle in the authentication process) will not be compromised.

More precisely, one generally assumes that both the encryption algorithms used in the protocol as well as the implementation of the protocol itself are correct, but that the communication media are non-trustable, i.e., one assumes the presence of an intruder able to eavesdrop messages, divert them, produce new messages from the already exchanged messages, and even impersonate some other participants. Indeed, following a pessimistic approach, we assume that each message exchange is intercepted by an intruder I and possibly modified. This means that the first step of our simplified protocol will be split into two substeps:

$$1A)\ A \to I : m = \langle A, B \rangle$$
$$1S)\ I \to S : m'$$

where m' is a fake message (possibly equal to m) produced by the intruder according to his *knowledge* about the previously exchanged (and intercepted) messages (and the protocol). The second step of the protocol is split in a similar way:

$$2S)\ S \to I : m = \{\langle K_B, B \rangle\}_{K_S^{-1}}$$
$$2A)\ I \to A : m'$$

Model-checking techniques (see, e.g., [3]) cannot be directly used for the verification of cryptographic protocols since these should necessarily be modeled by infinite state systems:

– the number of participants, keys, nonces[3] and more generally atomic data is unbounded,
– moreover, messages can be built from atomic data using encryption and pairing in an arbitrary way, so the set of the potential messages exchanged is also infinite (and even if the number of participants, keys, nonces, . . . , involved in the protocol is finite, the size of the exchanged messages is unbounded, at least for those possibly produced by the intruder),
– finally, several sessions of the protocol can be run (concurrently or sequentially, depending on the modeling).

---

[2] It is easy to see that this property may not hold for our simplified protocol in presence of malicious intruders.
[3] Nonces are random numbers incorporated into messages in order to avoid replay attacks.

A standard approach to overcome this problem is the use of abstraction techniques (see [6, 9, 8]) in order to reduce the infinite system modeling the cryptographic protocol into a finite (or into a still infinite, but simpler) one. The use of abstraction techniques is summarized by the following diagram:

$$
\begin{array}{ccc}
 & Abstraction & \\
M & \longrightarrow & M^a \\
& & \\
\overset{?}{\models} & & \overset{?}{\models} \\
& & \\
\phi & \longrightarrow & \phi^a
\end{array}
$$

Here, given a model M and a property ϕ, we want to know whether this property holds for M. Instead of working directly with M and ϕ, we use a function called *abstraction* that transforms M and ϕ into an abstract model M^a and an abstract property ϕ^a. Now, if the abstraction is *correct*, i.e., if we have:

$$
M^a \models \phi^a \quad \Rightarrow \quad M \models \phi
$$

then $M \models \phi$ can be deduced from $M^a \models \phi^a$. Using such an abstraction will solve our problem when $M^a \models \phi^a$ is decidable (in particular when the abstract model is finite). The abstraction will be called *adequate* if $M^a \models \phi^a$ holds. In general abstractions are introduced in a rather ad hoc way, that is, for each verification problem $M \models \phi$, one has to invent some (hopefully adequate) abstraction and to prove its correctness before using it.

The aim of this paper is to introduce a general setting where classes of abstractions, correct by construction, can be more easily defined. More precisely, we will define abstraction schemes and correctness criteria that will ensure that any abstraction defined according to these schemes and satisfying the corresponding criteria is correct. Then the adequacy problem will be made simpler by using the fact that we are not looking for arbitrary abstractions, but merely for abstractions suitable for the verification of cryptographic protocols. Our specific way of modeling these protocols will provide useful guidelines for choosing a relevant abstraction.

The rest of this paper is organized as follows. In Section 2 we sketch our modeling technique of cryptographic protocols by first-order structures. In Section 3 we detail abstraction schemes for the verification of first-order formula with respect to first-order structures, and we provide the corresponding correctness criteria. Then we prove in Section 4 that abstractions defined according to our method are (correct by construction and) optimal in a certain technical sense. In Section 5 we show how to apply our abstraction method in the context of cryptographic protocols by exhibiting some relevant special cases.

In the sequel we assume familiarity with the basic concepts of algebraic specification (see, e.g., [2]) and in particular with the usual institution of (many-sorted) first-order logic with equality **FOLeq**:

- An algebraic signature is a pair $\Theta = (S, \Omega)$, where S is a set of sorts and Ω is a set of function symbols (together with their arity). A model for such a signature is a Θ-algebra and the class of Θ-algebras is denoted by $\mathbf{Alg}(\Theta)$.
- A first-order signature is a triple $\Sigma = (S, \Omega, \Pi)$, where (S, Ω) is an algebraic signature called the algebraic signature underlying Σ and denoted by $[\Sigma]$, and Π is a set of predicate symbols (with their arity). First-order signatures form a category denoted by $\mathbf{Sign}$. A model for a first-order signature $\Sigma = (S, \Omega, \Pi)$ is a first-order structure $\mathcal{C}$ which is made of a $[\Sigma]$-algebra called the $[\Sigma]$-algebra underlying $\mathcal{C}$ and denoted by $[\mathcal{C}]$ together with an interpretation for each predicate symbol in Π. $\mathbf{Str}(\Sigma)$ denotes the class of all Σ-structures.
- The set of first-order formulas (with equality) built from Σ and a set of variables X is denoted by $\mathbf{Sen}(\Sigma)$.
- $\models$ denotes the usual satisfaction relation between Σ-structures and Σ-formulas.

In the following, by default "signature" means "first-order signature".

2   Modeling Cryptographic Protocols by First-Order Structures

As any protocol, a cryptographic protocol is naturally modeled by a labeled transition system. However, in the specific context of cryptographic protocols, the content and structure of the messages exchanged is especially important, since, as explained in the previous section, to model the *knowledge of the intruder*, we will use the set of messages exchanged (and therefore intercepted by the intruder). More precisely, assuming that the intruder "knows" a set of messages s, he will be able to produce a (possibly fake) message m if the ingredients necessary to build m can be "derived" from s by the intruder. This is formally captured by a predicate $\Vdash$ which states under which conditions a message m can be produced by the intruder from his current knowledge represented by a set of messages s. Following [11], this predicate $\Vdash$ is defined inductively by the following rules:

$$\frac{m \in s}{s \Vdash m} \quad (\xi)$$

$$\frac{s \Vdash m_1 \qquad s \Vdash m_2}{s \Vdash \langle m_1, m_2 \rangle} \quad (\pi)$$

$$\frac{s \Vdash \langle m_1, m_2 \rangle}{s \Vdash m_1} \quad (\pi_1') \qquad \frac{s \Vdash \langle m_1, m_2 \rangle}{s \Vdash m_2} \quad (\pi_2')$$

$$\frac{s \Vdash m \qquad s \Vdash k}{s \Vdash \{m\}_k} \quad (\gamma) \qquad \frac{s \Vdash \{m\}_k \qquad s \Vdash k^{-1}}{s \Vdash m} \quad (\gamma')$$

The first four rules are obvious whereas (γ) and (γ') express a strong hypothesis about the encryption algorithms used in the protocol. For instance, (γ') states that if the intruder wants to produce m from $\{m\}_k$, he must know the inverse of k (hence these rules reflect our assumption that the encryption algorithms are perfect, as generally done in the literature [4, 13, 10]).

Hence, a first part of the model is a first-order structure corresponding to:

- atomic data (keys, nonces, principals, etc.), together with some operations (e.g., obtaining the inverse of a given key),
- messages built from atomic data using two constructions:
  - the pair construction, denoted $\langle m_1, m_2 \rangle$, to put together messages m_1 and m_2,
  - the encrypt construction, denoted $\{m\}_k$, to encrypt the message m using the key k,
- sets of messages with the usual operations and membership predicate,
- the predicate $\Vdash$.

Then the labeled transition system itself is also described as a first-order structure (enriching the above one), where the transition relation between states is modeled by a predicate $q \xrightarrow{(l,m)} q'$ (see e.g. [1]).

To summarize, the whole cryptographic protocol is modeled by a first-order structure, which is itself described by a specification structured as follows:

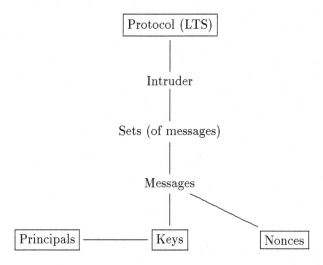

In the above diagram, the boxed pieces of specification correspond to the parts that are specific to a given cryptographic protocol, while the other parts are independent of the protocol considered. The specification of the intruder is merely the axiomatization of the predicate $\Vdash$ according to the rules given above. For the specification of the labeled transition system, one has to keep in mind that, as explained in the previous section, each step of the protocol is split into two sub-steps. So the predicate $q \xrightarrow{(l,m)} q'$ is axiomatized by a formula $q \xrightarrow{(l,m)} q' \Leftrightarrow r$, where r is a disjunction of formulas, one for each sub-step of the protocol (the full description of the transition relation corresponding to our toy cryptographic protocol is given in the following example).

Example 1 (A CASL specification of the toy cryptographic protocol).
spec PRINCIPAL =
    sort   *Princ*
    op $S : Princ$
end

spec KEY =
    PRINCIPAL
then **free type** $Key ::= k(Princ) \mid k^{-1}(Princ)$
    op $\_{}^{-1} : Key \to Key;$
    $\forall x : Princ$
    • $k(x)^{-1} = k^{-1}(x)$
    • $k^{-1}(x)^{-1} = k(x)$
end

spec MESSAGE =
    PRINCIPAL **and** KEY
then **free type** $Msg ::=$ *sort Princ* | *sort Key* %% Coercion of atomic data
                               | $\langle \_, \_ \rangle(Msg; Msg)$ %% Pairs
                                 | $\{\_\}\_(Msg; Key)$ %% Encryption
end

spec INTRUDER =
    MESSAGE **and** SET [**sort** *Msg*]
then **free {**
    pred $\_ \Vdash \_ : Set[Msg] \times Msg$
    $\forall m, m_1, m_2 : Msg; \; k : Key; \; s : Set[Msg]$
    • $m \in s \Rightarrow s \Vdash m$
    • $(s \Vdash m_1 \wedge s \Vdash m_2) \Rightarrow s \Vdash \langle m_1, m_2 \rangle$
    • $s \Vdash \langle m_1, m_2 \rangle \Rightarrow s \Vdash m_1$
    • $s \Vdash \langle m_1, m_2 \rangle \Rightarrow s \Vdash m_2$
    • $(s \Vdash \{m\}_k \wedge s \Vdash k^{-1}) \Rightarrow s \Vdash m$
    • $(s \Vdash m \wedge s \Vdash k) \Rightarrow s \Vdash \{m\}_k$ **}**
end

spec TOY_CRYPTOGRAPHIC_PROTOCOL =
    INTRUDER
then **free types**
             $Lbl ::= 1A \mid 1S \mid 2S \mid 2A;$
             %% As explained before, the two steps are split into four sub-steps
             $State ::= (\_.Adir : Set[Msg];$
                       $\_.Sid : Princ;$
                       $\_.I : Set[Msg])$
             %% Each state is a tuple containing :
             %% The directory where A records the public keys
             %% The identity of the participant received by the server S
             %% The set of messages known by the intruder

pred $\_\_\xrightarrow{(\_,\_)}\_\_ : State \times Lbl \times Msg \times State$
$\forall q, q' : State; \; m : Msg; \; l : Lbl$

- $q \xrightarrow{(l,m)} q' \Leftrightarrow$
  $(l = 1A$
  $\qquad \wedge \; \exists x : Princ \; \bullet \; \exists y : Princ \; \bullet$
  $\qquad\qquad m = \langle x, y \rangle \; \wedge \; q' = (q.Adir, q.Sid, q.I \cup m))$
  $\vee \; (l = 1S$
  $\qquad \wedge \; \exists x : Princ \; \bullet \; \exists y : Princ \; \bullet$
  $\qquad\qquad m = \langle x, y \rangle \; \wedge \; q.I \Vdash m \; \wedge \; q' = (q.Adir, y, q.I))$
  $\vee \; (l = 2S$
  $\qquad \wedge \; m = \{\langle k(q.Sid), q.Sid \rangle\}_{k^{-1}(S)}$
  $\qquad \wedge \; q' = (q.Adir, q.Sid, q.I \cup m))$
  $\vee \; (l = 2A$
  $\qquad \wedge \; (\exists x : Princ \; \bullet \; \exists y : Key \; \bullet$
  $\qquad\qquad m = \{\langle y, x \rangle\}_{k^{-1}(S)} \; \wedge \; q.I \Vdash m$
  $\qquad\qquad \wedge \; q' = (q.Adir \cup \langle x, y \rangle, q.Sid, q.I))$

then free {
  pred $Reach : State \times State$
  $\forall q, q', q'' : State; \; m : Msg; \; l : Lbl$
  - $Reach(q, q)$
  - $Reach(q, q') \wedge q \xrightarrow{(l,m)} q' \; \Rightarrow \; Reach(q, q'') \}$

end

Now, we can formalize the properties of our toy protocol.

- The following sentence states that the server's private key remains secret:

$$\forall q_0, q : State \; \bullet \; \neg(q_0.I \Vdash k^{-1}(S)) \wedge Reach(q_0, q) \; \Rightarrow \; \neg(q \Vdash k^{-1}(S))$$

- A's directory contains only valid pairs $\langle name, key \rangle$:

$$\forall q_0, q : State \; \bullet \; \forall b : Princ \; \bullet \; \forall k : Key \; \bullet$$
$$q_0.Adir = empty \wedge Reach(q_0, q) \wedge \langle b, k \rangle \in q.Adir \; \Rightarrow \; k = k(b)$$

3   Algebraic Abstractions

In this section, we describe our general setting for defining abstractions suitable for verification problems of the form $\mathcal{C} \models \phi$ (see Section 1), where $\mathcal{C}$ is a first-order structure and ϕ is a first-order formula (both w.r.t. a given signature Σ).[4] The abstraction problem can be split into the following subtasks:

---

[4] Abstractions are usually described by means of *Galois connections* (see [8]). This can indeed be done as well for our algebraic abstractions. However, in our case, a direct presentation without explicit Galois connections is much more easily to follow, and this is why we will adopt the later style here.

1. Abstract $\mathcal{C}$ into $\mathcal{A}$.
2. Abstract ϕ into ϕ^a.
3. Check the correctness of this abstraction.

Let us focus on the first subtask. A crucial remark is that any Σ-structure $\mathcal{C}$ can be considered as a $[\Sigma]$-algebra $[\mathcal{C}]$ together with an interpretation for each predicate symbol in Σ. Thus, abstracting $\mathcal{C}$ can itself be decomposed into, first, abstracting the underlying $[\Sigma]$-algebra $[\mathcal{C}]$, and then, finding an adequate abstraction for each predicate symbol interpretation.

To abstract algebras, we use, as usual, homomorphisms. Let $\alpha : [\mathcal{C}] \to \mathcal{B}$ be a $[\Sigma]$-morphism between the given $[\Sigma]$-algebra $[\mathcal{C}]$ and some $[\Sigma]$-algebra $\mathcal{B}$. To understand how to extend $\mathcal{B}$ into a Σ-structure $\mathcal{A}$ abstracting $\mathcal{C}$, i.e., to understand how to provide relevant interpretations for the predicate symbols, let us consider the following illustrative example.

Example 2. Consider the following CASL specification [7], specifying natural numbers and lists of natural numbers in the usual way:

spec BASICNATLIST =
    types *Nat* ::= *0* | *suc*(*Nat*);
           List ::= *empty* | *cons*(*Nat*; *List*)
    pred __ *is_in* __ : *Nat* × *List*
    $\forall x, y : Nat; \; l : List$
    • $\neg(x \; is\_in \; empty)$
    • $x \; is\_in \; cons(y, l) \Leftrightarrow (x = y \lor x \; is\_in \; l)$
end

spec NATLIST = **free** { BASICNATLIST }

The specification NATLIST has (up to isomorphism) only one model $\mathcal{C}$ defined by:

- $\mathcal{C}_{Nat} \overset{\text{def}}{=} \mathbb{N}$,
- $\mathcal{C}_{List} \overset{\text{def}}{=} \mathbb{N}^*$ (set of finite sequences of elements of $\mathbb{N}$),
- *0*, *suc*, *empty*, and *cons* are interpreted in the usual way.

Now consider the specification NATLIST' given by:

spec NATLIST' =
free { BASICNATLIST **then**
        $\forall x, y : Nat; \; l : List$
        • $suc(suc(x)) = suc(x)$
        • $cons(x, cons(y, l)) = cons(y, cons(x, l))$
        • $cons(x, cons(x, l)) = cons(x, l)$ }

This specification also has a unique model $\mathcal{A}$ up to isomorphism, given by:

- $\mathcal{A}_{Nat} \overset{\text{def}}{=} \{\mathtt{nul}, \mathtt{pos}\}$,
- $0^{\mathcal{A}} \overset{\text{def}}{=} \mathtt{nul}$, $suc^{\mathcal{A}}(\mathtt{nul}) \overset{\text{def}}{=} \mathtt{pos}$, and $suc^{\mathcal{A}}(\mathtt{pos}) \overset{\text{def}}{=} \mathtt{pos}$,
- $\mathcal{A}_{List}$ is the powerset of $\mathcal{A}_{Nat}$ and $is\_in^{\mathcal{A}}$ is the usual membership predicate.

Then, there is a unique homomorphism α from $[\mathcal{C}]$ to $\mathcal{B} = [\mathcal{A}]$. Given n in $\mathcal{C}_{Nat}$ and l in $\mathcal{C}_{List}$, we have n $is\_in^{\mathcal{C}}$ l implies $\alpha(n)$ $is\_in^{\mathcal{A}}$ $\alpha(l)$, because α is an homomorphism, so in order to prove that n is not in l, it is sufficient to prove that $\alpha(n)$ is not in $\alpha(l)$. However, n $is\_in^{\mathcal{C}}$ l cannot be deduced from $\alpha(n)$ $is\_in^{\mathcal{A}}$ $\alpha(l)$ (consider for instance $n = 2$ and $l = cons(1, empty)$). A stronger property, for instance $\alpha(n)$ $is\_in^{\mathcal{A}}$ $\alpha(l)$ and $\alpha(n) = \mathtt{nul}$, is required.

The above example illustrates that in general, two (distinct) interpretations in the abstract domain should be associated with each interpretation of a predicate symbol in $\mathcal{C}$, one for validation and one for refutation. Hence, the idea is to replace each predicate symbol P interpreted in the concrete domain by two predicate symbols $P_{\oplus}$ and $P_{\ominus}$, which means using a new signature Σ^* for the abstract structure.

Abstraction of a formula ϕ can then be done in a straightforward way: each positive (resp. negative) occurrence of a predicate symbol P in ϕ is replaced by $P_{\oplus}$ (resp. by $P_{\ominus}$). We can even construct a dual formula obtained by replacing the positive (resp. negative) occurrences of P in ϕ by $P_{\ominus}$ (resp. by $P_{\oplus}$). If these formulas are denoted $\phi_{\oplus}$ and $\phi_{\ominus}$ and if $\mathcal{B}$ is "correctly" embedded into a Σ^*-structure $\mathcal{A}$ (i.e., if some conditions on the interpretations of $P_{\oplus}$ and of $P_{\ominus}$ in $\mathcal{A}$ w.r.t. both the interpretation of P in $\mathcal{C}$ and the morphism α hold), then we will obtain the following properties:

$$\mathcal{A} \models \phi_{\oplus} \ \Rightarrow \ \mathcal{C} \models \phi$$
$$\text{and} \ \ \mathcal{A} \not\models \phi_{\ominus} \ \Rightarrow \ \mathcal{C} \not\models \phi$$

It may obviously happen that neither $\mathcal{A} \models \phi_{\oplus}$ nor $\mathcal{A} \not\models \phi_{\ominus}$. In such a case we cannot conclude whether $\mathcal{C} \models \phi$ or not, which means the chosen abstraction is not adequate.

Example 3. For instance, continuing Example 2 above, a "correct" Σ^*-structure with respect to the Σ-structure associated to the specification NATLIST is given by the specification:

spec NATLIST" $=$
    NATLIST' **with pred** $is\_in \mapsto is\_in_{\ominus}$
    %% $is\_in_{\ominus}$ is the same as is_in
    %% Recall that $\alpha(n)$ is_in $\alpha(l)$ $\Rightarrow$ n is_in l
then **pred** $\_\_$ $is\_in_{\oplus}$ $\_\_$: $Nat \times List$
    $\forall x : Nat; \ l : List \bullet x \ is\_in_{\oplus} \ l \Leftrightarrow (x \ is\_in_{\ominus} \ l \wedge x = 0)$

Remark 1. Related ideas (duplicating predicate symbols) have been used in some works on abstractions for reactive systems [9, 6, 12], using automata and some kind of temporal logic. Let us stress that here, due to the crucial rôle of the messages exchanged in a cryptographic protocol, it seems more convenient to model a cryptographic protocol by a first-order structure in order to obtain a uniform framework for modeling both the data part and the control part of the protocol.

Remark 2. Assume we want to validate some formula of the form $\phi \stackrel{\text{def}}{=} \exists x. \; \phi'$ and suppose that this formula is valid in the abstract domain (i.e., $\mathcal{A} \models \phi_\oplus$). In this case, there exists $a \in \mathcal{A}$ such that the valuation $x \mapsto a$ satisfy ϕ'. In order to transpose this result in $\mathcal{C}$, it is natural to intend to use an object $c \in \mathcal{C}$ such that $\alpha(c) = a$. This is why in the sequel, arbitrary homomorphisms α will not be used but only epimorphisms.

We are now ready to formalize the above intuitive principles.

Definition 1. *Let $\Sigma = (S, \Omega, \Pi)$ be a signature. The split signature Σ^* associated with Σ is defined by:*

$$\Sigma^* \stackrel{\text{def}}{=} (S, \Omega, \Pi^*)$$

where Π^* is the disjoint union of two copies of Π and of a new family of predicate symbols $\{=_\oplus : s \times s \mid s \in S^*\}$ and we note:

$$\begin{aligned} \Pi^* \stackrel{\text{def}}{=} \; & \{P_\oplus : s_1 \times \ldots \times s_n \mid P : s_1 \times \ldots \times s_n \in \Pi\} \\ & \cup \; \{P_\ominus : s_1 \times \ldots \times s_n \mid P : s_1 \times \ldots \times s_n \in \Pi\} \\ & \cup \; \{=_\oplus : s \times s \mid s \in S\}. \end{aligned}$$

Definition 2. *Given a signature $\Sigma = (S, \Omega, \Pi)$ and the associated split signature Σ^*, a Σ-structure $\mathcal{C}$, a Σ^*-structure $\mathcal{A}$ and a $[\Sigma]$-epimorphism $\alpha : [\mathcal{C}] \to [\mathcal{A}]$, $\mathcal{A}$ is correct with respect to $\mathcal{C}$ and α, denoted by $\mathcal{C} \sqsubset_\alpha \mathcal{A}$, if and only if:*

i) for each $P : s_1 \times \ldots \times s_n \in \Pi$, we have, for all $c_1 \in \mathcal{C}_{s_1}, \ldots, c_n \in \mathcal{C}_{s_n}$:

$$P_\oplus^{\mathcal{A}}(\alpha(c_1), \ldots, \alpha(c_n)) \Rightarrow P^{\mathcal{C}}(c_1, \ldots, c_n)$$

$$\text{and } P^{\mathcal{C}}(c_1, \ldots, c_n) \Rightarrow P_\ominus^{\mathcal{A}}(\alpha(c_1), \ldots, \alpha(c_n)),$$

ii) for each $s \in S$, we have, for all $c, c' \in \mathcal{C}_s$:

$$\alpha(c) =_\oplus^{\mathcal{A}} \alpha(c') \Rightarrow c = c'.$$

Given a Σ-structure $\mathcal{C}$ and a $[\Sigma]$-epimorphism $\alpha : [\mathcal{C}] \to \mathcal{B}$, there are in general several ways to extend the $[\Sigma]$-algebra $\mathcal{B}$ into a Σ-structure $\mathcal{A}$ (i.e., $[\mathcal{A}] = \mathcal{B}$) correct with respect to both $\mathcal{C}$ and α. This is why we introduce a notion of precision in the definition below in order to be able to compare such extensions. Of course, when two Σ^*-structures $\mathcal{A}$ and $\mathcal{A}'$ are correct with respect to $\mathcal{C}$ and α, we have $[\mathcal{A}] = [\mathcal{A}']$. Essentially, the notion of precision only concerns the respective interpretations of the predicate symbols on $\mathcal{A}$ and $\mathcal{A}'$. For the Σ^*-structure $\mathcal{A}$, to be more precise than $\mathcal{A}'$ intuitively means that $\mathcal{A}$ allows us to validate and refute at least all the formulas that can be validated or refuted using $\mathcal{A}'$.

Definition 3. *Let $\Sigma = (S, \Omega, \Pi)$ be a signature and $\mathcal{A}, \mathcal{A}' \in \mathbf{Str}(\Sigma^*)$. $\mathcal{A}$ is more precise than $\mathcal{A}'$ (denoted by $\mathcal{A} \preccurlyeq \mathcal{A}'$) if and only if $[\mathcal{A}] = [\mathcal{A}']$ and:*

i) for all $s \in S$ and $a_1, a_2 \in \mathcal{A}_s$, $a_1 =_\oplus^{\mathcal{A}'} a_2 \Rightarrow a_1 =_\oplus^{\mathcal{A}} a_2$,

ii) for all $P : s_1 \times \ldots \times s_n \in \Pi$, for all $a_1 \in \mathcal{A}_{s_1}, \ldots, a_n \in \mathcal{A}_{s_n}$:

$$P_{\oplus}^{\mathcal{A}'}(a_1, \ldots, a_n) \Rightarrow P_{\oplus}^{\mathcal{A}}(a_1, \ldots, a_n)$$

$$\text{and } P_{\ominus}^{\mathcal{A}}(a_1, \ldots, a_n) \Rightarrow P_{\ominus}^{\mathcal{A}'}(a_1, \ldots, a_n)$$

This relation clearly defines a partial order on $\mathbf{Str}(\Sigma^*)$.

Proposition 1 (Conservation of correctness with respect to precision).
Given a signature Σ, a Σ-structure $\mathcal{C}$, two Σ^*-structures $\mathcal{A}$ and $\mathcal{A}'$ and a $[\Sigma]$-epimorphism $\alpha : [\mathcal{C}] \to [\mathcal{A}]$, if $\mathcal{C} \sqsubset_\alpha \mathcal{A}$ and $\mathcal{A} \preccurlyeq \mathcal{A}'$ then we have $\mathcal{C} \sqsubset_\alpha \mathcal{A}'$.

Proof. Obvious since for each predicate symbol $P : s_1 \times \ldots \times s_n$ in Σ and any $c_1 \in \mathcal{C}_{s_1}, \ldots, c_n \in \mathcal{C}_{s_n}$, we have, using the above definitions of correctness and precision:

$$P_{\oplus}^{\mathcal{A}'}(\alpha(c_1), \ldots, \alpha(c_n)) \Rightarrow P_{\oplus}^{\mathcal{A}}(\alpha(c_1), \ldots, \alpha(c_n)) \Rightarrow P^{\mathcal{C}}(c_1, \ldots, c_n)$$
$$P^{\mathcal{C}}(c_1, \ldots, c_n) \Rightarrow P_{\ominus}^{\mathcal{A}}(\alpha(c_1), \ldots, \alpha(c_n)) \Rightarrow P_{\ominus}^{\mathcal{A}'}(\alpha(c_1), \ldots, \alpha(c_n))$$

and similarly for $=_\oplus$.   □

Definition 4 (Canonical abstraction). *Assume given a signature $\Sigma = (S, \Omega, \Pi)$, a Σ-structure $\mathcal{C}$ and $\alpha : [\mathcal{C}] \to \mathcal{B}$ an epimorphism of $[\Sigma]$-algebras. Then there is a canonical Σ^*-structure associated to $\mathcal{C}$ and α, denoted by $\mathcal{C}^\alpha$ and defined by:*

$-\ [\mathcal{C}^\alpha] \stackrel{\text{def}}{=} \mathcal{B}$,
$-$ *for all $P : s_1 \times \ldots \times s_n \in \Pi$, $a_1 \in \mathcal{C}_{s_1}^\alpha, \ldots, a_n \in \mathcal{C}_{s_n}^\alpha$:*
   $\bullet\ P_{\oplus}^{\mathcal{C}^\alpha}(a_1, \ldots, a_n)$ *if and only if for all $c_1 \in \alpha^{-1}(a_1), \ldots, c_n \in \alpha^{-1}(a_n)$, we have $P^{\mathcal{C}}(c_1, \ldots, c_n)$,*
   $\bullet\ P_{\ominus}^{\mathcal{C}^\alpha}(a_1, \ldots, a_n)$ *if and only if there exists $c_1 \in \alpha^{-1}(a_1), \ldots, c_n \in \alpha^{-1}(a_n)$ such that $P^{\mathcal{C}}(c_1, \ldots, c_n)$,*
$-$ *for all $s \in S$ and $a_1, a_2 \in \mathcal{C}_s^\alpha$, $a_1 =_{\oplus}^{\mathcal{C}^\alpha} a_2$ if and only if $a_1 = a_2$ and $\alpha^{-1}(a_1)$ and $\alpha^{-1}(a_2)$ both contain exactly one element.*

This construction allows us to reformulate the notion of correctness in terms of precision.

Proposition 2 (Characterization of correctness). *Let be given a signature Σ, a Σ-structure $\mathcal{C}$, a Σ^*-structure $\mathcal{A}$ and a $[\Sigma]$-epimorphism $\alpha : [\mathcal{C}] \to [\mathcal{A}]$. Then $\mathcal{C} \sqsubset_\alpha \mathcal{A}$ if and only if $\mathcal{C}^\alpha \preccurlyeq \mathcal{A}$ (and consequently, $\mathcal{C}^\alpha$ is the most precise Σ^*-structure correct with respect to $\mathcal{C}$ and α).*

Proof. See the Appendix.   □

Note that, using the same notations as above, if $\mathcal{A}$ is correct with respect to $\mathcal{C}$ and α, then $\mathcal{A}$ satisfies the following conditions:

i) for all $P : s_1 \times \ldots \times s_n \in \Pi$, $P_{\oplus}^{\mathcal{A}} \subseteq P_{\ominus}^{\mathcal{A}}$,
ii) $=_\oplus$ is a subset of the natural equality on $\mathcal{A}$.

Such a Σ^*-structure $\mathcal{A}$, correct with respect to $\mathcal{C}$ and α, allows us to validate (or refute) a subset of formulas on Σ, namely the atomic formulas. This property has now to be extended to more general formulas in order to obtain the abstract counterpart of the concrete properties considered.

Definition 5. *Given a Σ-formula $\phi \in \mathbf{Sen}(\Sigma)$, we define two formulas in $\mathbf{Sen}(\Sigma^*)$, denoted respectively by $\phi_\oplus$ and $\phi_\ominus$, inductively on the structure of ϕ as follows:*

– *for $P : s_1 \times \ldots \times s_n \in \Pi$, $t_1 \in T_\Sigma(X)_{s_1}, \ldots, t_n \in T_\Sigma(X)_{s_n}$:*

$$(Pt_1 \ldots t_n)_\oplus \stackrel{\text{def}}{=} P_\oplus t_1 \ldots t_n \quad and \quad (Pt_1 \ldots t_n)_\ominus \stackrel{\text{def}}{=} P_\ominus t_1 \ldots t_n,$$

– *for $s \in S$ and $t_1, t_1 \in T_\Sigma(X)_s$:*

$$(t_1 = t_2)_\oplus \stackrel{\text{def}}{=} t_1 =_\oplus t_2 \quad and \quad (t_1 = t_2)_\ominus \stackrel{\text{def}}{=} t_1 = t_2,$$

– *for $\phi_1, \phi_2 \in \mathbf{Sen}(\Sigma)$:*

$$(\phi_1 \wedge \phi_2)_\oplus \stackrel{\text{def}}{=} \phi_{1\oplus} \wedge \phi_{2\oplus},$$
$$(\phi_1 \wedge \phi_2)_\ominus \stackrel{\text{def}}{=} \phi_{1\ominus} \wedge \phi_{2\ominus},$$
$$(\phi_1 \vee \phi_2)_\oplus \stackrel{\text{def}}{=} \phi_{1\oplus} \vee \phi_{2\oplus},$$
$$(\phi_1 \vee \phi_2)_\ominus \stackrel{\text{def}}{=} \phi_{1\ominus} \vee \phi_{2\ominus},$$

– *for $\phi \in \mathbf{Sen}(\Sigma)$ and $x : s \in X$:*

$$(\forall x : s. \ \phi)_\oplus \stackrel{\text{def}}{=} \forall x : s. \ \phi_\oplus,$$
$$(\forall x : s. \ \phi)_\ominus \stackrel{\text{def}}{=} \forall x : s. \ \phi_\ominus,$$
$$(\exists x : s. \ \phi)_\oplus \stackrel{\text{def}}{=} \exists x : s. \ \phi_\oplus,$$
$$(\exists x : s. \ \phi)_\ominus \stackrel{\text{def}}{=} \exists x : s. \ \phi_\ominus,$$
$$(\neg \phi)_\oplus \stackrel{\text{def}}{=} \neg(\phi_\ominus),$$
$$(\neg \phi)_\ominus \stackrel{\text{def}}{=} \neg(\phi_\oplus).$$

We can now state our first main result.

Theorem 1 (Correctness result). *Assume given a signature Σ, a Σ-structure $\mathcal{C}$, and a Σ^*-structure $\mathcal{A}$ such that $\mathcal{C} \sqsubseteq_\alpha \mathcal{A}$, where $\alpha : [\mathcal{C}] \to [\mathcal{A}]$ is a $[\Sigma]$-epimorphism. Then for any $\phi \in \mathbf{Sen}(\Sigma)$ we have:*

$$\mathcal{A} \models \phi_\oplus \ \Rightarrow \ \mathcal{C} \models \phi$$
$$and \ \mathcal{C} \models \phi \ \Rightarrow \ \mathcal{A} \models \phi_\ominus.$$

Proof. See the Appendix. □

Proposition 3. *Assume given a signature Σ and two Σ^*-structures $\mathcal{A}$ and $\mathcal{A}'$ such that $\mathcal{A} \preccurlyeq \mathcal{A}'$. Then for any $\phi \in \mathbf{Sen}(\Sigma)$ we have:*

$$\mathcal{A}' \models \phi_\oplus \ \Rightarrow \ \mathcal{A} \models \phi_\oplus$$
$$and \ \mathcal{A} \models \phi_\ominus \ \Rightarrow \ \mathcal{A}' \models \phi_\ominus.$$

Proof. By induction on the structure of ϕ.   □

The framework developed above and the corresponding results mean that, to abstract a verification problem of the form $\mathcal{C} \models \phi$, where $\mathcal{C}$ is a first-order structure and ϕ is a first-order formula (both w.r.t. a given signature Σ), it is enough to, first, abstract the corresponding underlying algebra $[\mathcal{C}]$ by choosing some $[\Sigma]$-epimorphism $\alpha : [\mathcal{C}] \to \mathcal{B}$, and then, extend the $[\Sigma]$-algebra $\mathcal{B}$ into a Σ^*-structure $\mathcal{A}$ correct w.r.t. $\mathcal{C}$ and α, which amounts to choose a suitable interpretation for the duplicated predicate symbols $P_\oplus$ and $P_\ominus$ (and $\mathcal{C}^\alpha$ is a good candidate). Then we can use $\phi_\oplus$ and $\phi_\ominus$ which are syntactically derived from ϕ.

4   Optimality of Algebraic Abstractions

In the previous section, we have explained how to use algebraic abstractions which are, by construction, correct abstractions. However, a correct algebraic abstraction may turn out not to be adequate, and in this case one has to look for another better suited abstraction. We will show in this section that, given a verification problem of the form $\mathcal{C} \models \phi$, for each choice of an epimorphism $\alpha : [\mathcal{C}] \to \mathcal{B}$, there is a corresponding optimal algebraic abstraction: if it is not adequate, then any other abstraction built from the same epimorphism will not be adequate either. Hence, our optimality results will allow us to reduce the adequacy problem to the right choice of the underlying epimorphism of algebras. Intuitively, Proposition 3 shows that $\mathcal{C}^\alpha$ is the "best" structure we can use once α is chosen. It is however far from obvious whether our syntactic abstraction of formulas ϕ into $\phi_\oplus$, $\phi_\ominus$ is optimal or not. The main aim of this section is to formally establish that $(\phi_\oplus, \phi_\ominus)$ is indeed the best possible choice.

Hence, we will prove that, for each application:

$$\tau : \mathbf{Sen}(\Sigma) \to \mathbf{Sen}(\Sigma^*)^2$$
$$\phi \quad \mapsto (\tau^1(\phi), \tau^2(\phi))$$

satisfying for all $\phi \in \mathbf{Sen}(\Sigma)$ and for all $\mathcal{C} \in \mathbf{Str}(\Sigma)$ and $\mathcal{A} \in \mathbf{Str}(\Sigma^*)$ correct with respect to $\mathcal{C}$ and an epimorphism α:

$$\mathcal{A} \models \tau^1(\phi) \Rightarrow \mathcal{C} \models \phi \quad \text{and} \quad \mathcal{C} \models \phi \Rightarrow \mathcal{A} \models \tau^2(\phi)$$

we have:

$$\mathcal{A} \models \tau^1(\phi) \Rightarrow \mathcal{A} \models \phi_\oplus \quad \text{and} \quad \mathcal{A} \models \phi_\ominus \Rightarrow \mathcal{A} \models \tau^2(\phi)$$

for all $\phi \in \mathbf{Sen}(\Sigma)$. We will assume that these applications τ are compatible with signature morphisms, since in our framework it only makes sense to consider abstraction techniques generic w.r.t. the signatures of the structures to be abstracted.

To prove our optimality result, we will rely on a technical trick, namely we will abstract again $\mathcal{A}$, and we will therefore consider a further split signature Σ^{**}. Moreover, in a first step, we will work within the **FOL** institution (instead of the **FOLeq** institution), and then we will of course come back to **FOLeq**.

4.1   Optimality in FOL

The institution of (many-sorted) first-order logic is denoted **FOL**:

$$\mathbf{FOL} = (\mathbf{Sign}, \mathbf{Sen_{FOL}}(), \mathbf{Str}(), \models)$$

(where **Sign** and **Str**() are the same as in **FOLeq**). Abstractions can be done in **FOL** in the same way as in **FOLeq**, but since there is no predicate $=_{\oplus}$ to add, we use, instead of Σ^*, the signature:

$$\Sigma^{\bullet} = (S, \Omega, \{P_{\oplus} \mid P \in \Pi\} \cup \{P_{\ominus} \mid P \in \Pi\})$$

where $\Sigma = (S, \Omega, \Pi)$.

In order to state and prove the optimality result in **FOL**, we are going to proceed as follow:

- abstraction of signatures will be turned into a functor (then it will be possible to define a translation which is compatible with respect to signature morphisms),
- a $\Sigma^{\bullet\bullet}$-structure will be associated to each $\Sigma^{\bullet}$-structure,
- then it will be possible to prove the optimality result using abstractions.

With each signature morphism $\sigma : \Sigma \to \Sigma'$ is associated a morphism $\sigma^{\bullet} : \Sigma^{\bullet} \to \Sigma'^{\bullet}$ defined by:

- $\sigma^{\bullet}$ acts as σ on the sorts and function symbols of Σ,
- for each predicate symbol P of Σ:

$$\sigma^{\bullet}(P_{\oplus}) \overset{\text{def}}{=} (\sigma(P))_{\oplus} \quad \text{and} \quad \sigma^{\bullet}(P_{\ominus}) \overset{\text{def}}{=} (\sigma(P))_{\ominus}.$$

Moreover, for each signature Σ there is a canonical homomorphism:

$$|.|_{\Sigma} : \Sigma^{\bullet} \to \Sigma$$

defined by:

- $|.|_{\Sigma}$ is the identity on sorts and function symbols,
- for each predicate symbol P of Σ:

$$|P_{\oplus}|_{\Sigma} \overset{\text{def}}{=} P \text{ and } |P_{\ominus}|_{\Sigma} \overset{\text{def}}{=} P.$$

Then, the following diagram is commutative:

$$
\begin{array}{ccc}
\Sigma & \xrightarrow{\ \sigma\ } & \Sigma' \\
{\scriptstyle |.|_{\Sigma}} \uparrow & & \uparrow {\scriptstyle |.|_{\Sigma'}} \\
\Sigma^{\bullet} & \xrightarrow[\ \sigma^{\bullet}\ ]{} & \Sigma'^{\bullet}
\end{array}
$$

Finally, in the proof of optimality, we are going to use the satisfaction condition in first-order logic, i.e., given a signature morphism $\sigma : \Sigma \to \Sigma'$, $\phi \in \mathbf{Sen_{FOL}}(\Sigma)$ and $\mathcal{C} \in \mathbf{Str_{FOL}}(\Sigma')$ we have:

$$\mathcal{C} \models \sigma(\phi) \Leftrightarrow \mathcal{C}|_\sigma \models \phi$$

(see [2] for details).

So, there are two homomorphisms from $\Sigma^{\bullet\bullet}$ to $\Sigma^\bullet$:

- the first one is the canonical homomorphism $|.|_{\Sigma^\bullet} : \Sigma^{\bullet\bullet} \to \Sigma^\bullet$, which satisfies for each predicate P in Σ annotated by s_1 in $\Sigma^\bullet$ and s_2 in $\Sigma^{\bullet\bullet}$:

$$|(P_{s_1})_{s_2}|_{\Sigma^\bullet} = P_{s_1},$$

- the second one is the homomorphism associated to the canonical homomorphism $|.|_\Sigma : \Sigma^\bullet \to \Sigma$, denoted by $|.|_\Sigma^\bullet$, and we have:

$$|(P_{s_1})_{s_2}|_\Sigma^\bullet = (|P_{s_1}|_\Sigma)_{s_2} = P_{s_2}.$$

When we reduce a $\Sigma^\bullet$-structure $\mathcal{A}$ along $|.|_{\Sigma^\bullet}$, we obtain the structure defined in Section 3 (Definition 4) and denoted $\mathcal{A}^{id}$ (where id denotes the identity epimorphism). This structure is known to be correct with respect to $\mathcal{A}$, but, in the proof of optimality, we need to reduce $\mathcal{A}$ along the other homomorphism. So, we consider $\mathcal{A}|_{|.|_\Sigma^\bullet}$, and the first thing to do is to prove that this structure is also correct with respect to $\mathcal{A}$. This is the aim of the following lemma.

Lemma 1. *Let be given a signature Σ and a $\Sigma^\bullet$-structure $\mathcal{A} \in \mathbf{Str_{FOL}}(\Sigma^\bullet)$ such that for each predicate symbol $P : s_1 \times \ldots \times s_n$ in Σ:*

$$P_\oplus^\mathcal{A} \subseteq P_\ominus^\mathcal{A}.$$

Let $|.|_\Sigma^\bullet$ denote the signature homomorphism from $\Sigma^{\bullet\bullet}$ to $\Sigma^\bullet$ associated to the canonical homomorphism $|.|_\Sigma$ from $\Sigma^\bullet$ to Σ. Then $\mathcal{A} \sqsubseteq_{id} \mathcal{A}|_{|.|_\Sigma^\bullet}$, where id denote the identity from $[\mathcal{A}]$ to $[\mathcal{A}|_{|.|_\Sigma^\bullet}] = [\mathcal{A}]$.

Proof. See the Appendix. □

We now just have to define the class of translations considered and we can prove the optimality theorem.

Definition 6. *A translation (in **FOL**) is a family $\tau = (\tau_\Sigma)_{\Sigma \in \mathbf{Sign}}$ where, for each $\Sigma \in \mathbf{Sign}$, τ_Σ is an application:*

$$\tau_\Sigma : \mathbf{Sen_{FOL}}(\Sigma) \to \mathbf{Sen_{FOL}}(\Sigma^\bullet)^2$$
$$\phi \mapsto (\tau_\Sigma^1(\phi), \tau_\Sigma^2(\phi))$$

A translation is correct *if for all $\Sigma \in \mathbf{Sign}$, $\phi \in \mathbf{Sen_{FOL}}(\Sigma)$, $\mathcal{C} \in \mathbf{Str_{FOL}}(\Sigma)$ and $\mathcal{A} \in \mathbf{Str_{FOL}}(\Sigma^\bullet)$ correct with respect to $\mathcal{C}$ and an epimorphism $\alpha : [\mathcal{C}] \to [\mathcal{A}]$ we have:*

$$\mathcal{A} \models \tau_\Sigma^1(\phi) \Rightarrow \mathcal{C} \models \phi \quad and \quad \mathcal{C} \models \phi \Rightarrow \mathcal{A} \models \tau_\Sigma^2(\phi)$$

and natural *if for* $\Sigma, \Sigma' \in \mathbf{Sign}$, $\sigma : \Sigma \to \Sigma'$ *and* $\phi \in \mathbf{Sen_{FOL}}(\Sigma)$, *we have:*

$$\tau_{\Sigma'}(\sigma(\phi)) = \sigma^\bullet(\tau_\Sigma(\phi))$$

i.e., for each $\Sigma \in \mathbf{Sign}$ *the following diagram is commutative:*

$$
\begin{array}{ccc}
\mathbf{Sen_{FOL}}(\Sigma) & \xrightarrow{\;\;\tau_\Sigma\;\;} & \mathbf{Sen_{FOL}}(\Sigma^\bullet)^2 \\
\sigma \downarrow & & \downarrow \sigma^\bullet \times \sigma^\bullet \\
\mathbf{Sen_{FOL}}(\Sigma') & \xrightarrow[\;\;\tau_{\Sigma'}\;\;]{} & \mathbf{Sen_{FOL}}(\Sigma'^\bullet)^2
\end{array}
$$

Remark 3. Consider the functor $\mathbf{Sen}^\bullet_{\mathbf{FOL}}(.)$ from $\mathbf{Sign}$ to $\mathbf{Set}$ (the category of sets) defined by:

- for $\Sigma \in \mathbf{Sign}$, $\mathbf{Sen}^\bullet_{\mathbf{FOL}}(\Sigma) = \mathbf{Sen_{FOL}}(\Sigma^\bullet)^2$,
- for $\sigma : \Sigma \to \Sigma'$, $\mathbf{Sen}^\bullet_{\mathbf{FOL}}(\sigma) = \sigma^\bullet \times \sigma^\bullet$,

then a translation is natural if and only if it defines a natural transformation from $\mathbf{Sen_{FOL}}(.)$ to $\mathbf{Sen}^\bullet_{\mathbf{FOL}}(.)$.

Theorem 2 (Optimality in FOL). *Let be given a correct and natural translation* τ *and let* $\Sigma \in \mathbf{Sign}$ *and* $\mathcal{A} \in \mathbf{Str_{FOL}}(\Sigma^\bullet)$ *be such that for each predicate symbol* $P : s_1 \times \ldots \times s_n$ *in* Σ, *we have* $P^{\mathcal{A}}_\oplus \subseteq P^{\mathcal{A}}_\ominus$. *Then:*

$$\mathcal{A} \models \tau^1_\Sigma(\phi) \Rightarrow \mathcal{A} \models \phi_\oplus \quad and \quad \mathcal{A} \models \phi_\ominus \Rightarrow \mathcal{A} \models \tau^2_\Sigma(\phi).$$

Proof. See the Appendix.   □

Remark 4. This result holds in particular when $\mathcal{A}$ is correct with respect to $\mathcal{C} \in \mathbf{Str_{FOL}}(\Sigma)$ and an epimorphism $\alpha : [\mathcal{C}] \to [\mathcal{A}]$.

4.2   Optimality in FOLeq

The optimality result in **FOLeq** will be proved using the previous theorem in **FOL** and that is why we first define an encoding (following [14]) of **FOLeq** into **FOL**. Since sentences and structures in **FOL** can be considered as sentences and structures in **FOLeq**, we finally obtain an encoding of **FOLeq** into **FOLeq**. Then, the class of translations considered in **FOLeq** is the class of translations correct, natural (as above) and commuting with this encoding.

For each $\Sigma = (S, \Omega, \Pi) \in \mathbf{Sign}$, we consider the signature:

$$\overline{\Sigma} = (S, \Omega, \Pi \cup \{eq_\Sigma : s \times s \mid s \in S\})$$

(where $\{eq_\Sigma : s \times s \mid s \in S\}$ is a fresh family of predicate symbols). Using $\overline{\Sigma}$, we obtain a one to one translation of formulas:

$$
\begin{array}{ccc}
\mathbf{Sen}(\Sigma) & \to & \mathbf{Sen_{FOL}}(\overline{\Sigma}) \\
\phi & \mapsto & \overline{\phi}
\end{array}
$$

and a mapping on the structures:

$$\mathbf{Str}(\Sigma) \to \mathbf{Str}(\overline{\Sigma})$$
$$\mathcal{C} \quad \mapsto \quad \overline{\mathcal{C}}$$

satisfying the property:

$$\mathcal{C} \models \phi \Leftrightarrow \overline{\mathcal{C}} \models \overline{\phi}$$

These mappings define an encoding of **FOLeq** into **FOL**. Moreover, $\overline{\Sigma^*}$ and $\overline{\Sigma}^{\bullet}$ are clearly isomorphic and in the sequel we are going to identify them. Since $\overline{\Sigma}^{\bullet} \subseteq \overline{\Sigma}^*$, the following relations between sentences hold:

$$\mathbf{Sen_{FOL}}(\Sigma) \subseteq \mathbf{Sen}(\Sigma) \cong \mathbf{Sen_{FOL}}(\overline{\Sigma}),$$
$$\mathbf{Sen}(\overline{\Sigma^*}) \subseteq \mathbf{Sen}(\overline{\Sigma}^*).$$

In **FOLeq**, we need to consider translations commuting with the above encoding of **FOLeq** into itself. Such a translation is called homogeneous. The homogeneity condition states that the equality is processed like predicates, at the syntactic level.

Definition 7. *A translation (in **FOLeq**) is a family* $\tau = (\tau_\Sigma)_{\Sigma \in \mathbf{Sign}}$, *where for each* $\Sigma \in \mathbf{Sign}$, τ_Σ *is an application* $\tau_\Sigma : \mathbf{Sen}(\Sigma) \to \mathbf{Sen}(\Sigma^*)^2$. *Correctness and naturality of such a translation can be defined like in **FOL**. Moreover,* τ *is said to be homogeneous if for each* $\Sigma \in \mathbf{Sign}$ *the following diagram is commutative :*

$$
\begin{array}{ccc}
\mathbf{Sen}(\Sigma) \xrightarrow{\;\cong\;} \mathbf{Sen_{FOL}}(\overline{\Sigma}) \lhook\joinrel\longrightarrow \mathbf{Sen}(\overline{\Sigma}) \\
\Big\downarrow{\scriptstyle \tau_\Sigma} \qquad\qquad\qquad\qquad\qquad \Big\downarrow{\scriptstyle \tau_{\overline{\Sigma}}} \\
\mathbf{Sen}(\Sigma^*)^2 \xrightarrow{\;\cong\;} \mathbf{Sen}(\overline{\Sigma^*})^2 \lhook\joinrel\longrightarrow \mathbf{Sen}(\overline{\Sigma}^*)^2
\end{array}
$$

(i.e., for all $\phi \in \mathbf{Sen}(\Sigma)$, $\tau_{\overline{\Sigma}}(\overline{\phi}) = \overline{\tau_\Sigma(\phi)}$*).*

The following lemma states that a correct, natural and homogeneous translation in **FOLeq** induces (and is entirely defined by) a correct and natural translation in **FOL**.

Lemma 2. *Let* τ *be a natural and homogeneous translation in **FOLeq**, then the following properties hold:*

i) *for each* $\Sigma \in \mathbf{Sign}$, $\tau_\Sigma(\mathbf{Sen_{FOL}}(\Sigma)) \subseteq \mathbf{Sen_{FOL}}(\Sigma^*)^2$ *(hence* τ *induces a natural translation* $\tau^{\mathbf{FOL}}$ *in **FOL** which is correct when* τ *is correct),*

ii) *for each* $\Sigma \in \mathbf{Sign}$, *the following diagram is commutative:*

$$
\begin{array}{ccc}
\mathbf{Sen_{FOL}}(\Sigma) \lhook\joinrel\longrightarrow \mathbf{Sen}(\Sigma) \xrightarrow{\;\cong\;} \mathbf{Sen_{FOL}}(\overline{\Sigma}) \\
\Big\downarrow{\scriptstyle \tau_\Sigma^{\mathbf{FOL}}} \qquad\qquad \Big\downarrow{\scriptstyle \tau_\Sigma} \qquad\qquad\qquad \Big\downarrow{\scriptstyle \tau_{\overline{\Sigma}}^{\mathbf{FOL}}} \\
\mathbf{Sen_{FOL}}(\Sigma^*)^2 \lhook\joinrel\longrightarrow \mathbf{Sen}(\Sigma^*)^2 \xrightarrow{\;\cong\;} \mathbf{Sen_{FOL}}(\overline{\Sigma}^*)^2
\end{array}
$$

Proof. See the Appendix.   □

Theorem 3 (Optimality in FOLeq). *Let τ be a correct, natural and homogeneous translation and let $\Sigma \in$ **Sign** and $\mathcal{A} \in$ **Str**(Σ^*) be such that, for each predicate symbol $P : s_1 \times \ldots \times s_n$ in Σ, we have $P_\oplus^\mathcal{A} \subseteq P_\ominus^\mathcal{A}$, and such that $=_\oplus^\mathcal{A}$ is a subset of the natural equality on $\mathcal{A}$. Then:*

$$\mathcal{A} \models \tau_\Sigma^1(\phi) \Rightarrow \mathcal{A} \models \phi_\oplus \quad \text{and} \quad \mathcal{A} \models \phi_\ominus \Rightarrow \mathcal{A} \models \tau_\Sigma^2(\phi).$$

Proof. See the Appendix.   □

Combining this result and Proposition 3, we obtain the following result about optimality of algebraic abstractions.

Corollary 1. *Assume given a correct, natural and homogeneous translation τ, a signature Σ, a Σ-structure $\mathcal{C}$, a Σ^*-structure $\mathcal{A}$ correct with respect to $\mathcal{C}$ and an epimorphism $\alpha : [\mathcal{C}] \to [\mathcal{A}]$. Then for any $\phi \in$ **Sen**(Σ) we have:*

$$\mathcal{A} \models \tau_\Sigma^1(\phi) \Rightarrow \mathcal{C}^\alpha \models \phi_\oplus$$
$$\text{and } \mathcal{C}^\alpha \models \phi_\ominus \Rightarrow \mathcal{A} \models \tau_\Sigma^2(\phi).$$

Proof (of Corollary 1). Since $\mathcal{C} \sqsubset_\alpha \mathcal{A}$, we can deduce $\mathcal{A} \models \phi_\oplus$ from $\mathcal{A} \models \tau_\Sigma^1(\phi)$ using Theorem 3. Then, applying Proposition 3, we obtain $\mathcal{C}^\alpha \models \phi_\oplus$. The proof is the same for the second implication.   □

5   Algebraic Abstractions at Work

According to the previous section, the main task to design an abstraction for a verification problem $\mathcal{C} \models \phi$ is to abstract $\mathcal{C}$. For this, as already explained, the non trivial part is to find suitable interpretations for the predicate symbols $P_\oplus$ and $P_\ominus$. In this section, we consider specific important cases for which we provide relevant guidelines.

The first important case is when the predicate to be abstracted is specified by a formula not involving this predicate. Abstracting it is in that case pretty obvious.

Proposition 4 (Predicate defined by a formula). *Assume given a structure $\mathcal{C}$ defined on a signature Σ of the form $\Sigma = \Sigma' \cup \{$**pred** $P : s_1 \times \ldots \times s_n\}$, such that $\mathcal{C} \models Px_1 \ldots x_n \Leftrightarrow \phi$, with $\phi \in$ **Sen**(Σ'). Moreover, assume the abstraction problem solved for $\mathcal{C}' = \mathcal{C}|_{\Sigma'}$ (i.e., assume given a Σ'^*-structure $\mathcal{A}'$ and an epimorphism $\alpha : [\mathcal{C}'] \to [\mathcal{A}']$ such that $\mathcal{A}'$ is correct with respect to $\mathcal{A}$ and α). Then any Σ-structure $\mathcal{A}$ satisfying:*

i) $\mathcal{A}|_{\Sigma'^*} = \mathcal{A}'$,
ii) $\mathcal{A} \models P_\oplus x_1 \ldots x_n \Leftrightarrow \phi_\oplus$,
iii) $\mathcal{A} \models P_\ominus x_1 \ldots x_n \Leftrightarrow \phi_\ominus$.

is correct with respect to $\mathcal{C}$ and α.

Proof. Follows directly from Theorem 1.   □

Example 4. In our cryptographic protocol models, the transition relation is defined by an axiom of the form:

$$q \xrightarrow{(l,m)} q' \Leftrightarrow r$$

and consequently we can use $r_\oplus$ and $r_\ominus$ in order to define $\xrightarrow{(.)}_\oplus$ and $\xrightarrow{(.)}_\ominus$ (disregarding how the rest of the abstraction is defined).

Remark 5. The above result will prove useful for structures defined by a specification (Σ, Φ) which can be stratified as follows:

- the set of predicate symbols of Σ is a disjoint union $\Pi \cup \{P_1, \ldots, P_n\}$,
- Φ is a set of Σ-sentences of the form:

$$\Phi = \Phi_0 \cup \Phi_1 \cup \ldots \cup \Phi_n$$

where, for $i \in \{0, \ldots, n\}$, Φ_i contains no occurrence of a predicate symbol P_j for $j \geq i$ and for $i \geq 1$, Φ_i is of the form:

$$\Phi_i = \{P_i x_1 \ldots x_k \Leftrightarrow \phi_i\}.$$

The second special case is when the predicate to be abstracted is defined inductively by some set of rules.

Proposition 5 (Predicate defined by a set of rules). *Assume given a structure $\mathcal{C}$ defined on a signature Σ of the form $\Sigma = \Sigma' \cup \{\textbf{pred } P : s_1 \times \ldots \times s_n\}$, such that the interpretation of P in $\mathcal{C}$ is the least predicate satisfying a set $\mathcal{R}$ of rules of the form:*

$$R = \frac{Pt_1^1 \ldots t_n^1 \quad \ldots \quad Pt_1^m \ldots t_n^m \quad \phi}{Pt_1^0 \ldots t_n^0}$$

with $\phi \in \textbf{Sen}(\Sigma)$, $t_1^0, \ldots, t_1^m \in T_\Sigma(X)_{s_1}$ and $t_n^0, \ldots, t_n^m \in T_\Sigma(X)_{s_n}$. Moreover, assume the abstraction problem solved for $\mathcal{C}' = \mathcal{C}|_{\Sigma'}$ (i.e., assume given a Σ'^*-structure $\mathcal{A}'$ and an epimorphism $\alpha : [\mathcal{C}'] \to [\mathcal{A}']$ such that $\mathcal{A}'$ is correct with respect to $\mathcal{A}$ and α). Then any Σ-structure $\mathcal{A}$ satisfying:

i) $\mathcal{A}|_{\Sigma'^*} = \mathcal{A}'$,
ii) for all $c_1 \in \mathcal{C}_{s_1}, \ldots, c_n \in \mathcal{C}_{s_n}$, $P_\oplus^\mathcal{A}(\alpha(c_1), \ldots, \alpha(c_n)) \Rightarrow P^\mathcal{C}(c_1, \ldots, c_n)$,
iii) $P_\ominus$ is the least predicate on $\mathcal{A}$ satisfying each rule in $\mathcal{R}_\ominus$, obtained from $\mathcal{R}$ by replacing each rule R by:

$$R_\ominus \stackrel{\text{def}}{=} \frac{P_\ominus t_1^1 \ldots t_n^1 \quad \ldots \quad P_\ominus t_1^m \ldots t_n^m \quad \phi_\ominus}{P_\ominus t_1^0 \ldots t_n^0}$$

is correct with respect to $\mathcal{C}$ and α.

Proof. See the Appendix. □

Example 5. For cryptographic protocols, the capacities of the intruder are modeled by a predicate $\Vdash$ defined by a set of rules. So $\Vdash_\ominus$ can be defined by the following rules:

$$\frac{m \in_\ominus s}{s \Vdash_\ominus m} \quad (\xi)$$

$$\frac{s \Vdash_\ominus m_1 \qquad s \Vdash_\ominus m_2}{s \Vdash_\ominus \langle m_1, m_2 \rangle} \quad (\pi)$$

$$\frac{s \Vdash_\ominus \langle m_1, m_2 \rangle}{s \Vdash_\ominus m_1} \quad (\pi'_1) \qquad\qquad \frac{s \Vdash_\ominus \langle m_1, m_2 \rangle}{s \Vdash_\ominus m_2} \quad (\pi'_2)$$

$$\frac{s \Vdash_\ominus m \qquad s \Vdash_\ominus k}{s \Vdash_\ominus \{m\}_k} \quad (\gamma) \qquad\qquad \frac{s \Vdash_\ominus \{m\}_k \qquad s \Vdash_\ominus k^{-1}}{s \Vdash_\ominus m} \quad (\gamma')$$

Example 6 (An abstraction for the study of cryptographic protocols). Let $\Sigma = (S, \Omega, \Pi)$ denote the signature of the specification of our toy cryptographic protocol given in Example 1 and let $\mathcal{C}$ be a model of this specification. We are going to describe a possible abstraction, wich will be useful to prove security properties about $\mathcal{C}$. The abstract structure $\mathcal{A}$ and the epimorphism $\alpha : [\mathcal{C}] \to [\mathcal{A}]$ will be described by the mean of the carrier sets of $\mathcal{A}$ together with a function $\alpha_s : \mathcal{C}_s \, to \mathcal{A}_s$ for each $s \in S$ (recall that $S = \{Princ, Key, Msg, Set[Msg], Lbl, State\}$):

– $\mathcal{A}_{Princ}$ is a set containing two values, S and A, and α_{Princ} is defined by:

$$\begin{aligned} \alpha_{Princ} : \mathcal{C}_{Princ} &\to \mathcal{A}_{Princ} \\ S &\mapsto S \\ x &\mapsto A \text{ if } x \neq S \end{aligned}$$

– $\mathcal{A}_{Key}$ is a set containing four values, $k(S)$, $k^{-1}(S)$, $k(A)$ and $k^{-1}(A)$, and α_{Key} is defined by:

$$\begin{aligned} \alpha_{Key} : \mathcal{C}_{Key} &\to \mathcal{A}_{Key} \\ k(S) &\mapsto k(S) \\ k^{-1}(S) &\mapsto k^{-1}(S) \\ k(x) &\mapsto k(A) \text{ if } x \neq S \\ k^{-1}(x) &\mapsto k^{-1}(A) \text{ if } x \neq S \end{aligned}$$

– from an idea of D. Le Métayer, an element of $\mathcal{A}_{Msg}$ is a non-empty set of "rules" r of the form:

$$r = k_1, \ldots, k_n \vdash d$$

where $k_1, \ldots, k_n \in \mathcal{A}_{Key}$ and $d \in \mathcal{A}_{Princ} \cup \mathcal{A}_{Key}$. For each $m \in \mathcal{C}_{Msg}$, $\alpha_{Msg}(m)$ is defined inductively on the structure of m:
   • if $m = d$ is atomic, $\alpha_{Msg}(m) = \{\vdash \alpha(d)\}$,

- if $m = \langle m_1, m_2 \rangle$, $\alpha_{Msg}(m) = \alpha_{Msg}(m_1) \cup \alpha_{Msg}(m_2)$,
- if $m = \{m'\}_k$, $\alpha_{Msg}(m) = \{\alpha_{Key}(k).r \mid r \in \alpha_{Msg}(m')\}$ where, for each $r = k_1, \ldots, k_n \vdash d \in \alpha_{Msg}(m')$, $\alpha_{Key}(k).r$ denotes the rule $\alpha_{Key}(k), k_1, \ldots, k_n \vdash d$

For example:

$$\alpha_{Msg}(\langle \{d_0\}_k, \{\{d_1\}_{k_2}\}_{k_1} \rangle) = \{k \vdash d_0; \ k_1, k_2 \vdash d_1\}$$

Formally, this part of the abstract structure is described by the following specification:

spec MESSAGE' =
        PRINCIPAL' **and** KEY'
        and SET[**sort** Key] **and** NON-EMPTY-SET[**sort** $Rule$]
then **free types** $Atom ::= $ **sort** $Key \mid$ **sort** $Princ$;
                        $Rule \ ::= \ \_\_ \vdash \_\_(Set[Key]; \ Atom)$;
                        $Msg \ ::= \ Non - Empty - Set[Rule]$
        ops $\langle \_\_, \_\_ \rangle : Msg \times Msg \to Msg$;
                $\{\_\_\}\_\_ : Msg \times Key \to Msg$;
then **%def**
        $\forall m, m' : Msg; \ k : Key; \ s : Set[Key]; \ d : Atom$
        - $\langle m, m' \rangle \ = \ m \cup m'$
        - $\{\{s \vdash d\}\}_k \ = \ \{\{k\} \cup s \vdash d\}$
        - $\{m \cup \{s \vdash d\}\}_k \ = \ \{m\}_k \cup \{\{k\} \cup s \vdash d\}$
end

- $\mathcal{A}_{Set[Msg]}$ is the powerset of $\mathcal{A}_{Msg}$ and $\alpha_{Set[Msg]}$ is the map induced by α_{Msg}:

$$\alpha_{Set[Msg]} : \mathcal{C}_{Set[Msg]} \quad \to \mathcal{A}_{Set[Msg]}$$
$$\{m_1, \ldots, m_n\} \mapsto \{\alpha_{Msg}(m_1), \ldots, \alpha_{Msg}(m_n)\}$$

- $\mathcal{A}_{Lbl} = \mathcal{C}_{Lbl}$ and α_{Lbl} is the identity map,
- $\mathcal{A}_{State} = \mathcal{A}_{Set[Msg]} \times \mathcal{A}_{Princ} \times \mathcal{A}_{Set[Msg]}$ (recall that $\mathcal{C}_{State} = \mathcal{C}_{Set[Msg]} \times \mathcal{C}_{Princ} \times \mathcal{C}_{Set[Msg]}$) and α_{State} is defined by:

$$\alpha_{State} : \mathcal{C}_{State} \quad \to \mathcal{A}_{State}$$
$$(dir, id, I) \mapsto (\alpha_{Set[Msg]}(dir), \alpha_{Princ}(id), \alpha_{Set[Msg]}(I))$$

Now we would have to define interpretations for the function symbols of Σ in order to turn the family $(\mathcal{A}_s)_{s \in S}$ into a $[\Sigma]$-algebra denoted $[\mathcal{A}]$, but we are going to skip this step since it is pretty obvious. So, we just have to give the interpretations of the predicate symbols of Σ^*, which are:

$$
\begin{aligned}
&\text{\_\_}=_\oplus\text{\_\_} &&: s \times s \text{ (for each } s \in S), \\
&\text{\_\_}\in_\oplus\text{\_\_} &&: Msg \times Set[Msg], \\
&\text{\_\_}\in_\ominus\text{\_\_} &&: Msg \times Set[Msg], \\
&\text{\_\_} \Vdash_\oplus \text{\_\_} &&: Set[Msg] \times Msg, \\
&\text{\_\_} \Vdash_\ominus \text{\_\_} &&: Set[Msg] \times Msg, \\
&\text{\_\_}\xrightarrow{(-,-)}_\oplus\text{\_\_} &&: State \times Lbl \times Msg \times State, \\
&\text{\_\_}\xrightarrow{(-,-)}_\ominus\text{\_\_} &&: State \times Lbl \times Msg \times State, \\
&Reach_\oplus &&: State \times State, \\
&Reach_\ominus &&: State \times State
\end{aligned}
$$

in order to obtain a Σ^*-structure correct with respect to $\mathcal{C}$ and α.

- $=_\oplus$ is defined on $\mathcal{A}_{Princ}$ and $\mathcal{A}_{key}$ using the following axioms:

$$
\forall x, y : Princ.\ x =_\oplus y \Leftrightarrow x = y = S
$$
$$
\forall x, y : Key.\ x =_\oplus y \Leftrightarrow x = y = k(S) \vee x = y = k^{-1}(S)
$$

- concerning the sort Msg, for $m, m' \in \mathcal{A}_{Msg}$, $m =_\oplus m'$ is true iff m and m' are equal to the same atomic data $d \in \mathcal{A}_{Princ} \cup \mathcal{A}_{Key}$ and, as an atomic data, $d =_\oplus d$,
- for two sets of messages $s, s' \in \mathcal{A}_{Set[Msg]}$, $s =_\oplus s'$ is true iff s and s' are equal to the same set $\{m_1, \ldots, m_n\}$ with $m_1 =_\oplus m_1, \ldots, m_n =_\oplus m_n$,
- concerning the sorts Lbl and $State$, $=_\oplus$ is defined on $\mathcal{A}_{Lbl}$ and $\mathcal{A}_{State}$ using the following axioms:

$$
\forall x, y : Lbl.\ x =_\oplus y \Leftrightarrow x = y,
$$
$$
\forall x, y : State.\ x =_\oplus y \Leftrightarrow x.Adir =_\oplus y.Adir \wedge x.Sid =_\oplus y.Sid \wedge x.I =_\oplus y.I
$$

- $\in_\ominus$ is the usual membership predicate for sets of messages whereas $\in_\oplus$ is defined by:

$$
\forall m : Msg. \forall s : Set[Msg].\ m \in_\oplus s \Leftrightarrow m \in_\ominus s \wedge m =_\oplus m
$$

- $\Vdash_\ominus$ is defined by the same rules as $\Vdash$ (see Example 5) whereas $\Vdash_\oplus$ is defined using the rule:

$$
\frac{k_1, \ldots, k_n \vdash d \in_\oplus s \quad s \Vdash_\oplus k_1 \quad \ldots \quad s \Vdash_\oplus k_n}{s \Vdash_\oplus d}
$$

(this definition is the only one, among all the others, that needs a non trivial proof of correctness),

- $\xrightarrow{(,)}_\oplus$ and $\xrightarrow{(,)}_\ominus$ are defined as in Example 4,
- using Proposition 5, $Reach_\ominus$ is defined by the two rules:

$$
\frac{}{Reach_\ominus(q, q)} \qquad \frac{Reach_\ominus(q, q') \quad q' \xrightarrow{(l,m)}_\ominus q''}{Reach_\ominus(q, q'')}
$$

whereas the following definition of $Reach_\oplus$ can be proved to be correct:

$$\frac{q =_\oplus q}{Reach_\oplus(q,q)} \qquad \frac{Reach_\oplus(q,q') \quad q' \xrightarrow{(l,m)}_\oplus q''}{Reach_\oplus(q,q'')}$$

Finally, we obtain a Σ^*-structure $\mathcal{A}$ which is correct with respect to $\mathcal{C}$ and α and finite. So, if we want to prove a property of the protocol, for instance if we want to check that:

$$\mathcal{C} \models \forall q_0, q : State \bullet \ \neg(q_0.I \Vdash k^{-1}(S)) \wedge Reach(q_0, q) \Rightarrow \neg(q \Vdash k^{-1}(S))$$

we just have to show that:

$$\mathcal{A} \models \forall q_0, q : State \bullet \ \neg(q_0.I \Vdash_\oplus k^{-1}(S)) \wedge Reach_\ominus(q_0, q) \Rightarrow \neg(q \Vdash_\ominus k^{-1}(S))$$

and this can be done using model-checking techniques (recall that $\mathcal{A}$ is finite).

6   Conclusion

In this paper, we have provided a general framework for defining correct abstractions for verification problems of the form $\mathcal{C} \models \phi$, where $\mathcal{C}$ is a first-order structure and ϕ is a first-order formula (both w.r.t. a given signature Σ). We have proven that our algebraic abstractions are in some sense optimal, and moreover we have provided useful guidelines to design abstractions in practice, in particular for first-order structures modeling cryptographic protocols.

We now intend to improve our guidelines in order to automatize as much as possible the abstraction step and to build libraries of abstractions suited for some classes of cryptographic protocols.

Proofs

Proof (of Proposition 2). Concerning $=_\oplus$, for $s \in S$ and $a_1, a_2 \in \mathcal{A}_s$, if we have $a_1 =_\oplus^{\mathcal{A}} a_2$, then for any $c_1 \in \alpha^{-1}(a_1)$ and $c_2 \in \alpha^{-1}(a_2)$, we have $c_1 = c_2$, since $a_1 = a_2$ and $\alpha^{-1}(a_1) = \alpha^{-1}(a_2)$ contains exactly one element. For a predicate $P : s_1 \times \ldots \times s_n$ in Σ and $a_1 \in \mathcal{A}_{s_1}, \ldots, a_n \in \mathcal{A}_{s_n}$, we have:

$$P_\oplus^{\mathcal{A}}(a_1, \ldots, a_n) \Rightarrow P^{\mathcal{C}}(c_1, \ldots, c_n)$$

for any choice of $c_1 \in \alpha^{-1}(a_1), \ldots, c_n \in \alpha^{-1}(a_n)$, and then by definition of $P_\oplus^{\mathcal{C}^\alpha}$:

$$P_\oplus^{\mathcal{A}}(a_1, \ldots, a_n) \Rightarrow P_\oplus^{\mathcal{C}^\alpha}(a_1, \ldots, a_n).$$

Similarly, if we have:

$$P_\ominus^{\mathcal{C}^\alpha}(a_1, \ldots, a_n)$$

then there is a choice of $c_1 \in \alpha^{-1}(a_1), \ldots, c_n \in \alpha^{-1}(a_n)$ such that $P^{\mathcal{C}}(c_1, \ldots, c_n)$ which implies $P_\ominus^{\mathcal{A}}(a_1, \ldots, a_n)$.   □

Proof (of Theorem 1). First, we prove that for any $\phi \in \mathbf{Sen}(\Sigma)$ and for any valuation $\nu : X \to \mathcal{C}$:

$$\mathcal{A}, \alpha\nu \models \phi_\oplus \Rightarrow \mathcal{C}, \nu \models \phi \quad \text{and} \quad \mathcal{C}, \nu \models \phi \Rightarrow \mathcal{A}, \alpha\nu \models \phi_\ominus$$

by induction on the structure of ϕ. Wlog, we can consider only formulas containing no occurrence of $\vee$ nor of $\exists$.

- if ϕ is of the form $t_1 = t_2$ and if $\mathcal{A}, \alpha\nu \models \phi$, then $\alpha\nu(t_1) =_\oplus^\mathcal{A} \alpha\nu(t_2)$ and $\nu(t_1) = \nu(t_2)$. Now if $\mathcal{C}, \nu \models \phi$ then we have $\alpha\nu(t_1) = \alpha\nu(t_2)$ since $\nu(t_1) = \nu(t_2)$ and α is an homomorphism,
- if ϕ is of the form $P t_1 \ldots t_n$, and if $\mathcal{A}, \alpha\nu \models \phi_\oplus$, then:

$$P_\oplus^\mathcal{A}(\alpha(\nu(t_1)), \ldots, \alpha(\nu(t_n)))$$

hence $P^\mathcal{C}(\nu(t_1), \ldots, \nu(t_n))$ (because $\mathcal{C} \sqsubseteq_\alpha \mathcal{A}$) and $\mathcal{C}, \nu \models \phi$. Now if we suppose that $\mathcal{C}, \nu \models \phi$, then:

$$P^\mathcal{C}(\nu(t_1), \ldots, \nu(t_n))$$

and, because $\mathcal{C} \sqsubseteq_\alpha \mathcal{A}$, $P_\ominus^\mathcal{A}(\alpha(\nu(t_1)), \ldots, \alpha(\nu(t_n)))$ i.e., $\mathcal{A}, \alpha\nu \models \phi_\ominus$,

- if $\phi = \phi_1 \wedge \phi_2$ and $\mathcal{A}, \alpha\nu \models \phi_\oplus$ then $\mathcal{A}, \alpha\nu \models \phi_{1\oplus}$ and $\mathcal{A}, \alpha\nu \models \phi_{2\oplus}$ and by induction hypothesis $\mathcal{C}, \nu \models \phi_1$ and $\mathcal{C}, \nu \models \phi_2$, so $\mathcal{C}, \nu \models \phi$. Now if we suppose that $\mathcal{C}, \nu \models \phi$ then $\mathcal{C}, \nu \models \phi_1$ and $\mathcal{C}, \nu \models \phi_2$ and by induction hypothesis $\mathcal{A}, \alpha\nu \models \phi_{1\ominus}$ and $\mathcal{A}, \alpha\nu \models \phi_{2\ominus}$ and so $\mathcal{A}, \alpha\nu \models \phi_\ominus$,
- if $\phi = \neg\phi_1$ and $\mathcal{A}, \alpha\nu \models \phi_\oplus$ then by definition of $\neg\phi_{1\oplus}$ we have $\mathcal{A}, \alpha\nu \not\models \phi_{1\ominus}$. Then by induction hypothesis $\mathcal{C}, \alpha\nu \not\models \phi_1$ and $\mathcal{C}, \alpha\nu \models \phi$. Now if we suppose that $\mathcal{C}, \nu \models \phi$, then $\mathcal{C}, \nu \not\models \phi_1$ and by induction hypothesis $\mathcal{A}, \alpha\nu \not\models \phi_{1\oplus}$, so $\mathcal{A}, \alpha\nu \models \phi_\ominus$,
- if $\phi = \forall x : s. \phi_1$ and $\mathcal{A}, \alpha\nu \models \phi_\oplus$ then given any $c \in \mathcal{C}_s$ we have by induction hypothesis:

$$\mathcal{C}, \nu[c/x] \models \phi_1$$

and then $\mathcal{C}, \nu \models \phi$. Now if we suppose that $\mathcal{C}, \nu \models \phi$ and if $a \in A_s$ then there exists $c \in \mathcal{C}_s$ such that $\alpha(c) = a$ (because α is an epimorphism) and $\mathcal{A}, (\alpha\nu)[a/x] \models \phi_1$ is a consequence of $\mathcal{C}, \nu[c/x] \models \phi_1$ using the induction hypothesis, so $\mathcal{A}, \alpha\nu \models \phi_\ominus$.

Now, if $\mathcal{A} \models \phi_\oplus$ and if $\nu : X \to \mathcal{C}$ is a valuation, then $\mathcal{A}, \alpha\nu \models \phi_\oplus$ and we can conclude that $\mathcal{C}, \nu \models \phi$, so $\mathcal{C} \models \phi$. When $\mathcal{C} \models \phi$ and $\nu' : X \to \mathcal{A}$ is a valuation, there is a valuation ν such that $\nu' = \alpha\nu$ (because α is an epimorphism) and then $\mathcal{C}, \nu \models \phi$ so $\mathcal{A}, \nu' \models \phi_\ominus$ and $\mathcal{A} \models \phi_\ominus$. $\qquad\square$

Proof (of Lemma 1). Let $\mathcal{A}^\bullet$ denote $\mathcal{A}|_{|.|_\Sigma^\bullet}$. We recall that each predicate symbol P in Σ gives two predicate symbols $P_\oplus$ and $P_\ominus$ in $\Sigma^\bullet$ and these predicates give four predicate symbols $(P_\oplus)_\oplus$, $(P_\oplus)_\ominus$, $(P_\ominus)_\oplus$, and $(P_\ominus)_\ominus$ in $\Sigma^{\bullet\bullet}$. So, we just have to prove that:

$$(P_\oplus)_\oplus^{\mathcal{A}^\bullet} \subseteq P_\oplus^\mathcal{A} \subseteq (P_\oplus)_\ominus^{\mathcal{A}^\bullet} \quad (P_\ominus)_\oplus^{\mathcal{A}^\bullet} \subseteq P_\ominus^\mathcal{A} \subseteq (P_\ominus)_\ominus^{\mathcal{A}^\bullet}.$$

Using the definition of $|.|_\Sigma$ and $|.|_\Sigma^\bullet$ we have:

$$
\begin{aligned}
|(P_\oplus)_\oplus|_\Sigma^\bullet &= |P_\oplus|_{\Sigma_\oplus} = P_\oplus, \\
|(P_\oplus)_\ominus|_\Sigma^\bullet &= |P_\oplus|_{\Sigma_\ominus} = P_\ominus, \\
|(P_\ominus)_\oplus|_\Sigma^\bullet &= |P_\ominus|_{\Sigma_\oplus} = P_\oplus, \\
|(P_\ominus)_\ominus|_\Sigma^\bullet &= |P_\ominus|_{\Sigma_\ominus} = P_\ominus.
\end{aligned}
$$

Then, using the assumption $P_\oplus^\mathcal{A} \subseteq P_\ominus^\mathcal{A}$:

$$
\begin{aligned}
(P_\oplus)_\oplus^{\mathcal{A}^\bullet} &= P_\oplus^\mathcal{A} \subseteq P_\ominus^\mathcal{A} = (P_\oplus)_\ominus^{\mathcal{A}^\bullet}, \\
(P_\ominus)_\oplus^{\mathcal{A}^\bullet} &= P_\oplus^\mathcal{A} \subseteq P_\ominus^\mathcal{A} = (P_\ominus)_\ominus^{\mathcal{A}^\bullet}.
\end{aligned}
$$

$\square$

Proof (of Theorem 2). Let $|.|$ denote the canonical homomorphism from $\Sigma^\bullet$ to Σ and $|.|^\bullet$ the associated homomorphism from $\Sigma^{\bullet\bullet}$ to $\Sigma^\bullet$. Then:

$$
\begin{aligned}
\mathcal{A} \models \tau_\Sigma^1(\phi) &\Leftrightarrow \mathcal{A} \models \tau_\Sigma^1(|\phi_\oplus|) && \text{because } \phi = |\phi_\oplus| \\
&\Leftrightarrow \mathcal{A} \models |\tau_{\Sigma^\bullet}^1(\phi_\oplus)|^\bullet && \text{naturality} \\
&\Leftrightarrow \mathcal{A}|_{|.|^\bullet} \models \tau_{\Sigma^\bullet}^1(\phi_\oplus) \\
&\Rightarrow \mathcal{A} \models \phi_\oplus && \text{correctness, using the above lemma.}
\end{aligned}
$$

And for the second implication:

$$
\begin{aligned}
\mathcal{A} \models \phi_\ominus &\Rightarrow \mathcal{A}|_{|.|^\bullet} \models \tau_{\Sigma^\bullet}^2(\phi_\ominus) && \text{correctness} \\
&\Rightarrow \mathcal{A} \models |\tau_{\Sigma^\bullet}^2(\phi_\ominus)|^\bullet \\
&\Rightarrow \mathcal{A} \models \tau_\Sigma^2(|\phi_\ominus|) && \text{naturality} \\
&\Rightarrow \mathcal{A} \models \tau_\Sigma^2(\phi) && \text{because } |\phi_\ominus| = \phi.
\end{aligned}
$$

Proof (of Lemma 2). For i), it is sufficient to prove that if $\phi \in \mathbf{Sen}(\Sigma)$ is such that ϕ contains no occurrence of a predicate eq_Σ, then $\overline{\tau_\Sigma(\phi)}$ contains no occurrence of either $eq_{\Sigma^\ast}$ or $eq_{\Sigma_\oplus}$. If ι denotes the signature inclusion of Σ into $\overline\Sigma$ we have by naturality of τ:

$$
\overline{\tau_\Sigma(\phi)} = \tau_{\overline\Sigma}(\overline\phi) = \tau_{\overline\Sigma}(\iota(\phi)) = \iota^\ast(\tau_\Sigma(\phi))
$$

and $\iota^\ast(\tau_\Sigma(\phi))$ can contain no occurrence of $eq_{\Sigma^\ast}$ nor of $eq_{\Sigma_\oplus}$ since they are not in $\iota(\Sigma)$. ii) directly results from the definition of $\tau^{\mathbf{FOL}}$.

Proof (of Theorem 3). For the first implication, we have:

$$
\begin{aligned}
\mathcal{A} \models \tau_\Sigma(\phi)_\oplus &\Leftrightarrow \overline{\mathcal{A}} \models \overline{\tau_\Sigma(\phi)_\oplus} \\
&\Leftrightarrow \overline{\mathcal{A}} \models \tau_{\overline\Sigma}^{\mathbf{FOL}}(\overline\phi)_\oplus && \text{by homogeneity} \\
&\Rightarrow \overline{\mathcal{A}} \models \overline\phi_\oplus && \text{applying optimality in } \mathbf{FOL} \\
&\Rightarrow \mathcal{A} \models \phi_\oplus
\end{aligned}
$$

and for the second one:

$$
\begin{aligned}
\mathcal{A} \models \phi_\ominus &\Leftrightarrow \overline{\mathcal{A}} \models \overline{\phi_\ominus} \\
&\Leftrightarrow \overline{\mathcal{A}} \models \overline\phi_\ominus \\
&\Rightarrow \overline{\mathcal{A}} \models \tau_{\overline\Sigma}^{\mathbf{FOL}}(\overline\phi)_\ominus && \text{applying optimality in } \mathbf{FOL} \\
&\Rightarrow \overline{\mathcal{A}} \models \overline{\tau_\Sigma(\phi)_\ominus} && \text{using homogeneity} \\
&\Rightarrow \mathcal{A} \models \tau_\Sigma(\phi)_\ominus
\end{aligned}
$$

$\square$

Proof (of Proposition 5). We prove that for every $c_1 \in \mathcal{C'}_{s_1}, \ldots, c_n \in \mathcal{C'}_{s_n}$ we have:

$$P^{\mathcal{C'}}(c_1, \ldots, c_n) \Rightarrow P_{\ominus}^{\mathcal{C'}}(\alpha(c_1), \ldots, \alpha(c_n))$$

by induction on the derivation of $P^{\mathcal{C'}}(c_1, \ldots, c_n)$. So assume there is a rule:

$$R = \frac{Pt_1^1 \ldots t_n^1 \quad \cdots \quad Pt_1^m \ldots t_n^m \quad \phi}{Pt_1^0 \ldots t_n^0} \quad \in \mathcal{R}$$

and a valuation $\nu : X \to \mathcal{C'}$ such that:

$$P^{\mathcal{C'}}(\nu(t_1^1), \ldots, \nu(t_n^1))$$
$$\vdots$$
$$P^{\mathcal{C'}}(\nu(t_1^m), \ldots, \nu(t_n^m))$$
$$\mathcal{C}, \nu \models \phi$$

and by induction hypothesis:

$$P_{\ominus}^{\mathcal{A'}}(\alpha\nu(t_1^1), \ldots, \alpha\nu(t_n^1))$$
$$\vdots$$
$$P_{\ominus}^{\mathcal{A'}}(\alpha\nu(t_1^m), \ldots, \alpha\nu(t_n^m))$$

then by Theorem 1, $\mathcal{A}, \alpha\nu \models \phi_{\ominus}$, so we can apply the rule $R_{\ominus} \in \mathcal{R}_{\ominus}$ and obtain $P_{\ominus}^{\mathcal{A'}}(\alpha\nu(t_1^0), \ldots, \alpha\nu(t_n^0))$. □

References

1. E. Astesiano, M. Broy, and G. Reggio. Algebraic specification of concurrent systems. chapter 13 in [2], pages 467–520, 1999.
2. E. Astesiano, H.-J. Kreowski, and B. Krieg-Brückner, editors. *Algebraic Foundations of Systems Specification.* Springer, 1999.
3. B. Bérard, M. Bidoit, A. Finkel, F. Laroussinie, A. Petit, L. Petrucci and P. Schnoebelen. *Systems and Software Verification.* Springer, July 2001.
4. D. Bolignano. Towards a mechanization of cryptographic protocol verification. In *Proc. 9th Int. Conf. Computer Aided Verification (CAV'97), Haifa, Israel, June 1997*, volume 1254 of *Lecture Notes in Computer Science*, pages 131–142. Springer, 1997.
5. D. Bolignano. Using abstract interpretation for the safe verification of security protocols. In *MFPS XV Mathematical Foundations of Programming Semantics, Fifteenth Conference, Tulane University, New Orleans, LA, April 28-May 1, 1999*, volume 20 of *Electronic Notes in Theor. Comp. Sci.* Elsevier Science, 1999.
6. E. M. Clarke, O. Grumberg, and D. E. Long. Model checking and abstraction. In *Proceedings of the Nineteenth Annual ACM Symposium on Principles of Programming Languages (POPL'92), Albuquerque, New Mexico, January 19–22, 1992*, pages 343–354. ACM Press, 1992.
7. Cofi: The common framework initiative for algebraic specification and development. http://www.brics.dk/Projects/CoFI/.

8. P. Cousot and R. Cousot. Abstract interpretation and application to logic programs. *Journal of Logic Programming*, 13(2–3):103–179, 1992.
9. D. Dams, R. Gerth, and O. Grumberg. Abstract interpretation of reactive systems. In *ACM Transactions on Programming Languages ans Systems*, volume 19, 1997.
10. D. Dolev and A. C. Yao. On the security of public key protocols. *IEEE Transactions on Information Theory*, 29(2):198–208, 1983.
11. J. Goubault-Larrecq. A method for automatic cryptographic protocol verification (extended abstract). In *Proc. 15th IPDPS 2000 Workshops, Cancun, Mexico, May 2000*, volume 1800 of *Lecture Notes in Computer Science*, pages 977–984. Springer, 2000.
12. S. Merz. Rules for abstraction. In R. K. Shyamasundar and K. Ueda, editors, *Advances in Computing Science (ASIAN'97)*, volume 1345 of *LNCS*, pages 32–45, Berlin, Germany, 1997. Springer.
13. L. C. Paulson. The inductive approach to verifying cryptographic protocols. *J. Computer Security*, 6:85–128, 1998.
14. A. Tarlecki. Towards heterogeneous specifications. In *Proc. Workshop on Frontiers of Combining Systems FroCoS'98, Amsterdam, October 1998*, Applied Logic Series. Kluwer Academic Publishers, 1998. (To appear).

Combining Logics: Parchments Revisited

C. Caleiro, P. Mateus, J. Ramos, and A. Sernadas

CMA/CLC – Dep. de Matemática, IST
Av. Rovisco Pais, 1049–001 Lisboa, Portugal
{ccal|pmat|jabr|acs}@math.ist.utl.pt

Abstract. In the general context of the theory of institutions, several
notions of parchment and parchment morphism have been proposed as
the adequate setting for combining logics. However, so far, they seem to
lack one of the main advantages of the combination mechanism known
as fibring: general results of transference of important logical properties
from the logics being combined to the resulting fibred logic. Herein, in
order to bring fibring to the institutional setting, we propose to work with
the novel notion of **c**-*parchment*. We show how both free and constrained
fibring can be characterized as colimits of **c**-*parchments*, and illustrate
both the construction and its preservation capabilities by exploring the
idea of obtaining partial equational logic by fibring equational logic with
a suitable logic of partiality. Last but not least, in the restricted context
of propositional based, we state and prove a collection of meaningful
soundness and completeness preservation results for fibring, with respect
to Hilbert-like proof-calculi.

1 Introduction

Recently, the problem of combining logics has been deserving much attention.
The practical impact of a theory of logic combination is clear for anyone working in knowledge representation or in formal specification and verification. In the
fields of artificial intelligence and software engineering, the need for working with
several formalisms at the same time is widely recognized. Besides, combinations
of logics are also of great theoretical interest [4]. Among the different combination techniques, both *fibring* [12,13,22] and combinations of parchments [20,21]
deserve close attention. In fact, although the work on parchments has found its
way into practice, see for instance [19], it lacks a feature that we consider essential: transference results for relevant properties of logics, such as soundness and
completeness. For *fibring*, however, recent significant preservation results have
been obtained [29,7]. Our goal in this paper is to bring both *fibring* and these
transference results to the setting of institutions.

This leads us, first of all, to a revised notion of parchment. It shall be made
clear that the detail provided by early definitions [14,20,21] is not enough to
capture the finer structure of models. In particular, for a smooth characterization of *fibring*, we need a notion that promotes logical consequence as a whole,
rather than just validity. In previous work [22,29,7], a validity based consequence

M. Cerioli and G. Reggio (Eds.): WADT/CoFI 2001, LNCS 2267, pp. 48–70, 2002.

has also been considered and related to this more "internal" notion [7]. Herein, however, we shall not make explicit use of it. Still, the distinction is crucial to the full understanding of many logics, including first-order logic and modal logic, and plays an essential role in the process. So, we propose to work with **c**-*parchments*, that essentially extend the model-theoretic parchments of [21] by endowing the algebras of truth-values with more than just a set of designated values. Namely, we require the set of truth-values to be structured according to a Tarskian closure operation as in [7], thus recovering an early proposal of Smiley [25].

Besides showing how **c**-*parchments* can be seen as presentations of institutions, a suitable notion of *morphism* is also proposed and shown to present institution op-morphisms. The reason for this relationship to the dual of the category of institutions and institution morphisms is precisely our intention to follow the "old slogan" in its strict sense, and use colimits for combination. Therefore, building on the fact that **c**-*parchments* are essentially functors over a suitable category of **c**-*rooms*, we manage to characterize both *free* and *constrained fibring* as colimits of **c**-*parchments*. We illustrate *fibring* by providing a detailed construction of an equational logic dealing with partiality, by combining equational logic with a suitable logic of partiality. This example, when compared with the way partiality is dealt with using previous notions of parchment [20,21,19], is in fact paradigmatic of the modular power of *fibring*. Along with the fibred semantics of partial equational logic, we also show that by simultaneously combining Hilbert-like proof-calculi for the given logics, a sound and complete calculus for partial equational logic can also be obtained.

In fact, given that the right amount of *structurality* [6] is embodied in the deduction rules of proof-calculi, their *fibring* is well understood [7] and meaningful. As in previous treatments of this issue, we shall achieve this by using *schema variables* to write *schema rules* that can then be instantiated with arbitrary formulae while building deductions. In this context, although just for the particular case of propositional based logics, we then state and prove a collection of soundness and completeness transference results for *fibring*. Preservation of soundness is easily just a consequence of the construction underlying *fibring*, as shown in [7]. On the contrary, as should be expected, completeness preservation results are in general not so easy to obtain. The completeness transference results that we shall present are based on the fundamental notion of *fullness*, as a means of guaranteeing that we always have enough models, extending original ideas from [29], further worked out in [7]. We provide completeness proofs for several classes of interpretation structures, including partially-ordered ones, using standard techniques in logic and algebra, such as congruence and Lindenbaum-Tarski algebras. Rephrasing the main Theorem of [29], we also mention the case of powerset structures inspired by general models for modal logic (see for instance [16]) whose completeness proof uses a Henkin-style technique.

The rest of the paper is organized as follows. In Section 2 we introduce the novel notion of **c**-*parchment* and show how it relates to institutions. For the sake of illustration, we show how to represent two well known logics as **c**-parchments.

Section 3 is devoted to developing the categorial setting of c-parchments as indexed categories of c-*rooms* and to establishing the cocompleteness of the corresponding categories. In Section 4, we define *fibring* of c-rooms (and c-parchments) and provide its characterization as a categorial colimit. In order to bring the necessary insight on the construction, we then dwell on the interesting example of obtaining partial equational logic by fibring equational logic with a logic of partiality. Moreover, paving the way to the subsequent sections, we also show how a sound and complete calculus for the resulting logic can be obtained by putting together (*fibring*) the proof-calculus of equational logic with a suited calculus for the logic of partiality. Section 5 introduces the details of the appropriate notion of Hilbert-like *proof-calculi*, using schematic rules, and shows also that their *fibring* corresponds to a colimit in the appropriate category. The whole treatment is restricted to the particular case of propositional based logics, since that is sufficient for the forthcoming preservation results. Finally, in Section 6, we state and prove our soundness and completeness preservation results for fibring of propositional based logics. We conclude by discussing the limitations of the work presented, and how we expect to overcome them in future work.

2   The Notion of c-Parchment

We start by introducing some notation. In the sequel, $\mathbf{AlgSig}_\phi$ denotes the category of algebraic many-sorted signatures $\langle S, O \rangle$, where S is a set (of *sorts*) and $O = \{O_u\}_{u \in S^+}$ is a family of sets (of *operators*) indexed by their type, with a distinguished sort $\phi \in S$ (for formulae) and morphisms preserving it. Given such a signature $\langle S, O \rangle$, we denote by $\mathbf{Alg}(\langle S, O \rangle)$ the category of $\langle S, O \rangle$-algebras and homomorphisms, and by $\mathrm{cAlg}(\langle S, O \rangle)$ the class of all pairs $\langle \mathcal{A}, \mathbf{c} \rangle$ with $\mathcal{A} \in |\mathbf{Alg}(\langle S, O \rangle)|$ and $\mathbf{c}$ a closure operation on $|\mathcal{A}|_\phi$ (the carrier of sort ϕ, that we can see as the set of truth-values). Recall that $\mathbf{c} : \wp(|\mathcal{A}|_\phi) \to \wp(|\mathcal{A}|_\phi)$ is required to be extensive - $B \subseteq B^{\mathbf{c}}$, monotonous - $B^{\mathbf{c}} \subseteq (B \cup B')^{\mathbf{c}}$, and idempotent - $(B^{\mathbf{c}})^{\mathbf{c}} = B^{\mathbf{c}}$, for all $B, B' \subseteq |\mathcal{A}|_\phi$. We shall use $\mathcal{W}_{\langle S, O \rangle}$ to denote the free $\langle S, O \rangle$-algebra (the *word* algebra), and $[\![\_]\!]^{\mathcal{A}}$ (for word *interpretation*) to denote the unique $\mathbf{Alg}(\langle S, O \rangle)$-homomorphism from $\mathcal{W}_{\langle S, O \rangle}$ to any given $\langle S, O \rangle$-algebra $\mathcal{A}$. Also recall that every $\mathbf{AlgSig}_\phi$-morphism $h : \langle S_1, O_1 \rangle \to \langle S_2, O_2 \rangle$ has an associated reduct functor $\_|_h : \mathbf{Alg}(\langle S_2, O_2 \rangle) \to \mathbf{Alg}(\langle S_1, O_1 \rangle)$. As usual, we shall preferably write $\hat{h}$ (for word *translation*) instead of $[\![\_]\!]^{\mathcal{W}_{\langle S_2, O_2 \rangle}|_h}$ to denote the unique $\mathbf{Alg}(\langle S_1, O_1 \rangle)$-homomorphism from $\mathcal{W}_{\langle S_1, O_1 \rangle}$ to $\mathcal{W}_{\langle S_2, O_2 \rangle}|_h$.

Definition 1. A c-*parchment* is a tuple $P = \langle \mathbf{Sig}, L, \mathcal{M} \rangle$ where:

- $\mathbf{Sig}$ is a category (of *abstract signatures*);
- $L : \mathbf{Sig} \to \mathbf{AlgSig}_\phi$ is a functor (of *concrete syntax*);
- $\mathcal{M} = \{M_\Sigma\}_{\Sigma \in |\mathbf{Sig}|}$, with $M_\Sigma \subseteq \mathrm{cAlg}(L(\Sigma))$ (a class of *structured algebras*),

such that the following condition holds for every $\mathbf{Sig}$-morphism $\sigma : \Sigma_1 \to \Sigma_2$:

- $\langle \mathcal{A}|_{L(\sigma)}, \mathbf{c} \rangle \in M_{\Sigma_1}$ for every $\langle \mathcal{A}, \mathbf{c} \rangle \in M_{\Sigma_2}$.

Note that although this coherence condition is stricter than the one considered in [21], all the examples presented there indeed correspond to this particular case. For the present approach using **c**-parchments we believe that this strict version is the most appropriate. It is true, however, that the coherence condition of [21] plays an essential role in the context of combining model-theoretic parchments.

It is easy to see that a model-theoretic parchment can be extracted from a **c**-parchment. However, the idea here is to take advantage of the closure operation in each structured algebra $\langle \mathcal{A}, \mathbf{c} \rangle$ and to go beyond the obvious choice of the set $D = \emptyset^{\mathbf{C}} \subseteq |\mathcal{A}|_\phi$ of designated values. Instead, we shall allow the set of distinguished values to vary freely among all possible **c**-closed sets $T \subseteq |\mathcal{A}|_\phi$. Of course, given $\Sigma \in |\mathbf{Sig}|$, we can recognize $|\mathcal{W}_{L(\Sigma)}|_\phi$ (the carrier of sort ϕ in the word algebra) as the set $Sen(\Sigma)$ of formulae. We shall use φ, ψ with or without primes and subscripts to denote formulae, and Φ, Ψ to denote sets of formulae. Furthermore, we can set $Mod(\Sigma) = \{\langle \langle \mathcal{A}, \mathbf{c} \rangle, T \rangle : \langle \mathcal{A}, \mathbf{c} \rangle \in M_\Sigma, T^{\mathbf{C}} = T \subseteq |\mathcal{A}|_\phi\}$ to be the class of models, and define the *satisfaction* relation $\Vdash_\Sigma$ by:

– $\langle \langle \mathcal{A}, \mathbf{c} \rangle, T \rangle \Vdash_\Sigma \varphi$ if $[\![\varphi]\!]^{\mathcal{A}} \in T$.

As usual, we extend satisfaction to sets of sentences by letting $\langle \langle \mathcal{A}, \mathbf{c} \rangle, T \rangle \Vdash_\Sigma \Phi$ if and only if $\langle \langle \mathcal{A}, \mathbf{c} \rangle, T \rangle \Vdash_\Sigma \varphi$ for every $\varphi \in \Phi$. We denote the induced *semantic entailment* relation by $\vDash_\Sigma$. It is defined as usual, from satisfaction, by $\Phi \vDash_\Sigma \psi$ if $\langle \langle \mathcal{A}, \mathbf{c} \rangle, T \rangle \Vdash_\Sigma \psi$ whenever $\langle \langle \mathcal{A}, \mathbf{c} \rangle, T \rangle \Vdash_\Sigma \Phi$, for every $\langle \langle \mathcal{A}, \mathbf{c} \rangle, T \rangle \in Mod(\Sigma)$. However, easily, this definition can be seen to correspond precisely to:

– $\Phi \vDash_\Sigma \psi$ if $[\![\psi]\!]^{\mathcal{A}} \in \{[\![\varphi]\!]^{\mathcal{A}} : \varphi \in \Phi\}^{\mathbf{C}}$, for every $\langle \mathcal{A}, \mathbf{c} \rangle \in M_\Sigma$.

Note that validity is still represented by $\emptyset \vDash_\Sigma \varphi$. For the sake of illustration, let us develop two well known examples: classical first-order logic and propositional normal modal logic.

Example 1. Classical first-order logic.
Let X be a fixed denumerable set of variables.

– The category of abstract signatures is $\mathbf{Set}^{\mathbb{N}} \times \mathbf{Set}^{\mathbb{N}}$, the category of pairs of $\mathbb{N}$-ranked alphabets $\langle F, P \rangle$ (of function and predicate symbols) and rank preserving functions.
– The concrete syntax functor $L : \mathbf{Set}^{\mathbb{N}} \times \mathbf{Set}^{\mathbb{N}} \to \mathbf{AlgSig}_\phi$ is defined by:
  • $L(\langle F, P \rangle) = \langle \{\tau, \phi\}, O \rangle$ with $O_\tau = X \cup F_0$, $O_{\tau^n \tau} = F_n$ for $n > 0$, $O_{\tau^n \phi} = P_n$ for $n \in \mathbb{N}$, $O_{\phi\phi} = \{\neg\} \cup \{\forall x : x \in X\}$, $O_{\phi^2\phi} = \{\Rightarrow\}$ and $O_w = \emptyset$ otherwise;
  • $L(h : \langle F, P \rangle \to \langle F', P' \rangle)$ is the identity on the sorts τ, ϕ, the variables in X, the quantifiers $\forall x$ and the logical connectives $\neg, \Rightarrow$, and maps each n-ary function symbol $f \in F_n$ to $h(f) \in F'_n$ and each n-ary predicate symbol $p \in P_n$ to $h(p) \in P'_n$.
– Each $M_{\langle F, P \rangle}$ is the class of all structured algebras $\langle \mathcal{A}, \mathbf{c} \rangle$ obtained from $\langle F, P \rangle$-interpretations $I = \langle D, \_I \rangle$ with $D \neq \emptyset$ a set, $f_I : D^n \to D$ for $f \in F_n$, and $p_I \subseteq D^n$ for $p \in P_n$, as follows:

- $|\mathcal{A}|_\tau = D^{\mathbf{Asg}(X,D)}$ and $|\mathcal{A}|_\phi = \wp(\mathbf{Asg}(X,D))$, where $\mathbf{Asg}(X,D) = D^X$ is the set of all assignments ρ to variables in D;
- $x_\mathcal{A}(\rho) = \rho(x)$ for $x \in X$, $f_\mathcal{A}(e_1,\dots,e_n)(\rho) = f_I(e_1(\rho),\dots,e_n(\rho))$ for $f \in F_n$, $p_\mathcal{A}(e_1,\dots,e_n) = \{\rho \in \mathbf{Asg}(X,D) : \langle e_1(\rho),\dots,e_n(\rho)\rangle \in p_I\}$ for $p \in P_n$, $\neg_\mathcal{A}(r) = \mathbf{Asg}(X,D) \setminus r$, $\forall x_\mathcal{A}(r) = \{\rho \in \mathbf{Asg}(X,D) : \rho[x/d] \in r$ for every $d \in D\}$, and $\Rightarrow_\mathcal{A}(r_1,r_2) = (\mathbf{Asg}(X,D) \setminus r_1) \cup r_2$;
- $\mathbf{c} : \wp(|\mathcal{A}|_\phi) \to \wp(|\mathcal{A}|_\phi)$ is the cut closure operation induced by set inclusion, that is, for every $R \subseteq \wp(|\mathcal{A}|_\phi)$, $R^{\mathbf{c}} = \{r \subseteq |\mathcal{A}|_\phi : (\bigcap R) \subseteq r\}$ is the principal ideal determined by $(\bigcap R)$ on the complete lattice $\langle \wp(|\mathcal{A}|_\phi), \supseteq \rangle$.

$Sen(\langle F,P\rangle)$ is the set of all first-order formulae build using the predicate symbols in P from the terms build using the function symbols in F. Easily, in such interpretation structures, the denotation of a formula corresponds to the set of all assignments for which it holds, and $\vDash_{\langle F,P\rangle}$ reflects reasoning with a fixed assignment. Moreover, $\emptyset^{\mathbf{c}} = \{\mathbf{Asg}(X,D)\}$ and thus a formula holds in a model if it is true for all possible assignments, as usual. Of course, expectedly, $\vDash_{\langle F,P\rangle}$ does not comply with the rule of generalization. Namely, in general $\{\varphi\} \nvDash_{\langle F,P\rangle} (\forall x\, \varphi)$. Instead, however, as a rule for generating theorems out of theorems, generalization is such that $\langle\langle \mathcal{A}, \mathbf{c}\rangle, \emptyset^{\mathbf{c}}\rangle \Vdash_{\langle F,P\rangle} (\forall x\, \varphi)$ whenever $\langle\langle \mathcal{A}, \mathbf{c}\rangle, \emptyset^{\mathbf{c}}\rangle \Vdash_{\langle F,P\rangle} \varphi$.

Example 2. Propositional normal modal logic.
- The category of abstract signatures is **Set**.
- The concrete syntax functor $L : \mathbf{Set} \to \mathbf{AlgSig}_\phi$ is defined by:
  - $L(\mathrm{PS}) = \langle \{\phi\}, O\rangle$ with $O_\phi = \mathrm{PS}$, $O_{\phi\phi} = \{\Box, \neg\}$, $O_{\phi^2\phi} = \{\Rightarrow\}$ and $O_w = \emptyset$ otherwise;
  - $L(h : \mathrm{PS} \to \mathrm{PS}')$ is the identity on the sort ϕ, the modality $\Box$ and the connectives $\neg, \Rightarrow$, and maps each $p \in \mathrm{PS}$ to $h(p) \in \mathrm{PS}'$;
- Each M_{PS} is the class of all structured algebras $\langle \mathcal{A}, \mathbf{c}\rangle$ obtained from a Kripke frame $\langle W, R\rangle$ and a valuation $\vartheta : \mathrm{PS} \to \wp(W)$, where $W \neq \emptyset$ is a set and $R \subseteq W^2$, as follows:
  - $|\mathcal{A}|_\phi = \wp(W)$;
  - $p_\mathcal{A} = \vartheta(p)$ for $p \in \mathrm{PS}$, $\Box_\mathcal{A}(U) = \{w \in W : \{u \in W : wRu\} \subseteq U\}$, $\neg_\mathcal{A}(U) = W \setminus U$ and $\Rightarrow_\mathcal{A}(U_1, U_2) = (W \setminus U_1) \cup U_2$;
  - $\mathbf{c} : \wp(|\mathcal{A}|_\phi) \to \wp(|\mathcal{A}|_\phi)$ is, as before, the cut closure operation induced by set inclusion.

Easily, in such a structure, $\emptyset^{\mathbf{c}} = \{W\}$ and a modal formula holds in a model if it is true for all worlds in that model. In fact, the denotation of a formula corresponds precisely to the set of all worlds where it holds. Therefore, $\vDash_{\mathrm{PS}}$ reflects reasoning over a given fixed world. Again, it does not comply with the rule of necessitation. In general, $\{\varphi\} \nvDash_{\mathrm{PS}} (\Box\varphi)$. Instead, as a rule for theorem generation, necessitation is such that $\langle\langle \mathcal{A}, \mathbf{c}\rangle, \emptyset^{\mathbf{c}}\rangle \Vdash_{\mathrm{PS}} (\Box\varphi)$ whenever $\langle\langle \mathcal{A}, \mathbf{c}\rangle, \emptyset^{\mathbf{c}}\rangle \Vdash_{\mathrm{PS}} \varphi$.

3   The Category of c-Parchments

As for *morphisms of c-parchments* we propose a version specially tailored for *fibring*. It is essentially dual to the notion of morphism of model-theoretic parchments used in [21], although in our case the relation between algebras is a little stricter (in exactly the same sense of the previous coherence condition for c-parchments).

Definition 2. A *morphism of c-parchments* from $P_1 = \langle \mathbf{Sig}_1, L_1, \mathcal{M}_1 \rangle$ to $P_2 = \langle \mathbf{Sig}_2, L_2, \mathcal{M}_2 \rangle$ is a pair $\langle \Phi, \eta \rangle : P_1 \to P_2$ where:

- $\Phi : \mathbf{Sig}_2 \to \mathbf{Sig}_1$ is a functor;
- $\eta : L_1 \circ \Phi \to L_2$ is a natural transformation,

such that, for every $\Sigma \in |\mathbf{Sig}_2|$, the following condition holds:

- $\langle \mathcal{A}|_{\eta_\Sigma}, \mathbf{c} \rangle \in M_{1,\Phi(\Sigma)}$ for every $\langle \mathcal{A}, \mathbf{c} \rangle \in M_{2,\Sigma}$.

Clearly, c-parchments and their morphisms constitute a category **CPar**. Moreover, the construction of an institution out of a c-parchment easily extends to a functor from **CPar** to the dual of the category **Ins** of institutions and institution morphisms [15]. In fact, a c-parchment is no more than a functor from a signature category to the following category **CPRoom** of *rooms for c-parchments*, or just c-*rooms*.

Definition 3. A c-*room* is a pair $R = \langle \langle S, O \rangle, M \rangle$ with $\langle S, O \rangle \in |\mathbf{AlgSig}_\phi|$ and $M \subseteq \mathrm{cAlg}(\langle S, O \rangle)$. A *morphism of c-rooms* from $R_1 = \langle \langle S_1, O_1 \rangle, M_1 \rangle$ to $R_2 = \langle \langle S_2, O_2 \rangle, M_2 \rangle$ is an $\mathbf{AlgSig}_\phi$-morphism $h : \langle S_1, O_1 \rangle \to \langle S_2, O_2 \rangle$ such that $\langle \mathcal{A}|_h, \mathbf{c} \rangle \in M_1$ for every $\langle \mathcal{A}, \mathbf{c} \rangle \in M_2$.

A c-parchment $P = \langle \mathbf{Sig}, L, \mathcal{M} \rangle$ essentially corresponds to the functor $P : \mathbf{Sig} \to \mathbf{CPRoom}$ such that $P(\Sigma) = \langle L(\Sigma), M_\Sigma \rangle$ and $P(\sigma) = L(\sigma)$. It is straightforward to show that this correspondence is not only a bijection, but that it indeed extends to an isomorphism of categories. Namely, a morphism $\langle \Phi, \eta \rangle : P_1 \to P_2$ of c-parchments corresponds precisely to a functor $\Phi : \mathbf{Sig}_2 \to \mathbf{Sig}_1$ and a natural transformation $\eta : P_1 \circ \Phi \to P_2$. Thus, in exactly the same way that the category **Ins** of institutions and institution morphisms corresponds to a Grothendieck construction on categories of functors to the category **Room(2)** of [14], and the category **MPar** of model-theoretic parchments and morphisms corresponds to a Grothendieck construction on categories of functors to the category **MPRoom** of [21], the dual of our category **CPar** corresponds to a Grothendieck construction on categories of functors to the dual of **CPRoom**. As in the other cases, the cocompleteness of **CPar** follows immediately from the cocompleteness of **CPRoom**.

Proposition 1. **CPRoom** *is cocomplete.*

Proof. We just show what coproducts and coequalizers look like in **CPRoom**. For the purpose, let I be a set and $\{R_i = \langle \langle S_i, O_i \rangle, M_i \rangle\}_{i \in I}$ a family of c-rooms. Of course, we capitalize on the well known fact that $\mathbf{AlgSig}_\phi$ is cocomplete. (Coproducts) Let $\{h_j : \langle S_j, O_j \rangle \to \coprod_{i \in I} \langle S_i, O_i \rangle\}_{j \in I}$ be a coproduct in $\mathbf{AlgSig}_\phi$. A coproduct $\{h_j : R_j \to \langle \coprod_{i \in I} \langle S_i, O_i \rangle, M \rangle\}_{j \in I}$ in **CPRoom** can be obtained by taking $M \subseteq \mathrm{cAlg}(\coprod_{i \in I} \langle S_i, O_i \rangle)$ to be the class of all pairs $\langle \mathcal{A}, \mathbf{c} \rangle$ such that $\langle \mathcal{A}|_{h_j}, \mathbf{c} \rangle \in M_j$ for every $j \in I$.

(Coequalizers) Let $I = \{1, 2\}$, $h', h'' : R_1 \to R_2$ be **CPRoom**-morphisms and $h : \langle S_2, O_2 \rangle \to \langle S, O \rangle$ a coequalizer of $h', h'' : \langle S_1, O_1 \rangle \to \langle S_2, O_2 \rangle$ in **AlgSig**$_\phi$. A coequalizer $h : R_2 \to \langle \langle S, O \rangle, M \rangle$ of $h', h'' : R_1 \to R_2$ in **CPRoom** can be obtained by taking $M \subseteq \mathrm{cAlg}(\langle S, O \rangle)$ to be the class of all pairs $\langle \mathcal{A}, \mathbf{c} \rangle$ such that $\langle \mathcal{A}|_h, \mathbf{c} \rangle \in M_2$. □

The following result is a simple corollary of the corresponding Grothendieck construction, similar to the one in [26].

Proposition 2. CPar *is cocomplete.*

Proof. Again, we just show what coproducts and coequalizers look like in **CPar**. For the purpose, let I be a set and $\{P_i : \mathbf{Sig}_i \to \mathbf{CPRoom}\}_{i \in I}$ a family of c-parchments.
(Coproducts) A coproduct $\{\langle \Pi_i, \iota_i \rangle : P_i \to P\}_{i \in I}$ in **CPar** can be obtained by taking the functor $P = (\coprod_I \circ (\prod_{i \in I} P_i)) : \prod_{i \in I} \mathbf{Sig}_i \to \mathbf{CPRoom}$, where: each $\Pi_j : \prod_{i \in I} \mathbf{Sig}_i \to \mathbf{Sig}_j$ is the corresponding projection functor; $\coprod_I$ is the coproduct functor left adjoint to the diagonal functor $\Delta_I : \mathbf{CPRoom} \to \mathbf{CPRoom}^I$; $\prod_{i \in I} P_i : \prod_{i \in I} \mathbf{Sig}_i \to \mathbf{CPRoom}^I$ is the unique functor such that $(\prod_{i \in I} P_i)_j = P_j \circ \Pi_j$ for every $j \in I$, resulting from the universality of the product $\{\_i : \mathbf{CPRoom}^I \to \mathbf{CPRoom}\}_{i \in I}$ in **Cat**; and each $\iota_{j,\Sigma} : P_j(\Sigma_j) \to \coprod_{i \in I} P_i(\Sigma_i)$ is the corresponding injection on the coproduct, for every $\Sigma = \langle \Sigma_i \rangle_{i \in I} \in |\prod_{i \in I} \mathbf{Sig}_i|$.
(Coequalizers) Let $I = \{1, 2\}$. A coequalizer $\langle \Phi, \eta \rangle : P_2 \to P$ of a pair of morphisms $\langle \Phi', \eta' \rangle, \langle \Phi'', \eta'' \rangle : P_1 \to P_2$ in **CPar** can be obtained by taking the functor $P = \mathrm{coEq} \circ \langle \overline{\eta}'_\Phi, \overline{\eta}''_\Phi \rangle : \mathbf{Sig} \to \mathbf{CPRoom}$, where: $\Phi : \mathbf{Sig} \to \mathbf{Sig}_2$ is an equalizer of the functors $\Phi', \Phi'' : \mathbf{Sig}_2 \to \mathbf{Sig}_1$ in **Cat**; coEq is the coequalizer functor left adjoint to the diagonal functor $\Delta_{\downdownarrows} : \mathbf{CPRoom} \to \mathbf{CPRoom}^{\downdownarrows}$; $\langle \overline{\eta}'_\Phi, \overline{\eta}''_\Phi \rangle : \mathbf{Sig} \to \mathbf{CPRoom}^{\downdownarrows}$ is the unique functor such that $\langle \overline{\eta}'_\Phi, \overline{\eta}''_\Phi \rangle_1 = \overline{\eta}' \circ \Phi$ and $\langle \overline{\eta}'_\Phi, \overline{\eta}''_\Phi \rangle_2 = \overline{\eta}'' \circ \Phi$, with $\overline{\eta}', \overline{\eta}'' : \mathbf{Sig}_2 \to \mathbf{CPRoom}^{\downarrow}$ the functors corresponding to the natural transformations η' and η'', resulting from the universality of the pullback $\_{1}, \_{2} : \mathbf{CPRoom}^{\downdownarrows} \to \mathbf{CPRoom}^{\downarrow}$ of $\langle \mathrm{dom}, \mathrm{cod} \rangle, \langle \mathrm{dom}, \mathrm{cod} \rangle : \mathbf{CPRoom}^{\downarrow} \to A \times A$ in **Cat**; and $\eta_\Sigma : P_2(\Phi(\Sigma)) \to \mathrm{coEq}(\langle \eta'_{\Phi(\Sigma)}, \eta''_{\Phi(\Sigma)} \rangle)$ is the corresponding coequalizer, for $\Sigma \in |\mathbf{Sig}|$. □

4   Fibred Semantics

As we have said before, morphisms of c-parchments (as well as morphisms of c-rooms) have been set up having in mind the characterization of *fibring* via colimits. We now extend our previous characterizations of the construction [22,29,7] to the level of c-rooms and c-parchments, and concentrate on the particular cases of colimit defining *fibring constrained by sharing of symbols*. In the remainder of the paper, in fact, we restrict our attention to just c-rooms. As we have seen, colimits can be smoothly lifted to the level of c-parchments.
Let us consider fixed two arbitrary c-rooms $R_1 = \langle \langle S_1, O_1 \rangle, M_1 \rangle$ and $R_2 = \langle \langle S_2, O_2 \rangle, M_2 \rangle$. For simplicity, we shall assume that when fibring R_1 and R_2

the required sharing of syntax is specified by means of the largest common subsignature of $\langle S_1, O_1 \rangle$ and $\langle S_2, O_2 \rangle$. That is, by default, we shall assume to be sharing the signature $\langle S_0, O_0 \rangle$ with $S_0 = S_1 \cap S_2$ (it always includes at least the sort ϕ) and $O_{0,u} = O_{1,u} \cap O_{2,u}$ for $u \in S_0^+$ is shared according to the corresponding signature inclusion morphisms $h_1 : \langle S_0, O_0 \rangle \to \langle S_1, O_1 \rangle$ and $h_2 : \langle S_0, O_0 \rangle \to \langle S_2, O_2 \rangle$. We denote by R_0 the canonical c-room $\langle \langle S_0, O_0 \rangle, M_0 \rangle$ where $M_0 = \text{cAlg}(\langle S_0, O_0 \rangle)$. In the simplest possible case when $S_0 = \{\phi\}$ and $O_0 = \emptyset$ we say that the *fibring* is *free* or *unconstrained*.

Definition 4. The *fibring* of R_1 and R_2 (constrained by sharing $\langle S_0, O_0 \rangle$) is the c-room $R_1 \circledast R_2 = \langle \langle S, O \rangle, M \rangle$ such that:

- $S = S_1 \cup S_2$, with inclusions $f_i : S_i \to S$;
- $O_u = O_{1,u} \cup O_{2,u}$ if $u \in S_0^+$, $O_u = O_{i,u}$ if $u \in S_i^+ \setminus S_0^+$ and $O_u = \emptyset$ otherwise, with inclusions $g_i : O_i \to O$;
- $M \subseteq \text{cAlg}(\langle S, O \rangle)$ is the class of all pairs $\langle \mathcal{A}, \mathbf{c} \rangle$ such that $\langle \mathcal{A}|_{\langle f_1, g_1 \rangle}, \mathbf{c} \rangle \in M_1$ and $\langle \mathcal{A}|_{\langle f_2, g_2 \rangle}, \mathbf{c} \rangle \in M_2$.

Clearly, M consists precisely of all those $\langle \mathcal{A}, \mathbf{c} \rangle$ that can be obtained by joining together any two $\langle \mathcal{A}_1, \mathbf{c}_1 \rangle \in M_1$ and $\langle \mathcal{A}_2, \mathbf{c}_2 \rangle \in M_2$ such that $|\mathcal{A}_1|_s = |\mathcal{A}_2|_s = |\mathcal{A}|_s$ for every $s \in S_0$, $o_{\mathcal{A}_1} = o_{\mathcal{A}_2} = o_{\mathcal{A}}$ for every $o \in O_{0,u}$ with $u \in S_0^+$, and $\mathbf{c}_1 = \mathbf{c}_2 = \mathbf{c}$.

Proposition 3. *The fibring of c-rooms R_1 and R_2 (constrained by sharing $\langle S_0, O_0 \rangle$) is a pushout of $\{h_i : R_0 \to R_i\}_{i \in \{1,2\}}$ in* **CPRoom**.

As a simple corollary, when the fibring is free, $R_1 \circledast R_2$ is a coproduct of R_1 and R_2 in **CPRoom**.

Let us now analyze in some detail a new application of fibring, made possible in this setting. The example concerns partial equational logic and the way it can be obtained by fibring equational logic with a logic of partiality. A similar idea had already been proposed in [19,20,21], but for a different notion of parchment and without any concerns for preservation results. For this reason, we do think that the approach followed here is much more direct and modular. In fact, as we show, a nice proof-calculus for the fibred partial equational logic is obtained by putting together the proof-calculi for equational logic together with a calculus suited for the logic of partiality adopted.

Example 3. Partial equational logic.
We start by representing equational logic as a c-room, for a given $\mathbb{N}$-ranked alphabet F of function symbols. In order to keep the focus on partiality, we shall just consider an unsorted version of equational logic. Let X be a fixed denumerable set of variables and Eq an equational specification (set of equations) over F and X.

- The concrete syntax signature $\langle S, O \rangle$ is such that:
  - $S = \{\tau, \phi\}$;
  - $O_\tau = X \cup F_0$, $O_{\tau^n \tau} = F_n$ for $n > 0$, $O_{\tau^2 \phi} = \{=\}$ and $O_w = \emptyset$ otherwise.

- M is the class of all structured algebras $\langle \mathcal{A}, \mathbf{c} \rangle$ obtained from an F-algebra $\mathcal{F}$ that is a model[1] of Eq as follows:
  - $|\mathcal{A}|_\tau = |\mathcal{F}|^{\mathbf{Asg}(X, |\mathcal{F}|)}$ and $|\mathcal{A}|_\phi = \wp(\mathbf{Asg}(X, |\mathcal{F}|))$, where $\mathbf{Asg}(X, |\mathcal{F}|) = |\mathcal{F}|^X$ is the set of all assignments ρ to variables in $|\mathcal{F}|$;
  - $x_\mathcal{A}(\rho) = \rho(x)$ for $x \in X$, $f_\mathcal{A}(e_1, \ldots, e_n)(\rho) = f_\mathcal{F}(e_1(\rho), \ldots, e_n(\rho))$ for $f \in F_n$, and $=_\mathcal{A} (e_1, e_2) = \{ \rho \in \mathbf{Asg}(X, |\mathcal{F}|) : e_1(\rho) = e_2(\rho) \}$;
  - $\mathbf{c} : \wp(|\mathcal{A}|_\phi) \to \wp(|\mathcal{A}|_\phi)$ is defined as in Example 1 for first-order logic.

Clearly, as formulae, we have precisely the equations between terms build using the function symbols in F and the variables in X. Moreover, in all such interpretation structures, the denotation of an equation is precisely the set of all assignments where it holds. Moreover, $\emptyset^\mathbf{c} = \{ \mathbf{Asg}(X, |\mathcal{F}|) \}$ and an equation holds in a model if the values of the two terms coincide for all possible assignments. Again, $\vDash$ reflects reasoning with a fixed assignment and does not comply with the rule of substitution. In general, $\{t_1 = t_2\} \nvdash t_1 \sigma = t_2 \sigma$, where σ is some substitution of variables by terms. Instead, we have that if $\langle \langle \mathcal{A}, \mathbf{c} \rangle, \emptyset^\mathbf{c} \rangle \Vdash t_1 = t_2$ then $\langle \langle \mathcal{A}, \mathbf{c} \rangle, \emptyset^\mathbf{c} \rangle \Vdash t_1 \sigma = t_2 \sigma$.

We aim at obtaining a $\mathbf{c}$-room for unsorted partial equational logic by fibring the $\mathbf{c}$-room above with a suitable room $\langle \langle S, O \rangle, M \rangle$ for partiality. For the purpose, let $G = \{G_n\}_{n \in \mathbb{N}}$ be a ranked subalphabet of F. Operations in G shall be considered total, whereas all other operations in F can be partial. Here is a very simple possibility:

- The concrete syntax signature $\langle S, O \rangle$ is such that:
  - $S = \{\tau, \phi\}$;
  - $O_\tau = X \cup F_0$, $O_{\tau^n \tau} = F_n$ for $n > 0$, $O_{\tau\phi} = \{D\}$, $O_{\tau^2\phi} = \{=\}$ and $O_w = \emptyset$ otherwise;
- M is the class of all structured algebras $\langle \mathcal{A}, \mathbf{c} \rangle$ obtained from an F-algebra $\mathcal{F}$ with a distinguished element $* \in |\mathcal{F}|$ and satisfying:
  - $f_\mathcal{F}(a_1, \ldots, a_n) = *$ whenever some $a_i = *$, for every $f \in F_n$;
  - $g_\mathcal{F}(a_1, \ldots, a_n) \neq *$ whenever all $a_i \neq *$, for every $g \in G_n$,
  plus a given binary relation $R \subseteq |\mathcal{F}| \times |\mathcal{F}|$ satisfying:
  - $a R *$ or $* R a$ imply $a = *$,
  as follows:
  - $|\mathcal{A}|_\tau = |\mathcal{F}|^{\mathbf{Asg}(X, |\mathcal{F}|)}$ and $|\mathcal{A}|_\phi = \wp(\mathbf{Asg}(X, |\mathcal{F}|))$;
  - $x_\mathcal{A}(\rho) = \rho(x)$ for $x \in X$, $f_\mathcal{A}(e_1, \ldots, e_n)(\rho) = f_\mathcal{F}(e_1(\rho), \ldots, e_n(\rho))$ for $f \in F_n$, $D_\mathcal{A}(e) = \{ \rho \in \mathbf{Asg}(X, |\mathcal{F}|) : e(\rho) \neq * \}$, and $=_\mathcal{A} (e_1, e_2) = \{ \rho \in \mathbf{Asg}(X, |\mathcal{F}|) : e_1(\rho) R e_2(\rho) \}$;
  - $\mathbf{c} : \wp(|\mathcal{A}|_\phi) \to \wp(|\mathcal{A}|_\phi)$ is again the cut closure operation induced by set inclusion.

In this case we are only concerned with partiality. Hence, we impose the least possible constraints to the interpretation R of equality.

The desired $\mathbf{c}$-room for unsorted partial equational logic, obtained by constrained fibring and sharing both sorts τ and ϕ, the variables X, the equality symbol $=$, and the operations in F is as follows:

---

[1] This means that $[\![t_1]\!]_\rho^\mathcal{F} = [\![t_2]\!]_\rho^\mathcal{F}$ for every assignment $\alpha : X \to |\mathcal{F}|$ and every equation $t_1 = t_2 \in$ Eq.

– The concrete syntax signature $\langle S, O \rangle$ is such that:
  - $S = \{\tau, \phi\}$;
  - $O_\tau = X \cup F_0$, $O_{\tau^n \tau} = F_n$ for $n > 0$, $O_{\tau\phi} = \{D\}$, $O_{\tau^2\phi} = \{=\}$ and $O_w = \emptyset$ otherwise;
– M is the class of all structured algebras $\langle \mathcal{A}, \mathbf{c} \rangle$ obtained from an F-algebra $\mathcal{F}$ that is a model of Eq with a distinguished element $* \in |\mathcal{F}|$ and satisfying:
  - $f_{\mathcal{F}}(a_1, \ldots, a_n) = *$ whenever some $a_i = *$, for every $f \in F_n$;
  - $g_{\mathcal{F}}(a_1, \ldots, a_n) \neq *$ whenever all $a_i \neq *$, for every $g \in G_n$,
  as follows:
  - $|\mathcal{A}|_\tau = |\mathcal{F}|^{\mathbf{Asg}(X, |\mathcal{F}|)}$ and $|\mathcal{A}|_\phi = \wp(\mathbf{Asg}(X, |\mathcal{F}|))$;
  - $x_{\mathcal{A}}(\rho) = \rho(x)$ for $x \in X$, $f_{\mathcal{A}}(e_1, \ldots, e_n)(\rho) = f_{\mathcal{F}}(e_1(\rho), \ldots, e_n(\rho))$ for $f \in F_n$, $D_{\mathcal{A}}(e) = \{\rho \in \mathbf{Asg}(X, |\mathcal{F}|) : e(\rho) \neq *\}$, and $=_{\mathcal{A}} (e_1, e_2) = \{\rho \in \mathbf{Asg}(X, |\mathcal{F}|) : e_1(\rho) = e_2(\rho)\}$;
  - $\mathbf{c} : \wp(|\mathcal{A}|_\phi) \to \wp(|\mathcal{A}|_\phi)$ is again the cut closure operation induced by set inclusion.

Note that, in the resulting $\mathbf{c}$-room, equality is necessarily interpreted as *strong equality*. *Existential equality* can also be introduced, by adding the symbol $\overset{e}{=}$ to the partial logic room, where it should be interpreted as follows:

$$\overset{e}{=}_{\mathcal{A}} (e_1, e_2) = \{\rho \in \mathbf{Asg}(X, |\mathcal{F}|) : e_1(\rho) \neq * \text{ and } e_1(\rho) \, R \, e_2(\rho)\}.$$

One may ask what can be achieved at the proof-theoretic level. Can we expect to obtain a calculus for partial equational logic, by putting together a calculus for equational logic and a calculus for the logic of partiality? The answer is yes. Just consider the following deduction rules for partiality:

$$\frac{D(f(t_1, \ldots, t_n))}{D(t_i)} \; f \in F_n, i = 1, \ldots, n \qquad \frac{D(t_1) \; \ldots \; D(t_n)}{D(g(t_1, \ldots, t_n))} \; g \in G_n$$

$$\frac{D(t_1) \quad t_1 = t_2}{D(t_2)} \qquad \frac{D(t_2) \quad t_1 = t_2}{D(t_1)}$$

and for the partial logic with existential equality ($\overset{e}{=}$) we may consider the additional rules:

$$\frac{t_1 \overset{e}{=} t_2}{D(t_1)} \qquad \frac{t_1 \overset{e}{=} t_2}{D(t_2)} \qquad \frac{t_1 \overset{e}{=} t_2}{t_1 = t_2} \qquad \frac{D(t_1) \quad t_1 = t_2}{t_1 \overset{e}{=} t_2}.$$

It is easy to check that not only this calculus is sound and complete for the simple logic of partiality being adopted, but furthermore the calculus for partial equational logic obtained by adding these rules to the usual calculus for equational logic with specific axioms for each equation in Eq is also sound and complete, even with existential equality. The proof of soundness is straightforward. In what concerns completeness, we can easily adapt the usual techniques for partiality (see, for instance, [27,2]). We should also note that similar results could have been obtained by taking from the beginning a conditional equational specification instead of Eq.

Note that other approaches to partiality, namely using three-valued versions of equality, could also be considered. This remark, however, deserves some explanation. Of course, one may argue that by just fibring logics whose interpretation structures are based on different spaces of truth-values one obtains a logic without interpretation structures whatsoever. This is in fact a consequence of the strict form of the coherence conditions for **c**-parchments and their morphisms. Still, an important remark, already explored in previous work on combining logics (see for instance [20,21,22,29]), is that the process of combining logics depends as much on the particular logics being combined as on the specific way these logics are presented. In particular, the very same logic can be presented using several distinct alternative formulations of its class of interpretation structures (recall for instance the case of classical propositional logic, that can be equivalently presented with just two-valued structures or else with structures corresponding to all Boolean algebras). In fact, several ways of closing the class of interpretation structures of a logic have already been considered in connection with fibring [29,22], and we shall use similar ideas in the forthcoming notion of *fullness* in Section 6. For the moment, however, just note that adding *sound* interpretation structures to a logic equipped with a proof calculus is absolutely harmless with respect to the induced semantic entailment and that such an approach can be used to enhance radically the ability of combining that logic with others. In the present case, of obtaining an equational logic of partiality over a three-valued version of equality, the procedure might be similar to the following simplified but similar case.

Assume that we want to combine classical propositional logic with the following **c**-room for the three-valued propositional logic of Łukasiewicz (see [28]) over the set PS of propositional symbols:

- The concrete syntax signature $\langle S, O \rangle$ is such that:
  - $S = \{\phi\}$;
  - $O_\phi = \text{PS}$, $O_{\phi\phi} = \{\neg\}$, $O_{\phi^2\phi} = \{\Rightarrow\}$ and $O_w = \emptyset$ otherwise.
- M is the class of all structured algebras $\langle \mathcal{A}, \mathbf{c} \rangle$ such that:
  - $|\mathcal{A}|_\phi = \{0, \frac{1}{2}, 1\}$
  - $p_\mathcal{A} \in |\mathcal{A}|_\phi$ for $p \in \text{PS}$, $\neg_\mathcal{A}(v) = 1 - v$ and $\Rightarrow_\mathcal{A}(v_1, v_2) = \min(1, 1 - v_1 + v_2)$.
  - $\mathbf{c} : \wp(|\mathcal{A}|_\phi) \to \wp(|\mathcal{A}|_\phi)$ is the cut closure operation induced by the usual ordering $0 < \frac{1}{2} < 1$.

Let us also adopt the obvious two-valued **c**-room for classical logic over the set PS$'$ of propositional symbols:

- The concrete syntax signature $\langle S, O \rangle$ is such that:
  - $S = \{\phi\}$;
  - $O_\phi = \text{PS}'$, $O_{\phi\phi} = \{\neg\}$, $O_{\phi^2\phi} = \{\Rightarrow\}$ and $O_w = \emptyset$ otherwise.
- M is the class of all structured algebras $\langle \mathcal{A}, \mathbf{c} \rangle$ such that:
  - $|\mathcal{A}|_\phi = \{0, 1\}$
  - $p_\mathcal{A} \in |\mathcal{A}|_\phi$ for $p \in \text{PS}'$, $\neg_\mathcal{A}(v) = 1 - v$ and $\Rightarrow_\mathcal{A}(v_1, v_2) = \min(1, 1 - v_1 + v_2)$.
  - $\mathbf{c} : \wp(|\mathcal{A}|_\phi) \to \wp(|\mathcal{A}|_\phi)$ is the cut closure operation induced by the usual ordering $0 < 1$.

It is clear that, coherence condition *oblige*, the result of fibring the two rooms will have no interpretation structures. However, it is also possible to consider an extension of this room by also considering certain classically compatible three-valued structures, namely:

- all structured algebras $\langle \mathcal{A}, \mathbf{c} \rangle$ such that:
  - $|\mathcal{A}|_{\phi} = \{0, \frac{1}{2}, 1\}$
  - $p_{\mathcal{A}} \in \{0, 1\}$ for $p \in \mathrm{PS}'$, $\Rightarrow_{\mathcal{A}}(u_1, u_2) = \min(1, 1 - u_1 + u_2)$, $\neg_{\mathcal{A}}(u) = 1 - u$.
  - $\mathbf{c} : \wp(|\mathcal{A}|_{\phi}) \to \wp(|\mathcal{A}|_{\phi})$ is the cut closure operation induced by the usual ordering $0 < \frac{1}{2} < 1$.

It is not our objective to study precisely which kind of combined logic this might correspond to, but it is now clear that by using this enlarged **c**-room for classical propositional logic we do no longer get an empty class of interpretation structures after fibring. Indeed, if for instance there is no sharing of propositional symbols ($\mathrm{PS} \cap \mathrm{PS}' = \emptyset$), all of these new three-valued based classical structures clearly have counterparts on the Łukasiewicz side. If some propositional symbols are shared ($\mathrm{PS} \cap \mathrm{PS}' \neq \emptyset$) then all the three-valued interpretation structures corresponding to $\{0, 1\}$-valuations of the shared propositional symbols will still obtain after fibring.

5   Fibred Deduction

The most promising advantage of fibring concerns preservation results [12,13]. In the next section, we present a collection of soundness and completeness preservation results for fibring, within the context of a suitable notion of Hilbert-like proof-calculus that explicitly distinguishes theorem generating rules. We dedicate this section to presenting the details of such calculi and their fibring. However, we shall restrict ourselves to logics with a propositional base, namely, whose concrete syntax is given by a one-sorted signature whose only sort is the sort of formulae. Thus, from now on, we shall refer to ϕ-sorted signatures as those having ϕ as the unique sort. The forthcoming proof-theoretic definitions could of course be extended in order to encompass also logics without a propositional base, such as first-order logic or equational logic as in the preceding examples. However, we refrain from doing so, for simplicity, since the preservation results that we shall present are only applicable to the propositional base case.

The way the proof-calculi of equational logic and of the logic of partiality were put together in Example 3 to obtain the proof-calculus of partial equational logic, can in fact be systematized as a corresponding proof-theoretic form of fibring. The idea, originally used in [22], is to use schema variables in rules and schema variable substitutions in rule application. For the purpose, we consider fixed a denumerable set $\Xi = \{\xi_n : n \in \mathbb{N}\}$ of *schema formula variables*, to be used in writing *schema rules*. Letting $\langle \{\phi\}, O \rangle$ be a signature, we shall use $\mathcal{W}_{\langle \{\phi\}, O \rangle}(\Xi)$ to denote the free $\langle \{\phi\}, O \rangle$-algebra on the set Ξ of generators (the *schema word* algebra), and $[\![\_]\!]_{\alpha}^{\mathcal{A}}$ (for schema word *interpretation under* α) to denote the unique $\mathbf{Alg}(\langle \{\phi\}, O \rangle)$-homomorphism from $\mathcal{W}_{\langle \{\phi\}, O \rangle}(\Xi)$ to any given

$\langle\{\phi\}, O\rangle$-algebra $\mathcal{A}$ that extends a given *assignment* $\alpha : \Xi \to |\mathcal{A}|_\phi$. As before, given **AlgSig**$_\phi$-morphism $h : \langle\{\phi\}, O_1\rangle \to \langle\{\phi\}, O_2\rangle$, we shall preferably write $\widehat{h}$ (for schema word *translation*) instead of $[\![\_]\!]^{\mathcal{W}_{\langle\{\phi\}, O_2\rangle}|_\sigma}_{id_\Xi}$ to denote the unique **Alg**$(\langle\{\phi\}, O_1\rangle)$-homomorphism from $\mathcal{W}_{\langle\{\phi\}, O_1\rangle}(\Xi)$ to $\mathcal{W}_{\langle\{\phi\}, O_2\rangle}|_\sigma(\Xi)$ that identifies schema variables.

We can recognize $|\mathcal{W}_{\langle\{\phi\}, O\rangle}(\Xi)|_\phi$ (the carrier of sort ϕ in the schema word algebra) as the set of schema formulae over $\langle\{\phi\}, O\rangle$. We shall use γ, δ with or without primes and subscripts to denote schema formulae, and Γ, Δ to denote sets of schema formulae.

Definition 5. An *inference rule* over a signature $\langle\{\phi\}, O\rangle$ is a pair $r = \langle\Gamma, \delta\rangle$ where $\Gamma \cup \{\delta\} \subseteq |\mathcal{W}_{\langle\{\phi\}, O\rangle}(\Xi)|_\phi$ is a finite set.

Given such an inference rule r, we shall often use $\text{Prem}(r) = \Gamma$ and $\text{Conc}(r) = \delta$ to denote its premises and conclusion. As usual, we shall sometimes represent r simply by $\frac{\text{Prem}(r)}{\text{Conc}(r)}$, $\frac{\Gamma}{\delta}$, or even by $\frac{\gamma_1 \ldots \gamma_n}{\delta}$ if $\Gamma = \{\gamma_1 \ldots \gamma_n\}$. If the set of premises is empty the rule is sometimes also identified with its conclusion and referred to as an *axiom*. The following definition is similar to those in [22,29,7].

Definition 6. A *proof-calculus* is a triple $C = \langle\langle\{\phi\}, O\rangle, dR, gR\rangle$ where $\langle\{\phi\}, O\rangle$ is a signature and both dR and gR are sets of inference rules over $\langle\{\phi\}, O\rangle$ such that $\text{Prem}(r) \neq \emptyset$ for every $r \in gR$. A *morphism* between proof-calculi $C_1 = \langle\langle\{\phi\}, O_1\rangle, dR_1, gR_1\rangle$ and $C_2 = \langle\langle\{\phi\}, O_2\rangle, dR_2, gR_2\rangle$ is an **AlgSig**$_\phi$-morphism $h : \langle\{\phi\}, O_1\rangle \to \langle\{\phi\}, O_2\rangle$ such that $\widehat{h}(r) \in dR_2$ for every $r \in dR_1$, and $\widehat{h}(r) \in gR_2$ for every $r \in gR_1$.

In the definition above, dR denotes the set of rules allowed for building deductions in the calculus, while gR stands for the set of rules that are to be used just for building theorems out of theorems. Thus, the requirement that gR rules must have premises is quite natural. In the context of a signature morphism $h : \langle\{\phi\}, O_1\rangle \to \langle\{\phi\}, O_2\rangle$, given a rule $r = \frac{\gamma_1 \ldots \gamma_n}{\delta}$ over $\langle\{\phi\}, O_1\rangle$, we are of course writing $\widehat{h}(r)$ to denote the translated rule $\frac{\widehat{h}(\gamma_1) \ldots \widehat{h}(\gamma_n)}{\widehat{h}(\delta)}$ over $\langle\{\phi\}, O_2\rangle$. It is straightforward that proof-calculi constitute a category **PCalc**. Moreover, exactly as the category of c-parchments can be build over the category of c-rooms, we can use a similar Grothendieck construction to set up our proof-theoretic notion of logic as a category **Prf** of proof-systems built over **PCalc**.

A proof-calculus $C = \langle\langle\{\phi\}, O\rangle, dR, gR\rangle$ presents, as expected, a *deducibility* relation. However, this relation is build on top of a notion of *theoremhood*. For convenience, we define both notions over schema formulae. We say that a schema formula δ is a *theorem schema generated* from the set of schema formulae Γ if there exists a finite sequence $\gamma_1 \ldots \gamma_n$ of schema formulae such that:

- γ_n is δ;
- for each $i = 1, \ldots, n$ either $\gamma_i \in \Gamma$, or there exist a rule $r \in dR \cup gR$ and a schema variable substitution $\sigma : \Xi \to |\mathcal{W}_{\langle\{\phi\}, O\rangle}(\Xi)|_\phi$ such that $\text{Prem}(r)\sigma \subseteq \{\gamma_j : j < i\}$ and $\text{Conc}(r)\sigma$ is γ_i.

In the sequel, we shall refer to theorem schemata generated from $\emptyset$ simply by *theorem schemata*. In deductions now, only instances of rules in dR are allowed, together with theorem schemata. We say that δ is *deducible* from Γ, $\Gamma \vdash \delta$ if there exists a finite sequence $\gamma_1 \ldots \gamma_n$ such that:

- γ_n is δ;
- for each $i = 1, \ldots, n$ either $\gamma_i \in \Gamma$, or γ_i is a theorem schema, or there exist a rule $r \in dR$ and a schema variable substitution σ such that $\mathrm{Prem}(r)\sigma \subseteq \{\gamma_j : j < i\}$ and $\mathrm{Conc}(r)\sigma$ is γ_i.

Easily, $\emptyset \vdash \gamma$ if and only if γ is a theorem schema. Of course, since formulae are special cases of schema formulae, the notions of theorem and deduction also apply. The convenience for having introduced schema formulae is that both theoremhood and deducibility are *structural* [22] in the following sense: for every schema variable substitution σ, if δ is a theorem schema generated from Γ then $\delta\sigma$ is a theorem schema generated from $\Gamma\sigma$, and if $\Gamma \vdash \delta$ then $\Gamma\sigma \vdash \delta\sigma$.

Often, we shall work with a **c**-room $R = \langle\langle\{\phi\}, O\rangle, M\rangle$ equipped with a proof-calculus $C = \langle\langle\{\phi\}, O\rangle, dR, gR\rangle$. In this context, since R defines a semantic entailment relation $\vDash$ and C defines a deducibility relation $\vdash$, we shall use the following usual definitions:

- C is *sound* for R if $\Phi \vdash \psi$ implies $\Phi \vDash \psi$;
- C is *weak complete* for R if $\emptyset \vDash \varphi$ implies $\emptyset \vdash \varphi$;
- C is *finite complete* for R if Φ finite and $\Phi \vDash \psi$ imply $\Phi \vdash \psi$;
- C is *complete* for R if $\Phi \vDash \psi$ implies $\Phi \vdash \psi$,

for every set of formulae Φ and formula ψ.

The usual way of proving the soundness of a proof-calculus consists in establishing the *soundness* of each of its rules. An inference rule r is said to be *d-sound* for R if for every $\langle\mathcal{A}, \mathbf{c}\rangle \in M$ and every assignment $\alpha : \Xi \to |\mathcal{A}|_\phi$ we have that $[\![\mathrm{Conc}(r)]\!]^{\mathcal{A}}_\alpha \in \{[\![\gamma]\!]^{\mathcal{A}}_\alpha : \gamma \in \mathrm{Prem}(r)\}^{\mathbf{C}}$. On the other hand, r is said to be *g-sound* for R if for every $\langle\mathcal{A}, \mathbf{c}\rangle \in M$ and every assignment $\alpha : \Xi \to |\mathcal{A}|_\phi$, we have that $[\![\mathrm{Conc}(r)]\!]^{\mathcal{A}}_\alpha \in \emptyset^{\mathbf{C}}$ whenever $\{[\![\gamma]\!]^{\mathcal{A}}_\alpha : \gamma \in \mathrm{Prem}(r)\} \subseteq \emptyset^{\mathbf{C}}$. Clearly, if all the rules of C are sound for R, that is the rules in dR are d-sound and the rules in gR are g-sound, then C is sound for R [22,29,7].

For the sake of illustration, let us now present the well known calculus for propositional normal modal logic.

Example 4. Propositional normal modal logic.

- Consider any concrete syntax signature defined as in Example 2;
- The set dR of deduction rules is composed of the schema axioms

$$(\xi_1 \Rightarrow (\xi_2 \Rightarrow \xi_1))$$

$$((\xi_1 \Rightarrow (\xi_2 \Rightarrow \xi_3)) \Rightarrow ((\xi_1 \Rightarrow \xi_2) \Rightarrow (\xi_1 \Rightarrow \xi_3)))$$

$$(((\neg\xi_1) \Rightarrow (\neg\xi_2)) \Rightarrow (\xi_2 \Rightarrow \xi_1))$$

$$((\Box(\xi_1 \Rightarrow \xi_2)) \Rightarrow ((\Box\xi_1) \Rightarrow (\Box\xi_2)))$$

and the schema rule

$$\frac{\xi_1 \ (\xi_1 \Rightarrow \xi_2)}{\xi_2}$$

– The set gR of theorem generating rules contains just the schema rule

$$\frac{\xi_1}{(\Box\xi_1)}$$

It is well known (see, for instance, [16]) that this proof-calculus is both sound and complete for the **c**-room of propositional modal logic in Example 2.

As in the case of rooms, fibring of proof-calculi is to be characterized as a colimit. And in fact the corresponding category **PCalc** is cocomplete.

Proposition 4. PCalc *is cocomplete.*

Proof. We just show what coproducts and coequalizers look like in **PCalc**. For the purpose, let I be a set and $\{C_i = \langle\langle\{\phi\}, O_i\rangle, dR_i, gR_i\rangle\}_{i \in I}$ a family of proof-calculi. Again, we capitalize on the fact that $\mathbf{AlgSig}_\phi$ is cocomplete.
(Coproducts) Let $\{h_j : \langle\{\phi\}, O_j\rangle \to \coprod_{i \in I}\langle\{\phi\}, O_i\rangle\}_{j \in I}$ be a coproduct in the category $\mathbf{AlgSig}_\phi$. A coproduct $\{h_j : C_j \to \langle\coprod_{i \in I}\langle\{\phi\}, O_i\rangle, dR, gR\rangle\}_{j \in I}$ in **PCalc** can be obtained by taking $dR = \bigcup_{i \in I} \widehat{h}_i(dR_i)$ and $gR = \bigcup_{i \in I} \widehat{h}_i(gR_i)$.
(Coequalizers) Let $I = \{1, 2\}$, $h', h'' : C_1 \to C_2$ be **PCalc**-morphisms and $h : \langle\{\phi\}, O_2\rangle \to \langle\{\phi\}, O\rangle$ a coequalizer of $h', h'' : \langle\{\phi\}, O_1\rangle \to \langle\{\phi\}, O_2\rangle$ in $\mathbf{AlgSig}_\phi$. A coequalizer $h : C_2 \to \langle\langle\{\phi\}, O\rangle, dR, gR\rangle$ of $h', h'' : C_1 \to C_2$ in **PCalc** can be obtained by simply taking $dR = \widehat{h}(dR_2)$ and $gR = \widehat{h}(gR_2)$. $\square$

As a consequence of the Grothendieck construction, **Prf** is also cocomplete. The same applies to the following definition of fibring of proof-calculi. Let us consider fixed two arbitrary proof-calculi $C_1 = \langle\langle\{\phi\}, O_1\rangle, dR_1, gR_1\rangle$ and $C_2 = \langle\langle\{\phi\}, O_2\rangle, dR_2, gR_2\rangle$, and recall from the previous section the definition of the sharing common subsignature $\langle\{\phi\}, O_0\rangle$ and corresponding signature inclusion morphisms $h_1 : \langle\{\phi\}, O_0\rangle \to \langle\{\phi\}, O_1\rangle$ and $h_2 : \langle\{\phi\}, O_0\rangle \to \langle\{\phi\}, O_2\rangle$. We shall denote by C_0 the canonical proof-calculus $\langle\langle\{\phi\}, O_0\rangle, dR_0, gR_0\rangle$ where $dR_0 = gR_0 = \emptyset$.

Definition 7. The *fibring* of C_1 and C_2 (constrained by sharing $\langle\{\phi\}, O_0\rangle$) is the proof-calculus $C_1 \circledast C_2 = \langle\langle\{\phi\}, O\rangle, dR, gR\rangle$ such that:

– $O_u = O_{1,u} \cup O_{2,u}$ if $u \in S_0^+$, $O_u = O_{i,u}$ if $u \in S_i^+ \setminus S_0^+$ and $O_u = \emptyset$ otherwise, with inclusions $g_i : O_i \to O$;
– $dR = \langle\widehat{id_{\{\phi\}}, g_1}\rangle(dR_1) \cup \langle\widehat{id_{\{\phi\}}, g_2}\rangle(dR_2)$;
– $gR = \langle\widehat{id_{\{\phi\}}, g_1}\rangle(gR_1) \cup \langle\widehat{id_{\{\phi\}}, g_2}\rangle(gR_2)$.

Thus, the schema inference rules in the fibring are just the translation of the rules of each given proof-calculus translated to the fibred language.

Proposition 5. *The fibring of proof-calculi C_1 and C_2 (constrained by sharing $\langle S_0, O_0\rangle$) is a pushout of $\{h_i : C_0 \to C_i\}_{i \in \{1, 2\}}$ in* **PCalc.**

As a simple corollary, if the fibring is free, $C_1 \circledast C_2$ also corresponds to a coproduct of C_1 and C_2 in **PCalc**.

6   Preservation Results

We are now in the position of stating and proving our soundness and completeness preservation results. Excluding the very last completeness transference result, that is an instance of the one in [29], all the other results are suitably adapted from [7]. Recall that these are, for the moment, only concerned with propositional based logics. We shall discuss in the conclusions how these results can be extended to the general case. In the remainder of this section we shall consider fixed two **c**-rooms $R_1 = \langle\langle\{\phi\}, O_1\rangle, M_1\rangle$ and $R_2 = \langle\langle\{\phi\}, O_2\rangle, M_2\rangle$, and two corresponding proof-calculi $C_1 = \langle\langle\{\phi\}, O_1\rangle, dR_1, gR_1\rangle$ and $C_2 = \langle\langle\{\phi\}, O_2\rangle, dR_2, gR_2\rangle$.

To prove preservation of soundness we shall rely on guaranteeing the soundness of all inference rules.

Theorem 1. Soundness Preservation.
*Assume that all the rules of C_1 are sound for R_1 and all the rules of C_2 are sound for R_2. Then, the fibred proof-calculus $C_1 \circledast C_2$ is sound for the fibred **c**-room $R_1 \circledast R_2$.*

Proof. It is immediate, by definition of fibring, that in the conditions of the theorem all the rules of $C_1 \circledast C_2$ are sound for $R_1 \circledast R_2$.   □

Completeness preservation is, as expected, much harder to obtain. Our completeness preservation results are based on the following notion of *fullness*.

Definition 8. Let $\mathcal{P}$ be a class of closure operations. A **c**-room $\langle\langle\{\phi\}, O\rangle, M\rangle$ equipped with a proof-calculus C is said to be *full with respect to* $\mathcal{P}$ if M contains every structure $\langle\mathcal{A}, \mathbf{c}\rangle \in \mathrm{cAlg}(\langle\langle\{\phi\}, O\rangle\rangle)$ with $\langle|\mathcal{A}|_\phi, \mathbf{c}\rangle \in \mathcal{P}$ that makes all the rules in C sound.

Although fullness may seem to be a fairly strong requirement note that, as we have discussed before in the end of Section 4, the operation of turning a **c**-room full with respect to some given class $\mathcal{P}$ of closure operations in the context of a proof-calculus does not change its semantic entailment [29,7]. Moreover, the essential bit is that fullness is always preserved by fibring.

Proposition 6. *Let the **c**-rooms R_1 and R_2 equipped with the proof-calculi C_1 and C_2 be full with respect to a class $\mathcal{P}$ of closure operations. Then, the fibred **c**-room $R_1 \circledast R_2$ equipped with the fibred proof-calculus $C_1 \circledast C_2$ is full with respect to $\mathcal{P}$.*

Proof. Suppose that $\langle\mathcal{A}, \mathbf{c}\rangle$ makes all the rules in $C_1 \circledast C_2$ sound and $\langle|\mathcal{A}|_\phi, \mathbf{c}\rangle \in \mathcal{P}$. Easily, then, being each $\langle id_{\{\phi\}}, g_i\rangle : \langle\{\phi\}, O_i\rangle \rightarrow \langle\{\phi\}, O\rangle$ the signature morphism underlying both the morphisms from R_i to $R_1 \circledast R_2$ and from C_i to $C_1 \circledast C_2$ as defined in the fibring, and given that the rules of $C_1 \circledast C_2$ are corresponding translations of rules of C_1 or C_2, it easily follows that $\langle\mathcal{A}|_{\langle id_{\{\phi\}}, g_i\rangle}, \mathbf{c}\rangle$ makes all the rules of C_i sound. Moreover, $|(\mathcal{A}|_{\langle id_{\{\phi\}}, g_i\rangle})|_\phi = |\mathcal{A}|_\phi$ and therefore, by the fullness of each R_i equipped with C_i, it follows that $\langle\mathcal{A}|_{\langle id_{\{\phi\}}, g_i\rangle}, \mathbf{c}\rangle \in M_i$. Thus, by definition of fibring, $\langle\mathcal{A}, \mathbf{c}\rangle$ is a structure of the fibred room $R_1 \circledast R_2$.   □

We shall now present completeness preservation results for several choices of $\mathcal{P}$.

Theorem 2. Completeness Preservation – All Structures.
*Let the **c**-rooms R_1 and R_2 equipped with the proof-calculi C_1 and C_2 be full with respect to the class of all closure operations. Then, the fibred proof-calculus $C_1 \circledast C_2$ is complete for the fibred **c**-room $R_1 \circledast R_2$.*

Proof. We know that $R_1 \circledast R_2$ equipped with $C_1 \circledast C_2$ is full with respect to the class of all closure operations. It is very easy to see that the structure $\langle \mathcal{W}_{\langle \{\phi\}, O \rangle}, \mathbf{c} \rangle$ such that $\mathbf{c} = \vdash$, where $\langle \{\phi\}, O \rangle$ stands for the fibred signature, makes all the rules in $C_1 \circledast C_2$ sound. Therefore, the structure belongs to the fibred room $R_1 \circledast R_2$ as a consequence of fullness. Suppose now that, in the fibring, $\Phi \nvdash \varphi$. To show that $\Phi \nVdash \varphi$ it is enough to note that $[\![\_]\!]^{\mathcal{W}_{\langle \{\phi\}, O \rangle}}$ is the identity on formulae. □

Due to the plain use of the free word algebra, this result is, so to say, a little too much syntactic. Let us try and avoid such structures. A closure operation $\langle A, \mathbf{c} \rangle$ is said to be *elementary* if for every $a_1, a_2 \in A$, $a_1 \in \{a_2\}^{\mathbf{c}}$ and $a_2 \in \{a_1\}^{\mathbf{c}}$ together imply $a_1 = a_2$. In most of the interesting cases, the structure used in the previous proof clearly fails to be elementary.

In this case, however, we need a few further assumptions about the systems being fibred. A proof-calculus $C = \langle \langle \{\phi\}, O \rangle, dR, gR \rangle$ is said to be *congruent* if for every operator $o \in O_{\phi^n \phi}$, every set Γ of schema formulae closed for theorem generation, and all schema formulae $\gamma_1 \ldots, \gamma_n, \delta_1, \ldots, \delta_n$, it is the case that

$$\Gamma \cup \{o(\gamma_1 \ldots, \gamma_n)\} \vdash o(\delta_1, \ldots, \delta_n)$$

whenever

$$\Gamma \cup \{\gamma_i\} \vdash \delta_i \text{ and } \Gamma \cup \{\delta_i\} \vdash \gamma_i \text{ for } i = 1, \ldots, n.$$

Many proof-calculi, including the one for modal logic presented in Example 4, are indeed congruent. However, there are exceptions, as for instance the paraconsistent systems of [11].

Moreover, C is said to have an *implication* connective if there exists $\Rightarrow \in O_{\phi^2 \phi}$ such that, for every set Γ of schema formulae and all schema formulae γ, δ it is the case that

$$\Gamma \vdash (\gamma \Rightarrow \delta) \text{ if and only if } \Gamma \cup \{\gamma\} \vdash \delta.$$

This condition can be easily shown [29] to be equivalent to requiring

$$\emptyset \vdash (\xi_1 \Rightarrow \xi_1),$$

$$\{\xi_1, (\xi_1 \Rightarrow \xi_2)\} \vdash \xi_2,$$

$$\{\xi_2\} \vdash (\xi_1 \Rightarrow \xi_2)$$

and

$$\{(\xi_r \Rightarrow \gamma) : \gamma \in \text{Prem}(r)\} \vdash (\xi_r \Rightarrow \text{Conc}(r))$$

for each $r \in dR$ and some ξ_r not occurring in r.

Furthermore, in the presence of an implication, it is known [29] that congruence is equivalent to the condition that, for each operator $o \in O_{\phi^n \phi}$, the schema formula

$$(o(\xi_1 \ldots, \xi_n) \Rightarrow o(\xi_1', \ldots, \xi_n'))$$

is a theorem schema generated from

$$\{(\xi_1 \Rightarrow \xi_1'), (\xi_1' \Rightarrow \xi_1), \ldots, (\xi_n \Rightarrow \xi_n'), (\xi_n' \Rightarrow \xi_n)\}.$$

Given these characterizations of implication and congruence in the presence of an implication, it is very easy to see that the fibring of two congruent proof-calculi sharing an implication connective is also a congruent proof-calculus with implication. The proof of this property can be found in [29,7] and just uses the obvious fact that both theorem generation and deducibility are preserved by morphisms of proof-calculi.

Theorem 3. Completeness Preservation – Elementary Structures.
*Let the **c**-rooms R_1 and R_2 equipped with the proof-calculi C_1 and C_2 be full with respect to the class of all elementary closure operations. If both C_1 and C_2 are congruent and there is a shared implication connective, then the fibred proof-calculus $C_1 \circledast C_2$ is complete for the fibred **c**-room $R_1 \circledast R_2$.*

Proof. We know that $R_1 \circledast R_2$ equipped with $C_1 \circledast C_2$ is full with respect to the class of all elementary closure operations. Moreover, we also know that $C_1 \circledast C_2$ is congruent and has an implication connective $\Rightarrow$. It is easy to see that the binary relation $\equiv$ defined on formulae by $\varphi_1 \equiv \varphi_2$ if both $\{\varphi_1\} \vdash \varphi_2$ and $\{\varphi_2\} \vdash \varphi_1$ is an equivalence relation. Moreover, if $\langle \{\phi\}, O \rangle$ is the fibred signature, the congruence of $C_1 \circledast C_2$ immediately implies that $\equiv$ is a congruence relation on the word algebra $\mathcal{W}_{\langle \{\phi\}, O \rangle}$. Thus, let us consider the structure $\langle \mathcal{W}_{\langle \{\phi\}, O \rangle} / \equiv, \mathbf{c} \rangle$, corresponding to the Lindenbaum-Tarski quotient algebra together with the closure defined by $\{[\psi] : \psi \in \Psi\}^{\mathbf{c}} = \{[\psi'] : \Psi \vdash^l \psi'\}$, where $[\_]$ applied to a formula denotes its equivalence class under $\equiv$. Once again, it is straightforward to show that this structure makes all the rules in $C_1 \circledast C_2$ sound. In fact, it is clear that $[\_]^{\mathcal{W}_{\langle \{\phi\}, O \rangle} / \equiv} = [\_]$. Therefore, the structure belongs to the fibred room $R_1 \circledast R_2$ as a consequence of fullness. Suppose that, in the fibring, $\Phi \nvdash \varphi$. The structure just built clearly shows that $\Phi \nvDash \psi$. $\square$

Let us try and improve even more on this result, by requiring that the algebra of truth-values has the usual property of being ordered. Every partial-order $\langle A, \leq \rangle$ easily induces two polarities $\mathrm{Upp}(B) = \{a \in A : b \leq a \text{ for every } b \in B\}$ and $\mathrm{Low}(B) = \{a \in A : a \leq b \text{ for every } b \in B\}$, and a cut closure $\mathbf{c}$ on A defined by $B^{\mathbf{c}} = \mathrm{Upp}(\mathrm{Low}(B))$, where $B \subseteq A$, as in [3]. In such case, $\langle A, \mathbf{c} \rangle$ is said to be a *partially ordered* closure operation.

Theorem 4. Completeness Preservation – Partially Ordered Structures I.
*Let the **c**-rooms R_1 and R_2 equipped with the proof-calculi C_1 and C_2 be full with respect to the class of all partial-order closure operations. If both C_1 and C_2 are congruent and there is a shared implication connective, then the fibred proof-calculus $C_1 \circledast C_2$ is weak complete for the fibred **c**-room $R_1 \circledast R_2$.*

Proof. The proof of this result is similar to the previous one, but now considering a different closure operation on the Lindenbaum-Tarski algebra. Just note that the binary relation $\leq$ defined by $[\varphi_1] \leq [\varphi_2]$ if $\emptyset \vdash (\varphi_1 \Rightarrow \varphi_2)$ is a partial order on the quotient of the set of formulae. Therefore, let us consider the structure $\langle \mathcal{W}_{\langle\{\phi\},O\rangle}/\equiv, \mathbf{c}\rangle$ where $\mathbf{c}$ is the cut closure induced by $\leq$ as explained above. Let us check, in this less trivial case, that this structure indeed makes all the rules of $C_1 \circledast C_2$ sound. Just note that an assignment α in the quotient algebra sends each schema variable to an equivalence class of formulae. Therefore, if we set a schema variable substitution σ such that $[\sigma(\xi)] = \alpha(\xi)$, it immediately follows that $[\![\_]\!]_\alpha^{\mathcal{W}_{\langle\{\phi\},O\rangle}/\equiv} = [\_\sigma]$. Consider a d-rule $r = \frac{\gamma_1...\gamma_n}{\delta}$ and fix an assignment α. We need to show that $[\delta\sigma] \in \{[\gamma_1\sigma],\ldots,[\gamma_n\sigma]\}^{\mathbf{c}}$. Therefore, let φ be a sentence such that $[\varphi] \leq [\gamma_i\sigma]$ for $i = 1,\ldots,n$. By definition, this means that $\emptyset \vdash (\varphi \Rightarrow \gamma_i\sigma)$ for each i. Therefore, by using the last requirement (concerning d-rules) in the characterization of implication together with the structurality of deducibility, we can chose σ as above such that $\sigma(\xi_r) = \varphi$ and therefore conclude that also $\emptyset \vdash (\varphi \Rightarrow \delta\sigma)$. Equivalently, this means that $[\varphi] \leq [\delta\sigma]$ and the d-rule is sound. Assume now that r is a g-rule. We need to show that if $\{[\gamma_1\sigma],\ldots,[\gamma_n\sigma]\} \in \emptyset^{\mathbf{c}}$ then also $[\delta\sigma] \in \emptyset^{\mathbf{c}}$. In this case, it is easy to see that $\emptyset^{\mathbf{c}}$ has precisely one element, corresponding to the equivalence class of all theorems of $C_1 \circledast C_2$. Thus, if all $\gamma_i\sigma$ are theorems, then by using the dR rule r we can conclude that also $\delta\sigma$ is a theorem, and the g-rule is sound. Therefore, as a consequence of fullness, the structure belongs to the fibred room $R_1 \circledast R_2$. Now, if $\emptyset \nvdash \psi$ this structure clearly shows that $\emptyset \nVdash \psi$. □

One may wonder why the construction fails if we want to go beyond weak completeness. If $\Phi \nvdash \psi$, in the structure built above, we would need to find a sentence that would imply all the formulae in Φ but not ψ. This, in general, can only be solved in the presence of some form of infinitary conjunction, since Φ may very well be an infinite set. However, we can still improve the result a little bit by also dealing with the finite case.

A proof-calculus $C = \langle \langle \{\phi\}, O\rangle, dR, gR\rangle$ is said to have a *conjunction* connective if there exists $\wedge \in O_{\phi^2\phi}$ such that, for every set of schema formulae Γ and all schema formulae $\delta_1, \delta_2, \gamma$ it is the case that

$$\Gamma \cup \{\delta_1, \delta_2\} \vdash \gamma \text{ if and only if } \Gamma \cup \{(\delta_1 \wedge \delta_2)\} \vdash \gamma.$$

This condition can be easily shown to be equivalent to requiring

$$\{(\xi_1 \wedge \xi_2)\} \vdash \xi_1$$
$$\{(\xi_1 \wedge \xi_2)\} \vdash \xi_2$$
$$\text{and}$$
$$\{\xi_1, \xi_2\} \vdash (\xi_1 \wedge \xi_2).$$

Theorem 5. Completeness Preservation – Partially Ordered Structures II.
Let the c-rooms R_1 and R_2 equipped with the proof-calculi C_1 and C_2 be full with

*respect to the class of all partial-order closure operations. If both C_1 and C_2 are congruent, there is a shared implication connective and at least one of them has a conjunction connective, then the fibred proof-calculus $C_1 \circledast C_2$ is finite complete for the fibred **c**-room $R_1 \circledast R_2$.*

Proof. Consider exactly the same structure of the previous proof, and suppose that $\{\varphi_1, \ldots, \varphi_n\} \not\vdash \psi$. Considering the sentence $\varphi = (\varphi_1 \wedge \ldots \wedge \varphi_n)$, it is trivial to check that $\emptyset \vdash (\varphi \Rightarrow \varphi_i)$ for each $i = 1, \ldots, n$. However, easily, it is also the case that $\emptyset \not\vdash (\varphi \Rightarrow \psi)$ and the structure shows that $\{\varphi_1, \ldots, \varphi_n\} \not\vdash \psi$. $\square$

Being unfeasible to require an infinitary conjunction connective and therefore obtain completeness, we now just point out to an alternative. The following result, that we just state, is an instance of the one proved in [29] and concerns algebras of sets in the style of general frames for modal logic (see [16]) and uses a Henkin-style construction. For a powerset lattice $\langle \wp(U), \subseteq \rangle$, the cut closure **c** induced by the polarities as above is such that $B^{\mathbf{c}} = \{b \subseteq \wp(U) : (\bigcap B) \subseteq b\}$. We have already used such closures in the examples. A closure operation $\langle A, \mathfrak{e} \rangle$ is said to be a *general powerset* closure operation if $A \subseteq \wp(U)$, $U \in A$ and $\mathfrak{e}$ is the closure induced by **c** on A, i.e., $B^{\mathfrak{e}} = B^{\mathbf{c}} \cap A$ for each $B \subseteq A$.

Theorem 6. Completeness Preservation – General Powerset Structures.
*Let the **c**-rooms R_1 and R_2 equipped with the proof-calculi C_1 and C_2 be full with respect to the class of all general powerset closure operations. If both C_1 and C_2 are congruent and there is a shared implication connective, then the fibred proof-calculus $C_1 \circledast C_2$ is complete for the fibred **c**-room $R_1 \circledast R_2$.*

Other approaches to completeness preservation, namely via encoding in a suitable meta-logic have already been employed in the specific context of para-consistent non-truth-functional logics [8] but should of course be also workable in this setting. Some forms of conditional equational logic, namely rewriting logic [17], seem to be good candidate meta-logics for this process of algebraization.

7   Conclusion

We have introduced the novel notion of **c**-parchment, by enriching the structure of interpretation algebras, as a means of presenting an institution, and we have shown how it can be used to bring the combination mechanism of fibring to the institutional setting. Moreover, **c**-parchments were shown to correspond to indexed categories of **c**-rooms and their fibring characterized as a colimit. Furthermore, we have put the construction into practice by exploring partial equational logic as a fibred logic. The example is even more interesting since we also obtain a proof-calculus for partial equational logic by fibring calculi for equational logic and the logic of partiality adopted. In fact, although in the simplest case of propositional based logics, we have established a collection of soundness and completeness preservation results for fibring, thus showing that the successfulness of the example is not just a mere coincidence.

Of course, this line work is far from over. The most important topic to be followed is precisely the extension of these preservation results to a more general context. Namely, the work reported in [24] seems to bring ideas that can also be explored in this context. This comment applies not only to preservation results but, first of all, also with the way we should deal with more complex notions of proof-calculus than the one used here. Work in this direction has already been done in [23], including ways of representing constrained schematic rules such as those used in many logics with terms and quantifiers. Two well known such examples appear in classical first-order logic's axioms $((\forall x(\varphi \Rightarrow \psi)) \Rightarrow (\varphi \Rightarrow (\forall x \, \psi)))$ and $((\forall x \, \varphi) \Rightarrow \varphi_t^x)$. However, in both cases there are well known constraints. Namely, in the first case, there is the requirement that x must not occur free in φ and, in the second, the requirement that t must be free for x in φ.

Another interesting possibility to be pursued is to explore a process of algebraization using a suitable meta-logic, as mentioned above, also putting in context the notion of fullness and its importance not only for completeness purposes but even more for shaping up the whole process of fibring logics, specially if they have interpretation structures with different bases. This alternative does not only bring us closer to the realm of algebraic logic [10,5,1], but it also has the advantage of being able to deal, at an adequate level, with non-truth-functional logics. In fact, several interesting paraconsistent and paracomplete logics, with meaningful applications in computer science, fail to be truth-functional in the sense that some of its operators are not congruent. Some preliminary work also in the setting up of a meaningful notion of non-truth-functional parchment can be found in [9]. However, in this context, there is still a lot of research work to be done in order to extend the usual algebraic techniques of logic to cope with the possible absence of congruence. We are also interested in studying the representation of fibring in logical frameworks. Namely, capitalizing on Meseguer's theory of general logics [18], we aim at characterizing the mechanism of fibring of logics within rewriting logic [17]. In particular, general representation preservation results are envisaged, that may determine the exact extent to which representations of fibred logics can be obtained out of representations of the logics being fibred. Last but not least, future work should also cover the characterization of fibring of logics presented by other means, either model or proof-theoretic, as well as the search for transference results for other interesting properties, such as decidability, the finite model property, interpolation and amalgamation.

Acknowledgments

The authors are grateful to Alberto Zanardo and Cristina Sernadas for many valuable discussions on these and related topics. Thanks are also due to the anonymous referees for suggesting several improvements on an earlier version of the paper. This work is part of the undergoing FibLog initiative and was partially supported by *Fundação para a Ciência e a Tecnologia* via the PRAXIS XXI Project ProbLog (PRAXIS/P/MAT/10002/1998).

References

1. H. Andréka, Á. Kurucz, I. Németi, and I. Sain. Applying algebraic logic: a general methodology. Preprint, Mathematical Institute of the Hungarian Academy of Sciences, Budapest, Hungary, 1994.
2. E. Astesiano and M. Cerioli. Free objects and equational deduction for partial conditional specifications. *Theoretical Computer Science*, 152(1):91–138, 95.
3. G. Birkhoff. *Lattice Theory*. AMS Colloquium Publications, 1967.
4. P. Blackburn and M. de Rijke. Why combine logics? *Studia Logica*, 59(1):5–27, 1997.
5. W. Blok and D. Pigozzi. *Algebraizable Logics*, volume 77 of *Memoires of the AMS*. American Mathematical Society, 1989.
6. D. Brown and R. Suszko. Abstract logics. *Dissertationes Mathematicae*, 102:9–41, 1973.
7. C. Caleiro. *Combining Logics*. PhD thesis, IST, TU Lisbon, 2000.
8. C. Caleiro, W.A. Carnielli, M.E. Coniglio, A. Sernadas, and C. Sernadas. Fibring non-truth-functional logics: Completeness preservation. Preprint, Dep. Mathematics, IST, Lisbon, Portugal, 2000. Submitted.
9. C. Caleiro and J. Marcos. Non-truth-functional fibred semantics. In H. R. Arabnia, editor, *Proceedings of the International Conference on Artificial Intelligence (IC-AI'2001)*, volume II, pages 841–847. CSREA Press, Athens GA, USA, 2001.
10. J. Czelakowski. Equivalential logics (I and II). *Studia Logica*, 40:227–236 and 355–372, 1981.
11. N.C.A. da Costa. On the theory of inconsistent formal systems. *Notre Dame Journal of Formal Logic*, 15(4):497–510, 1974.
12. D. Gabbay. Fibred semantics and the weaving of logics: part 1. *Journal of Symbolic Logic*, 61(4):1057–1120, 1996.
13. D. Gabbay. *Fibring Logics*. Clarendon Press - Oxford, 1999.
14. J. Goguen and R. Burstall. A study in the foundations of programming methodology: specifications, institutions, charters and parchments. In *Category Theory and Computer Programming*, volume 240 of *LNCS*, pages 313–333. Springer-Verlag, 1986.
15. J. Goguen and R. Burstall. Institutions: abstract model theory for specification and programming. *Journal of the ACM*, 39(1):95–146, 1992.
16. G. Hughes and M. Cresswell. *A New Introduction to Modal Logic*. Routledge, London, 1996.
17. N. Martí-Oliet and J. Meseguer. Rewriting logic as a logical and semantic framework. *Electronic Notes in Theoretical Computer Science*, 4, 1996.
18. J. Meseguer. General logics. In H.-D. Ebbinghaus *et al*, editor, *Proceedings of the Logic Colloquium'87*, pages 275–329. North-Holland, 1989.
19. T. Mossakowski. Using limits of parchments to systematically construct institutions of partial algebras. In *Recent Trends in Data Type Specification*, volume 1130 of *LNCS*, pages 379–393. Springer-Verlag, 1996.
20. T. Mossakowski, A. Tarlecki, and W. Pawłowski. Combining and representing logical systems. In *Category Theory and Computer Science 97*, volume 1290 of *LNCS*, pages 177–196. Springer-Verlag, 1997.
21. T. Mossakowski, A. Tarlecki, and W. Pawłowski. Combining and representing logical systems using model-theoretic parchments. In *Recent Trends in Algebraic Development Techniques*, volume 1376 of *LNCS*, pages 349–364. Springer-Verlag, 1998.

22. A. Sernadas, C. Sernadas, and C. Caleiro. Fibring of logics as a categorial construction. *Journal of Logic and Computation*, 9(2):149–179, 1999.

23. A. Sernadas, C. Sernadas, C. Caleiro, and T. Mossakowski. Categorial fibring of logics with terms and binding operators. In D. Gabbay and M. de Rijke, editors, *Frontiers of Combining Systems 2*, pages 295–316. Research Studies Press, 2000.

24. A. Sernadas, C. Sernadas, and A. Zanardo. Fibring modal first-order logics: Completeness preservation. Preprint, Section of Computer Science, Department of Mathematics, Instituto Superior Técnico, 1049-001 Lisboa, Portugal, 2001. Submitted for publication.

25. T. Smiley. The independence of connectives. *Journal of Symbolic Logic*, 27(4):426–436, 1962.

26. A. Tarlecki, R. Burstall, and J. Goguen. Some fundamental algebraic tools for the semantics of computation. Part 3: indexed categories. *Theoretical Computer Science*, 91:239–264, 1991.

27. M. Wirsing. Algebraic specification. In J. van Leeuwen, editor, *Handbook of Theoretical Computer Science*, pages 675–787. Elsevier Science Publishers, 1990.

28. R. Wójcicki. *Theory of Logical Calculi*. Synthese Library. Kluwer Academic Publishers, 1988.

29. A. Zanardo, A. Sernadas, and C. Sernadas. Fibring: Completeness preservation. *Journal of Symbolic Logic*, 66(1):414–439, 2001.

Canonical Institutions of Behaviour

J. Félix Costa and H. Lourenço

Departamento de Matemática, I.S.T., Universidade Técnica de Lisboa
Av. Rovisco Pais, 1049-001 Lisboa, Portugal
{fgc,hlouren}@math.ist.utl.pt

Abstract. The concept of behaviour plays a central role in the specification of a considerable number of different kinds of systems. In these settings a "behaviour" is seen as a possible evolution (or life-cycle) of the system, whereas the system itself is considered to be defined by the set of all its possible behaviours.

Examples of this kind of situation are common. Maybe the most well known and studied is that of concurrency theory: a behaviour is e.g. a stream of actions and the system is a process (in this case, a set of streams of actions).

If institutions are used as the way for specifying the systems, then it is customary to start by creating an institution for individual behaviours (where each model corresponds to a possible behaviour) from which the "system institution" – or "institution of behaviour", in our terminology – where each model is a set of behaviours is built.

The new institution is tightly bound to the base institution, sharing signatures and languages. Also, because the models are obtained from the base institution's models, the satisfaction relation is defined in terms of the base satisfaction relation.

In this paper it is shown that the construction of these institutions of behaviour can be carried out in a canonical way. Indeed, the construction does not depend in any way at all on the particular base institution chosen. It is also shown that several institutions presented since the 90's in WADT workshops and elsewhere arise as particular cases of this canonical construction [4, 2, 3, 6–8].

It is hoped that the proposed construction can be used as a shortcut for defining new useful institutions of behaviour.

1 Introduction

This paper emerged from the necessity of abstracting the methodology used in the past when creating a considerable number of different institutions.

More concretely, one of the authors followed an already recurrent pattern (started by the seminal paper [4] and continued in [2,3,7,8]) when trying to create an institution for hybrid systems[1] [6]. As it had always been the case, the construction heavily relied upon the specific case at hand.

The papers went roughly like this: in the first place a suitable notion of behaviour was considered, such as a stream of actions (in the case of concurrency

---

[1] A *hybrid system* is a system exhibiting both discrete and continuous behaviour.

M. Cerioli and G. Reggio (Eds.): WADT/CoFI 2001, LNCS 2267, pp. 71–84, 2002.

theory), after which an institution for talking about individual behaviours was defined (for instance, temporal logic). Then, and since the initial goal was to talk about systems consisting of sets of behaviours, another institution was defined based upon the initial one.

Later, J. Félix Costa directed him to some previous work that he had done when studying the relation between temporal theories and models of concurrency. We had resorted to a set of categorial tools that included topological theories.

Although the initial focus was in the search for a categorial formalization of the relation between temporal theories and concurrency models, the present work proves to be more general.

In the one hand, it provides a straightforward way of creating what we call "institutions of behaviour". These arise when specifying a system in terms of its individual behaviours (as was the case in [6]).

In the other hand, it is not restricted to temporal theories, although they indeed seem the right candidate when talking about "behaviours".

Another aspect that initially was somewhat overlooked and is now addressed is denotational semantics. In the abstract setting of institutions of behaviour, the relation between specifications (and, later on, theories) and models is investigated. In particular, it is shown that it is possible to give denotational semantics to the usual constructions of unconstrained and constrained parallel composition (provided these constructions can be represented at the signature level).

We assume that the reader has a working knowledge of the basic notions of category theory, logic and institutions. For the sake of completeness, most of the categorial constructions used are defined in the paper (a very complete reference can be found in [1]; another good source is [5]).

In Section 2 we present the plan for our journey. We start by defining both the starting and the destination points - behaviour structures and institutions of behaviour, respectively.

The journey itself is conducted in Section 3. Here all the steps to follow in order to transform a behaviour structure into an institution are detailed. The transformation resorts to established categorial tools, which happens to be quite enlightening.

Finally, in Section 4 we tackle the denotational semantics aspects of our quest. We start by relating specifications and models, and then go on to relate theories both with specifications and models. As we'll see, in this framework theories and models appear unified (the unification being materialized by an adjunction).

2   Roadmap

2.1   The Ingredients ...

The starting point in our journey towards an institution of behaviour is a precise definition of what a behaviour is in the considered context. In our setting, behaviours must be presented via *behaviour structures*.

Definition 1. A *behaviour structure* is a tuple $\langle Sig, Sen, Bh, \Vdash^B \rangle$ where

- *Sig* is a category whose objects are called *signatures*;
- $Sen : Sig \to Set$ is a functor associating to each signature a set of *sentences*;
- $Bh : Sig^{op} \to Set$ is a functor associating to each signature a set of *behaviours*;
- $\Vdash^B = \{ \Vdash^B_\Sigma \}_{\Sigma \in |Sig|}$ is a family of *satisfaction relations* where each $\Vdash^B_\Sigma$ is a subset of $Bh(\Sigma) \times Sen(\Sigma)$

and such that for all $\sigma : \Sigma \to \Sigma'$ in *Sig*, $\varphi \in Sen(\Sigma)$ and $b' \in Bh(\Sigma')$ the following *satisfaction condition* is met:

$$Bh(\sigma)(b') \Vdash^B_\Sigma \varphi \quad \text{iff} \quad b' \Vdash^B_{\Sigma'} Sen(\sigma)(\varphi)$$

Note that behaviour structures closely resemble institutions (indeed they are institutions). We adopted a lighter definition because in our usual setting the added structure of an institution (namely the functor $Int : Sig^{op} \to Cat$ for talking about interpretation structures) is not relevant for describing behaviours.

As an example of a behaviour structure, consider an example from concurrency theory.[2]

Example 2. The behaviour interpretation structure $\mathcal{P}$ of traces consists of the following:

- $Sig = Set$. Each $\Sigma \in |Sig|$ is a set of action symbols;
- for each $\Sigma \in |Sig|$, $Sen(\Sigma)$ is inductively defined as follows:
  - $\Sigma \subseteq Sen(\Sigma)$;
  - $(\neg\varphi), (\mathbf{X}\varphi) \in Sen(\Sigma)$ p.t. $\varphi \in Sen(\Sigma)$;
  - $(\varphi \Rightarrow \psi), (\varphi\mathbf{U}\psi) \in Sen(\Sigma)$ p.t. $\varphi, \psi \in Sen(\Sigma)$.
- for each $\sigma : \Sigma \to \Sigma'$ in *Sig*, $Sen(\sigma) = \underline{\sigma}$ where
  - $\underline{\sigma}(a) = \sigma(a)$ for each $a \in \Sigma$;
  - $\underline{\sigma}((\neg\varphi)) = (\neg\underline{\sigma}(\varphi))$;
  - $\underline{\sigma}((\varphi \Rightarrow \psi)) = (\underline{\sigma}(\varphi) \Rightarrow \underline{\sigma}(\psi))$;
  - etc.
- $Bh : Sig^{op} \to Set$ is the functor defined by:
  - $Bh(\Sigma) = (2^\Sigma)^{\mathbb{N}_0}$; each behaviour (or stream) is a sequence containing at each point in time the snapshot of actions that occurred at that point;
  - for $\sigma : \Sigma' \to \Sigma$ in Sig^{op} and $b' \in Bh(\Sigma')$,
    $Bh(\sigma)(b') = \lambda n.\sigma^{-1}(b'(n))$.
- $\Vdash^B$ is defined as usual: for $\Sigma \in |Sig|$, $b \in Bh(\Sigma)$ and $i \in \mathbb{N}_0$,
  - $b \Vdash^B_\Sigma \varphi$ iff $b, 0 \Vdash^B_\Sigma \varphi$;
  - $b, i \Vdash^B_\Sigma a$ iff $a \in b(i)$;
  - $b, i \Vdash^B_\Sigma (\neg\varphi)$ iff not $b, i \Vdash^B_\Sigma \varphi$;
  - etc.

---

[2]   As we'll see later, from this structure the process institution as presented in [3], for instance, arises.

2.2   ... and the Cake

Our aim is to obtain an institution from a behaviour structure. First of all, we recall what an institution is.

Definition 3. An *institution* is a tuple $\langle Sig, Sen, Int, \Vdash \rangle$ where

- *Sig* is a category whose objects are called *signatures*;
- *Sen* : *Sig* → *Set* is a functor associating to each signature a set of *sentences*;
- *Int* : *Sig*^{op} → *Cat* is a functor associating to each signature a category of *interpretation structures*;
- $\Vdash = \{\Vdash_\Sigma\}_{\Sigma \in |Sig|}$ is a family of *satisfaction relations* where each $\Vdash_\Sigma$ is a subset of $|Int(\Sigma)| \times Sen(\Sigma)$

and such that for all signature morphism $\sigma : \Sigma \to \Sigma'$, sentence $\varphi \in Sen(\Sigma)$, and interpretation structure $I' \in |Int(\Sigma')|$, the following *satisfaction condition* is met:

$$Int(\sigma)(I') \Vdash_\Sigma \varphi \quad \text{iff} \quad I' \Vdash_{\Sigma'} Sen(\sigma)(\varphi)$$

We'll refer to institutions arising from behaviour structures as *institutions of behaviour*.

2.3   What About the Recipe?

In a behaviour institution we're interested in keeping signatures and languages and having sets of behaviours as interpretation structures. Also, we want to extend the behaviour satisfaction relation from behaviours to sets of behaviours in such a way that the satisfaction condition is met.

How should we then go from a structure $\langle Sig, Sen, Bh, \Vdash^B \rangle$ to an institution $\langle Sig, Sen, Int, \Vdash \rangle$? As we'll see in the following section, there is a canonical way of doing this.

Along the way, some categorial tools are needed to ease the baking process. Most of these tools are presented in [1].

3   Canonical Institutions of Behaviour

It was referred earlier that interpretation structures are to be sets of behaviours. There is a categorial construction, *Spa*, that is well suited for representing this.

Definition 4. A *concrete category* over $\mathcal{X}$ is a pair $\langle \mathcal{A}, U \rangle$ where $\mathcal{A}$ is a category and $U : \mathcal{A} \to \mathcal{X}$ is a faithful functor (the underlying or forgetful functor).

Definition 5. Let $T : \mathcal{X} \to Set$ be a functor. $Spa(T)$ is the concrete category over $\mathcal{X}$ in which:

- the objects are pairs $\langle X, \alpha \rangle$ with $X \in |\mathcal{X}|$ and $\alpha \subseteq T(X)$;

- a morphism $f : \langle X, \alpha \rangle \to \langle Y, \beta \rangle$ is a morphism $f : X \to Y$ in $\mathcal{X}$ such that $Tf(\alpha) \subseteq \beta$;
- the underlying functor $U : Spa(T) \to \mathcal{X}$ is defined by $U(f : \langle X, \alpha \rangle \to \langle Y, \beta \rangle) = f : X \to Y$.

In the particular case of $Spa(Bh)$, objects are pairs $\langle \Sigma, \beta \rangle$ with $\beta \subseteq Bh(\Sigma)$ (that is, β is a set of behaviours) and each morphism $\sigma : \langle \Sigma, \beta \rangle \to \langle \Sigma', \beta' \rangle$ is a Sig^{op} morphism $\sigma : \Sigma \to \Sigma'$ s.t. $Bh(\sigma)(\beta) \subseteq \beta'$ (that is, each behaviour in β must be transformed by $Bh(\sigma)$ into a behaviour of β'). The underlying functor is $bS : Spa(Bh) \to Sig^{op}$.

In order to obtain not only interpretation structures but the complete functor Int, a bit more of work is necessary.

Definition 6. Let $F : \mathcal{A} \to \mathcal{X}$ be a functor, $f : X \to Y$ in $\mathcal{X}$ and A in $\mathcal{A}$ s.t. $F(A) = X$. A morphism $\hat{f} : A \to B$ in $\mathcal{A}$ is said to be *cocartesian* by F for f on A if

- $F(\hat{f}) = f$;
- for each pair or morphisms $g : A \to C$ in $\mathcal{A}$ and $h : Y \to Z$ in $\mathcal{X}$ such that $Z = F(C)$ and $h \circ f = F(g)$ there is a unique morphism $\hat{h} : B \to C$ such that $F(\hat{h}) = h$ and $\hat{h} \circ \hat{f} = g$.

Dually, we define cartesian morphisms.

Definition 7. A functor $F : \mathcal{A} \to \mathcal{X}$ is said to be a *cofibration* if for every $f : X \to Y$ in $\mathcal{X}$ and $A \in \mathcal{A}$ such that $F(A) = X$ there is a cocartesian morphism by F for f on A. Dually, we define fibrations.

Definition 8. A *bifibration* is a functor that is simultaneously a fibration and a cofibration.

Proposition 9. In $Spa(T)$, given $f : X \to Y$:

- $f : \langle X, \alpha \rangle \to \langle Y, Tf(\alpha) \rangle$ is cocartesian by U for f on $\langle X, \alpha \rangle$;
- $f : \langle X, (Tf)^{-1}(\beta) \rangle \to \langle Y, \beta \rangle$ is cartesian by U for f on $\langle Y, \beta \rangle$

and so the underlying functor is always a bifibration.

Bifibrations are quite interesting: under some additional conditions, a bifibration $\mathcal{A} \to \mathcal{X}$ induces a functor $\mathcal{X} \to Cat$.

Definition 10. A *cocleavage* κ for a cofibration $F : \mathcal{A} \to \mathcal{X}$ maps each pair $\langle f : F(A) \to Y, A \rangle$ to a cocartesian morphism by F for f on A. κ is said to be a *splitting* if it preserves identity and composition.

Example 11. Let $T : \mathcal{X} \to Set$ be a functor and $U : Spa(T) \to \mathcal{X}$ the underlying functor. ¿From Proposition 9 we obtain a cocleavage κ by defining that for each $f : X \to Y$ and $\langle X, \alpha \rangle \in |Spa(T)|$, $\kappa(f, \langle X, \alpha \rangle) = f : \langle X, \alpha \rangle \to \langle Y, T f(\alpha) \rangle$.

Definition 12. Let $F : \mathcal{A} \to \mathcal{X}$ be a functor and $X \in |\mathcal{X}|$. The *fiber* $\mathcal{A}_X$ of $\mathcal{A}$ over X is the subcategory of $\mathcal{A}$ whose objects are mapped by F to X and whose arrows are mapped to id_X.

We are now ready to go from a functor $\mathcal{A} \to \mathcal{X}$ to another one $\mathcal{X} \to Cat$.

Proposition 13. A cofibration $F : \mathcal{A} \to \mathcal{X}$ with splitting cocleavage κ induces a functor $G : \mathcal{X} \to Cat$ defined by:

– $G(X) = \mathcal{A}_X$;
– for $f : X \to Y$ in $\mathcal{X}$, $G(f) : \mathcal{A}_X \to \mathcal{A}_Y$ is the functor defined by
  • $Gf(A) = cod(\kappa(f, A))$;
  • $Gf(g : A \to A')$ is the only $h : Gf(A) \to Gf(A')$ s.t.
    $h \circ \kappa(f, A) = \kappa(f, A') \circ g$.

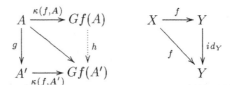

3.1   Interpretation Structures

Applying this proposition to the particular case of $Spa(T)$ categories provides us with the desired path from Bh to Int.

Corollary 14. In $Spa(T)$ (with $T : \mathcal{X} \to Set$), the underlying functor $U : Spa(T) \to \mathcal{X}$ induces $G : \mathcal{X} \to Cat$ s.t.

– $G(X) = Spa(T)_X$:
  • the objects of $G(X)$ are pairs $\langle X, \alpha \rangle$ with $\alpha \subseteq T(X)$;
  • id_X is a morphism $id_X : \langle X, \alpha \rangle \to \langle X, \beta \rangle$ iff $\alpha \subseteq \beta$.
– for $f : X \to Y$ in $\mathcal{X}$:
  • $Gf(\langle X, \alpha \rangle) = \langle Y, T f(\alpha) \rangle$;
  • $Gf(id_X) = id_Y$

We thus obtain $Int : Sig^{op} \to Cat$ from $bS : Spa(Bh) \to Sig^{op}$. An interpretation structure is a pair $\langle \Sigma, \beta \rangle$ with $\beta \subseteq Bh(\Sigma)$, and for $\sigma : \Sigma \to \Sigma'$ in Sig^{op},

$$Int(\sigma)(\langle \Sigma, \beta \rangle) = \langle \Sigma', Bh(\sigma)(\beta) \rangle.$$

3.2   Satisfaction

The satisfaction relation for institutions of behaviour must be obtained from the satisfaction relation of the underlying behaviour structure. This is easily accomplished by extending such relation to sets.

Definition 15. For each Σ, $\Vdash_\Sigma \subseteq |Int(\Sigma)| \times Sen(\Sigma)$ is defined by

$$\langle \Sigma, \beta \rangle \Vdash_\Sigma \varphi \quad \text{iff} \quad b \Vdash_\Sigma^B \varphi \text{ for every } b \in \beta$$

The satisfaction condition for the institution of behaviour is now a consequence of the satisfaction condition for the underlying behaviour structure.

Proposition 16. For every $\sigma : \Sigma \to \Sigma'$ in *Sig*, $\varphi \in Sen(\Sigma)$ and $I' \in |Int(\Sigma')|$:

$$Int(\sigma)(I') \Vdash_\Sigma \varphi \quad \text{iff} \quad I' \Vdash_{\Sigma'} Sen(\sigma)(\varphi)$$

Proof. Let $\sigma : \Sigma \to \Sigma'$ be a morphism in *Sig*, $\varphi \in Sen(\Sigma)$ a formula and $I' = \langle \Sigma', \beta' \rangle$ be a Σ'-interpretation structure. We then have that:

$$
\begin{aligned}
I' \Vdash_{\Sigma'} Sen(\sigma)(\varphi) \quad &\text{iff} \quad b' \Vdash_{\Sigma'}^B Sen(\sigma)(\varphi) \text{ for every } b' \in \beta' \\
&\text{iff} \quad Bh(\sigma)(b') \Vdash_\Sigma^B \varphi \text{ for every } b' \in \beta' \\
&\text{iff} \quad Int(\sigma)(I') \Vdash_\Sigma \varphi.
\end{aligned}
$$

$\square$

3.3   Institutions of Behaviour

The preceding sections have shown how to go from a behaviour structure to an institution (of behaviour) in a canonical way.

In the particular case of the behaviour structure $\mathcal{P}$, the corresponding institution of behaviour coincides with the institution for processes presented, among others, in [3].

4   Denotational Semantics

Now that we have a canonical way of constructing institutions of behaviour, we want to go a step further and study some of their properties.

This study is heavily biased towards denotational semantics aspects. We are mainly interested in the relation between specifications and interpretation structures, both of each arise from *Spa* constructions: both specifications and interpretation structures are sets (of formulae and behaviours, respectively).

The usefulness of denotational semantics whould be clear: it provides a way of relating constructions in the syntactical and semantic domains. For instance, it is desirable to have the property that putting together specifications corresponds in some way to putting together models of such specifications.

Since (co)products and (co)cartesian liftings are our preferred denotational semantics tools, it is natural to start by determining if (co)products and (co)cartesian liftings are inherent in behaviour institutions. If not, appropriate sufficiente conditions must be sought.

Again, the experience in past papers ([4,2,3,6,7,8]) is helpful. In all these papers, the existence of (co)products and (co)cartesian liftings is established in an ad-hoc way that can however be extrapolated to a fully general proof: as will be seen, the existence of these constructions is an "almost" intrinsic property of behaviour institutions, depending only on the existence of (co)products in Sig.

In the following we'll consider and relate the concrete category of specifications $\langle Spa(Sen), sS : Spa(Sen) \to Sig \rangle$ with the concrete category of behaviours $\langle Spa(Bh), bS : Spa(Bh) \to Sig^{op} \rangle$.

Again, most of the categorial tools come from [1].

4.1   (Co)completeness et al

Definition 17. Let $G : \mathcal{A} \to \mathcal{B}$ be a functor and $\mathcal{S} = (A \xrightarrow{f_i} A_i)_I$ a source in $\mathcal{A}$. $\mathcal{S}$ is called G-initial if for each source $\mathcal{T} = (B \xrightarrow{g_i} A_i)_I$ in $\mathcal{A}$ and $h : G(B) \to G(A)$ in $\mathcal{B}$ such that $G\mathcal{T} = G\mathcal{S} \circ h$ there exists a unique $\hat{h} : B \to A$ s.t. $\mathcal{T} = \mathcal{S} \circ \hat{h}$ and $h = G(\hat{h})$.

$$
\begin{array}{ccc}
A \xrightarrow{f_i} A_i & \qquad & GA \xrightarrow{Gf_i} GA_i \\
\hat{h} \uparrow \quad \nearrow g_i & & G\hat{h}=h \uparrow \quad \nearrow Gg_i \\
B & & GB
\end{array}
$$

G is called *topological* if every source $(B \xrightarrow{f_i} GA_i)_I$ has a unique G-initial lift.

Definition 18. *A concrete category* $\langle \mathcal{A}, U \rangle$ *over* $\mathcal{X}$ *is said* topological *if* U *is* topological.

The underlying functor in topological concrete categories has a very nice property, as the following proposition shows.

Proposition 19. Let $\langle \mathcal{A}, U \rangle$ be a concrete category over $\mathcal{X}$. If this concrete category is topological, then U uniquely lifts limits and colimits and preserves limits and colimits.

Proposition 20. Let $T : \mathcal{X} \to Set$ be a functor. In the concrete category $\langle Spa(T), U \rangle$, a source $(\langle X, \alpha \rangle \xrightarrow{f_i} \langle X_i, \alpha_i \rangle)_I$ is U-initial iff $\alpha = \bigcap_I T(f_i)^{-1}(\alpha_i)$, and so each source has a unique U-initial lift.

Additionally, if you consider the concrete category $\langle Spa(T)^{op}, U^{op} \rangle$, then a source $(\langle X, \alpha \rangle \xrightarrow{f_i} \langle X_i, \alpha_i \rangle)_I$ is U^{op}-initial iff $\alpha = \bigcup_I T(f_i)(\alpha_i)$.

The preceding proposition shows that in particular every Spa category is topological. A deeper result, however, is the following.

Corollary 21. The concrete category $\langle Spa(Bh), bS \rangle$ is (co)complete iff Sig^{op} is (co)complete. Similarly, the category $\langle Spa(Sen), sS \rangle$ is (co)complete iff Sig is (co)complete.

4.2   (Co)products

In the light of our denotational semantics study, the Proposition 19 implies that if (co)products exist in Sig^{op} (resp. Sig), then we have (co)products in $Spa(Bh)$ (resp. $Spa(Sen)$).

The following proposition characterizes (co)products in $Spa(T)$ categories.

Proposition 22. In the concrete category $\langle Spa(T), U \rangle$ over $\mathcal{X}$:

- if $\langle X, i', i'' \rangle$ is a coproduct of X' and X'' in $\mathcal{X}$, then

$$\langle \langle X, Ti'(\alpha') \cup Ti''(\alpha'') \rangle, i', i'' \rangle$$

  is a coproduct of $\langle X', \alpha' \rangle$ and $\langle X'', \alpha'' \rangle$ in $Spa(T)$;
- if $\langle X, p', p'' \rangle$ is a product of X' and X'' in $\mathcal{X}$, then

$$\langle \langle X, (Tp')^{-1}(\alpha') \cap (Tp'')^{-1}(\alpha'') \rangle, p', p'' \rangle$$

  is a product of $\langle X', \alpha' \rangle$ and $\langle X'', \alpha'' \rangle$ in $Spa(T)$.

(Co)products are frequently used to model unconstrained parallel composition. Proposition 22 shows how (co)products in the category of signatures can be lifted to (co)products in the concrete categories of specifications and behaviours.

Moreover, the fact that the underlying functors in these two categories are bifibrations enables us to obtain constrained parallel composition by (co)cartesian lifting of an appropriate signature morphism provided the constraints can be expressed at the signature level.[3]

The diagram below illustrates this. Consider two signatures Σ', Σ'', two specifications S', S'' over Σ', Σ'' (respectively) and two interpretation structures M', M'' over Σ', Σ'' (respectively).

Consider also that we want to obtain the constrained parallel composition S (resp. M) of the two specifications (resp. models) using a third signature Σ as the desired final signature. Usually this signature represents a restriction over the signature of the unconstrained parallel composition.

We can do this first by calculating $\Sigma' + \Sigma''$. Then we must provide a morphism $\sigma : \Sigma' + \Sigma'' \to \Sigma$ expressing the constraints to be observed. Finally, we obtain S and M as (co)cartesian lifting of σ.

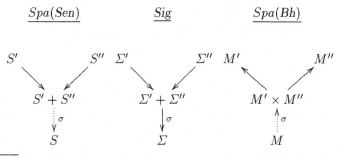

---

[3] As an example using the behaviour structure $\mathcal{P}$, we can represent action sharing or synchronization at the signature level.

Of course, in order to be able to talk about denotational semantics, it is important that the operations carried at the specification level can be transported to the model level. As we'll see in the next section, this is always the case in institutions of behaviour.

4.3   Specifications vs Interpretation Structures

In this section we answer the question: "Is there any relation between $Spa(Sen)^{op}$ and $Spa(Bh)$?". Apart from the trivial observation that both are concrete categories over Sig^{op}, there is a deeper relation: an adjunction

$$[\![\,\_\,]\!] : Spa(Sen)^{op} \to Spa(Bh)$$

Before we proceed we present an adapted version of the "Taut Lift" theorem introduced in [1].

Theorem 23 (Taut lift). Let $\langle \mathcal{A}, U \rangle$ and $\langle \mathcal{B}, V \rangle$ be concrete categories over $\mathcal{X}$, $G : \langle \mathcal{A}, U \rangle \to \langle \mathcal{B}, V \rangle$ a concrete functor and suppose that $\langle \mathcal{A}, U \rangle$ is topological. Then G is (right-)adjoint iff it sends U-initial sources into V-initial sources.

Def./Prop. 24. The functor $[\![\,\_\,]\!] : Spa(Sen)^{op} \to Spa(Bh)$ defined by

- $[\![\langle \Sigma, \Phi \rangle]\!] = \langle \Sigma, [\![\Phi]\!]_{\Sigma} \rangle = \langle \Sigma, \{b \in Bh(\Sigma) \mid b \Vdash^{B}_{\Sigma} \Phi\} \rangle^{4}$;
- $[\![\sigma : \langle \Sigma, \Phi \rangle \to \langle \Sigma', \Phi' \rangle]\!] = \sigma : \langle \Sigma, [\![\Phi]\!]_{\Sigma} \rangle \to \langle \Sigma', [\![\Phi']\!]_{\Sigma'} \rangle$.

is (right-)adjoint.

Proof. This is a simple consequence of the Taut lift theorem.

It suffices to prove that given a sS^{op}-initial source $(\langle \Sigma, \Phi \rangle \xrightarrow{\sigma_i} \langle X_i, \Phi_i \rangle)_I$ in $Spa(Sen)^{op}$, $([\![\langle \Sigma, \Phi \rangle]\!]_{\Sigma} \xrightarrow{\sigma_i} [\![\langle X_i, \Phi_i \rangle]\!]_{\Sigma})_I$ is bS-initial. In the light of Proposition 20, this amounts to proving that $[\![\bigcup_{i \in I} Sen(\sigma_i)(\Phi_i)]\!]_{\Sigma} = \bigcap_{i \in I} Bh(\sigma_i)^{-1}([\![\Phi_i]\!]_{\Sigma_i})$. Now:

$$
\begin{aligned}
&b \in [\![\textstyle\bigcup_{i \in I} Sen(\sigma_i)(\Phi_i)]\!]_{\Sigma} \\
\text{iff} \quad &b \Vdash^{B}_{\Sigma} \textstyle\bigcup_{i \in I} Sen(\sigma_i)(\Phi_i) \\
\text{iff} \quad &b \Vdash^{B}_{\Sigma} Sen(\sigma_i)(\Phi_i) && \text{for all } i \in I \\
\text{iff}^{5} \quad &Bh(\sigma_i)(b) \Vdash^{B}_{\Sigma_i} \Phi_i && \text{for all } i \in I \\
\text{iff} \quad &Bh(\sigma_i)(b) \in [\![\Phi_i]\!]_{\Sigma_i} && \text{for all } i \in I \\
\text{iff} \quad &b \in Bh(\sigma_i)^{-1}([\![\Phi_i]\!]_{\Sigma_i}) && \text{for all } i \in I \\
\text{iff} \quad &b \in \textstyle\bigcap_{i \in I} Bh(\sigma_i)^{-1}([\![\Phi_i]\!]_{\Sigma_i})
\end{aligned}
$$

□

Thus $[\![\,\_\,]\!]$ preserves limits: colimits in $Spa(Sen)$ are mapped to limits in $Spa(Bh)$. The functor $[\![\,\_\,]\!]$ has an extra useful property: it preserves cartesian morphisms.

Proposition 25. The functor $[\![\,\_\,]\!]$ preserves cartesian morphisms.

---

4   As usual, $b \Vdash^{B}_{\Sigma} \Phi$ means $b \Vdash^{B}_{\Sigma} \varphi$ for every $\varphi \in \Phi$

Proof. Consider a morphism $\sigma : \langle \Sigma, Sen(\sigma)(\Phi') \rangle \to \langle \Sigma', \Phi' \rangle$ in the category $Spa(Sen)^{op}$.

According to Proposition 9, we have that σ is cartesian by sS for $\sigma : \Sigma \to \Sigma'$ in $\langle \Sigma', \Phi' \rangle$. We want to prove that $[\![\sigma]\!] : [\![\langle \Sigma, Sen(\sigma)(\Phi') \rangle]\!] \to [\![\langle \Sigma', \Phi' \rangle]\!]$ is cartesian by bS for σ in $[\![\langle \Sigma', \Phi' \rangle]\!]$.

Note that $[\![\sigma]\!] : \langle \Sigma, [\![Sen(\sigma)(\Phi')]\!]_\Sigma \rangle \to \langle \Sigma', [\![\Phi']\!]_{\Sigma'} \rangle$. By Proposition 9, we have that $\sigma : \langle \Sigma, Bh(\sigma)^{-1}([\![\Phi']\!]_{\Sigma'}) \rangle \to \langle \Sigma', [\![\Phi']\!]_{\Sigma'} \rangle$ is cartesian by bS for σ in $\langle \Sigma', [\![\Phi']\!]_{\Sigma'} \rangle$, and so it is sufficient to prove that $[\![Sen(\sigma)(\Phi')]\!]_\Sigma = Bh(\sigma)^{-1}([\![\Phi']\!]_{\Sigma'})$, which can be done in a similar way to the proof of Proposition 24   □

For the fibre-inclined guy, this can be summarized as saying that $[\![\_]\!]$ is a fibred adjoint.

We can now complete the denotational semantics picture by noting that all operations may be carried at the specification level: the $[\![\_]\!]$ functor guarantees that they may be transported to the model level.

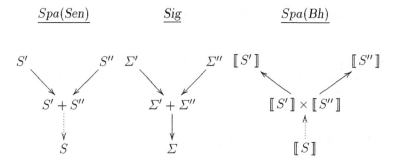

4.4   Theories

We now go a step further and study the relation between theories and interpretation structures. First of all, lets recall what a theory is.

Definition 26. The full concrete subcategory $\langle cSpa(Sen), tS \rangle$ of $\langle Spa(Sen), sS \rangle$ whose objects are closed for the entailment relation $\models^6$ is calles the category of theories.

We already know some properties of the relation between specifications and interpretation structures in institutions of behaviour. We prove that this relation is maintained if we replace "specifications" by "theories". Again we borrow some results from [1].

Definition 27. A concrete subcategory $\langle \mathcal{A}, U \rangle$ of $\langle \mathcal{B}, V \rangle$ is called *concretely reflective* in $\langle \mathcal{B}, V \rangle$ if for each object B of $\mathcal{B}$ there exists an identity-carried

---

[6] As usual, we define that $\Phi \models_\Sigma \varphi$ iff $I \vdash_\Sigma \Phi$ implies $I \vdash_\Sigma \varphi$ for each $I \in |Int(\Sigma)|$.

$\mathcal{A}$-reflection arrow $r_B : B \to A_B$.

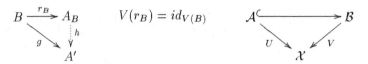

Proposition 28. A full concrete subcategory of a topological category $\mathcal{X}$ is topological provided it is concretely reflective in $\mathcal{X}$.

Proposition 29. The concrete category $\langle cSpa(Sen), tS \rangle$ is concretely reflective in $\langle Spa(Sen), sS \rangle$.

Proof. For each $\langle \Sigma, \Phi \rangle \in |Spa(Sen)|$, $id_\Sigma : \langle \Sigma, \Phi \rangle \to \langle \Sigma, \Phi^{\models_\Sigma} \rangle$ is a $cSpa(Sen)$-reflection for $\langle \Sigma, \Phi \rangle$. $\square$

We thus have again the result that $\langle cSpa(Sen), tS \rangle$ is (co)complete iff *Sig* is (co)complete.

Proposition 30. Consider a concrete category $\langle \mathcal{A}, U \rangle$ which is concretely reflective in $\langle \mathcal{B}, V \rangle$, and $r_B : B \to A_B$ be an identity-carried $\mathcal{A}$-reflection arrow for each $B \in |\mathcal{B}|$. Then:

- if $\hat{f} : A \to B'$ is cocartesian by V for $f : X \to X'$ on A then $r_{B'} \circ \hat{f} : A \to A_{B'}$ is cocartesian by U for f on A.

$$
\begin{array}{ccc}
A \xrightarrow{r_{B'} \circ \hat{f}} A_{B'} & A \xrightarrow{\hat{f}} B' & X \xrightarrow{f} X' \\
\searrow_g \quad \downarrow^{\hat{h}'} & \searrow_g \quad \downarrow^{\hat{h}} & \searrow_{V(g)} \quad \downarrow^h \\
A'' & A'' & X''
\end{array}
$$

$$
\begin{array}{c}
B' \xrightarrow{r_{B'}} A_{B'} \\
\searrow_{\hat{h}} \quad \downarrow^{\hat{h}'} \\
A''
\end{array}
$$

- if $\hat{f} : B \to A'$ is cartesian by V for $f : X \to X'$ on A then $\hat{f}' : A_B \to A'$ is cartesian by U for $f : X \to X'$ on A' where $\hat{f}' : A_{B'} \to A''$ is the unique morphism s.t. $\hat{f}' \circ r_B = \hat{f}$.

Corollary 31. The functor tS is a bifibration. In particular, given $\sigma : \Sigma \to \Sigma'$ (and using Proposition 9):

- $\sigma : \langle \Sigma, \Phi \rangle \to \langle \Sigma', Sen(\sigma)(\Phi)^{\models_{\Sigma'}} \rangle$ is cocartesian by tS for σ on $\langle \Sigma, \Phi \rangle$;
- $\sigma : \langle \Sigma, Sen(\sigma)^{-1}(\Phi') \rangle \to \langle \Sigma', \Phi' \rangle$ is cartesian by tS for σ on $\langle \Sigma', \Phi' \rangle$.

Also, the restriction of $[\![\,_-\,]\!]$ to $cSpa(Sen)^{op}$ is still an adjunction and preserves cartesian morphisms.

The diagram below relates the several categories and functors presented in the paper.

$$
\begin{array}{ccc}
cSpa(Sen) \hookrightarrow Spa(Sen) & \qquad & cSpa(Sen)^{op} \xrightarrow{\ [\,_-\,]\ } Spa(Bh) \\
{\scriptstyle tS}\Big\downarrow \qquad \Big\downarrow {\scriptstyle sS} & & {\scriptstyle sS^{op}}\Big\downarrow \qquad \Big\downarrow {\scriptstyle bS} \\
Sig \xleftrightarrow[\ id\ ] Sig & & Sig^{op} \xleftrightarrow[\ id\ ] Sig^{op}
\end{array}
$$

5 Concluding Remarks

In this paper we addressed a task that has been carried out in the past in a number of situations: the construction of a certain kind of institutions, which we named "behaviour institutions". This kind of institution arises whenever we want to talk about a system in terms of the set of its behaviours and already have a clear understanding and a formalization of individual behaviours.

We noted that, although in the past the construction of the institution has been done in a per-case basis, there is a canonical way of obtaining such institutions.

Furthermore, there is a set of properties shared by institutions of behaviour, regardless of the starting point chosen. In this paper we addressed the particular case of denotational semantics. In particular, the relation between specifications and models in institutions of behaviour was investigated.

This paper casts light on previous work [4,2,3,6,7,8] by providing a unified and abstract methodology for creating new useful institutions.

Acknowledgements

We thank J. L. Fiadeiro for many useful discussions on this topic. Our thanks go also to the Laboratório de Modelos e Arquitecturas Computacionais (Faculdade de Ciências da Universidade de Lisboa) which made possible the presentation in WADT.

References

1. J. Adámek, H. Herrlich, and G. Strecker. *Abstract and Concrete Categories*. John Wiley & Sons, 1990.
2. J. Fiadeiro and J. F. Costa. Institutions for behaviour specification. In E. Astesiano, G. Reggio, and A. Tarlecki, editors, *Recent Trends in Data Type Specification*, volume 906 of *Lecture Notes in Computer Science*, pages 273–289. Springer-Verlag, 1995.
3. J. Fiadeiro and J. F. Costa. Mirror, mirror in my hand... a duality between specifications and models of process behaviour. *Mathematical Structures in Computer Science*, 6:353–373, 1996.

4. J. Fiadeiro, J. F. Costa, A. Sernadas, and T. Maibaum. Process semantics of temporal logic specification. In M. Bidoit and C. Choppy, editors, *Recent Trends in Data Type Specification: 8th Workshop on Specification of Abstract Data Types - Selected Papers*, volume 655 of *Lecture Notes in Computer Science*, pages 236–253. Springer-Verlag, 1993.
5. Jacobs. *Categorical Logic and Type Theory*. Elsevier, 1999.
6. H. Lourenço and A. Sernadas. An institution of hybrid systems. In D. Bert and C. Choppy, editors, *Recent Trends in Algebraic Development Techniques - Selected Papers*, volume 1827 of *Lecture Notes in Computer Science*, pages 219–236. Springer-Verlag, 2000.
7. A. Sernadas, J. F. Costa, and C. Sernadas. An institution of object behaviour. In H. Ehrig and F. Orejas, editors, *Recent Trends in Data Type Specification*, volume 785 of *Lecture Notes in Computer Science*, pages 337–350. Springer-Verlag, 1994.
8. A. Sernadas, C. Sernadas, and C. Caleiro. Denotational semantics of object specification. *Acta Informatica*, 35:729–773, 1998.

The Lübeck Transformation System: A Transformation System for Equational Higher Order Algebraic Specifications

Walter Dosch and Sönke Magnussen

Institute for Software Technology and Programming Languages
Medical University of Lübeck
Lübeck, Germany
http://www.isp.mu-luebeck.de

Abstract. The Lübeck Transformation System supports the refinement of higher order algebraic specifications following sound transformation rules. We discuss the system requirements, describe the specification language and explain the life cycle of a specification in the transformation process. The system analyses various properties of the specification providing user guidance for further design decisions. The refinement relation is implemented by two refinement modes covering the different transformation rules for entire specifications and single axioms. Finally we describe the architecture and the implementation of the system. Throughout the paper, we accompany the presentation with a running example.

1 Introduction

Transformational programming offers a secure development process for the systematic design of software and hardware systems. The derivation starts with the specification of the required properties. The formal development process must preserve the behavioural properties of the specification. This can be guaranteed by a top-down approach favouring the "correct by construction" paradigm, or a bottom-up approach following the "invent and verify" style.

In the top-down approach, the derivation, also called synthesis, consists of a sequence of refinement steps starting from the specification. The refinement steps manifest design decisions wrt. the algorithmic structure and/or the data structure. The refinement is based on sound transformation rules [4,8] which can be combined into complex strategies. During the development, the system description passes through a series of abstraction levels starting from purely property-oriented specifications to efficient programs. As soon as the specification reaches an algorithmic level, it can be compiled into an executable program.

The single refinement steps are mechanizable whenever they follow syntactic transformation rules. As the basic requirement, a transformation system must support the safe manipulation of specifications wrt. a catalogue of transformation rules [2,3]. In general, a derivation employs many trivial steps, but also a few decisive steps which constitute proper design decisions [19]. An advanced

M. Cerioli and G. Reggio (Eds.): WADT/CoFI 2001, LNCS 2267, pp. 85–108, 2002.
© Springer-Verlag Berlin Heidelberg 2002

transformation system should support the automation of the simple steps as far as possible while leaving enough freedom for proper design decisions. The system should warn the programmer of critical properties initiating a possible revision of the derivation. More significant, in each stage the transformation system should inform the user of semantic properties guiding the further development.

In this paper, we present the first comprehensive description of the Lübeck Transformation System LTS — a new transformation system for higher order algebraic specifications. The system incorporates many advanced features for the analysis of algebraic specifications and includes some novel concepts for their manipulation and transformation. The implementation of the prototype, completely based on a strict functional language, is modular and remarkably slim. First experiences including complex derivations are encouraging — despite the currently poor user interface. We conceive of the prototype as an experimental transformation system for studying the mechanization of advanced programming techniques.

Section 2 starts with an informal overview of the system requirements and the specification language. We outline the transformation process by explaining the life cycle of a specification. Section 3 lays the formal basis of the specification language defining a loose constructor generated semantics. Section 4 presents the analysis of specifications where the system searches for interesting properties of the different constituents. We illustrate the analysis algorithms with an accompanying example. Section 5 introduces the refinement relation, characterizes the two refinement modes of LTS, and describes the administration of the transformation nodes. Section 6 enumerates the available refinements for entire specifications, viz. elementary steps and meta strategies like fold/unfold or fusion. In Section 7 we collect the single step commands and treat the induction principle in more detail. Then we expose a simple language for formulating transformation strategies. Section 8 describes the architecture and the implementation of the system. Finally, in Section 9 we discuss related work and outline future extensions.

We assume that the reader is familiar with the basic notions of algebraic specifications, rewriting, and program transformations. We include formal definitions only to the extent necessary for understanding the specific approach implemented in LTS .

2   Overview

This section presents an informal overview of the Lübeck Transformation System. We summarize the system requirements and outline the implementation. We sketch the specification language by a motivating example and describe the life cycle of specifications during the refinement procedure.

2.1   System Requirements and Implementation

LTS is a transformation system for the stepwise refinement of algebraic specifications. The derivations head for algorithmic specifications which can be compiled into Standard ML code. LTS is a prototype system which has been developed at the Institute for Software Technology and Programming Languages at the Medical University of Lübeck during the past two years. For the design of the kernel system, the emphasis was laid on clear semantics concepts supporting sound transformation rules which preserve the behavioural properties of the initial specification. The kernel system will gradually be extended towards a powerful program development tool by enriching the rule basis with more elaborate strategies encapsulating wide-spanned transformation steps.

LTS is completely implemented in Standard ML ; the top level environment of SML serves as user interface. During the start of the system, a collection of prelude specifications like Boolean values, natural numbers and lists are loaded. An autostart file is executed containing the user preferences and user-owned rule catalogues. After the system start, specifications can be loaded and manipulated by invoking transformation commands implemented as SML functions.

2.2   Specification Language

The Lübeck Transformation System handles algebraic specifications with higher order sorts having a constructor generated loose semantics. The notation of specifications is similar to the notation of CASL .

We illustrate the notation by a motivating example. The following specification Clist defines lists of elements of a parameter sort T . Lists are constructed from the empty list nil by forming singletons sglt and by concatenation conc. The sort list of lists forms a non-free data type, since the constructor equations ass, ntl and ntr constrain the data type: we have indeed linear lists and not binary trees. A loaded specification is displayed on the screen by the command specOutput .

```
- specOutput("Clist");

spec Clist =
sorts
    list = nil | sglt(T) | conc(list,list),
     T
ops
    cons : (T,list)list,
    snoc : (list,T)list
vars k:list,l:list,m:list,x:T
axioms
    ass:   conc(k,conc(l,m)) = conc(conc(k,l),m),
    ntl:   conc(nil,l) = l,
    ntr:   conc(l,nil) = l,
    cons:  cons(x,l) = conc(sglt x,l),
    snoc:  snoc(l,x) = conc(l,sglt x)
end
> val it = () : unit
-
```

2.3   Combining Specifications

New specifications can be constructed by giving an expression for an imported specification together with a new signature and new axioms. The importing expression is formed by the extension (;) and union (+) of existing specifications which are possibly renamed by a signature morphism.

The following specification Sumsq extends the union of Nat and the specification obtained from instantiating the parameter sort T of Clist by the sort nat, by a new sort op for mappings between natural numbers, and four operators sq, sum, map, and sumsq together with the corresponding axioms.

```
spec Sumsq = Nat + nL as Clist[T/nat] ;
sorts
        op = (nat)->nat
ops
    sq : (nat)nat,
    sum : (list)nat,
    map : (op,list)list,
    sumsq : (list)nat
vars k,l:list, x:nat, f:op
axioms
    sq1:   sq x = mult(x,x),
    sum1:  sum nil = zero,
    sum2:  sum (sglt x) = x,
    sum3:  sum (conc(k,l)) = add(sum k,sum l),
    map1:  map(f,nil) = nil,
    map2:  map(f,sglt x) = sglt (f x),
    map3:  map(f,conc(k,l)) = conc(map(f,k),map(f,l)),
    sumsq: sumsq l = (sum ~ (@map sq)) l
end
```

In the axiom sumsq, the symbol ~ denotes functional composition, and @ indicates a curried use of the functional map. Functional composition and currying allow building higher order terms.

When a specification is imported and modified by a signature mapping, an alias has to be provided. The elements of the signature can then be qualified by prefixing them with the alias. In this way, the elements resulting from different instantiations of the same specification can be distinguished.

For example, in the specification Sumsq the empty list of the instantiation nL of the specification Clist is qualified by nL.nil. Since there is a unique instantiation of Clist, the prefixes of the alias can be omitted in this case.

2.4   Life Cycle of Specifications during Refinement

Specifications show a characteristic life cycle when they are transformed with LTS. After a specification has been loaded, it resides in the system and is ready for transformation and code generation.

The life cycle of a specification residing in the system is illustrated in Fig. 1. After parsing and context checking, the specification is analysed in order to

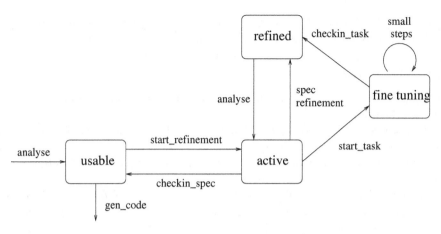

Fig. 1. Life cycle of a specification

inform the user about desirable or critical properties. These results assist the programmer in making further design decisions or in revising previous steps.

The user selects one of the loaded specifications and starts a refinement process. The selected specification becomes active, and the system enters the specification mode. This mode allows refinements tackling the entire specification. Additionally a specification can be transformed in fine tuning mode. Here a single axiom is picked as start axiom and then transformed under various logic and algebraic rules. A refinement in fine tuning mode leads to a couple of new equations logically implying the start axiom. The refinement of the overall specification consists in replacing the start axiom by the newly generated axioms. After a refinement step, the specification is analysed to update its semantic properties.

When the user decides to finish the transformation process, the refined specification can be inserted into the collection of loaded specifications overwriting the original specification. When a specification is detected to be algorithmic, it can be compiled into Standard ML code.

3   The Specification Language

The Lübeck Transformation System represents a general purpose system covering a wide range of different applications. As foundation, we use the theory of algebraic specifications. This framework captures different levels of abstraction in a coherent setting starting from property oriented specifications to executable functional programs over concrete data structures.

The first order restriction of the specification language forms a subset of CASL [14]. The syntax of the specification widely follows the notational conventions of CASL . In the sequel, we will mainly discuss the non-standard concepts of the specification language, in particular the higher order features. The interested reader is refered to [21,22] for the foundations of algebraic specifications. The higher order extension of algebraic specifications is similar to [13].

3.1   Signatures

A *signature* $\Sigma = (S, F, C)$ provides names for the sorts and functions occurring in a specification.

The set $S = SC \,\dot{\cup}\, SP$ of *basic sorts* is the disjoint union of the set SC of *constructor sorts* and the set SP of *parameter sorts*. The set $S^{\times, \rightarrow}$ of *derived sorts* comprises S, all *tuple sorts* $(s_1, \ldots, s_n)$ $(n \geq 2)$ and all *function sorts* $(s_1 \rightarrow s_2)$ defined inductively over S.

The $S^{\times, \rightarrow}$-indexed family F comprises the sets F_s of *function symbols* of *functionality* s. For a function sort $s = (s_1 \rightarrow s_2)$, we call s_1 the *argument sort* and s_2 the *result sort*.

The subfamily $C \subseteq F$ of *constructors* C_s comprises either constants with $s \in SC$ or function symbols with $s = (s_1 \rightarrow s_2)$ and $s_2 \in SC$. The remaining function symbols in $F \backslash C$ are called *operations*.

In the specification language, we distinguish between constructors and operations. This separation supports the analysis of specifications, since it enables the definition of sufficient syntactical criteria for desirable or critical properties. When generating SML code, the constructors lead to data type definitions, whereas the operations yield function definitions.

In the specification language, we also distinguish between constructor sorts, parameter sorts, and derived sorts. For each constructor sort, the collection of constructors determines the generation of the respective carrier. This generation principle confines the semantics of algebraic specifications to generated algebras. Constructor sorts can be constrained to non-free sorts by imposing equations between constructor terms. For the parameter sorts, there is no generation principle. The derived sorts support the declaration of the higher order functions. The interpretation of a derived sort is inferred from the interpretations of the basic sorts.

3.2   Algebraic Specifications

Terms are well-formed expressions formed over a signature; logic formulae describe properties of models. An algebraic specification constitutes an axiomatic definition of a software or hardware component.

Let X be an $S^{\times, \rightarrow}$-indexed family of sets X_s of *free variables* of sort s. The $S^{\times, \rightarrow}$-indexed family $T(\Sigma, X)$ of *terms* with free variables from X is the wrt. set inclusion least set validating

- $X_s \subseteq T_s(\Sigma, X)$ (variables)
- $F_s \subseteq T_s(\Sigma, X)$ (function symbols)
- If $f \in F_{(s_1, \ldots, s_n) \rightarrow s_{n+1}}$,
  then $(@f) \in T_{(s_1 \rightarrow \ldots \rightarrow (s_n \rightarrow s_{n+1}) \ldots)}(\Sigma, X)$.   (currying)
- If $n \geq 2$ and $t_i \in T_{s_i}(\Sigma, X)$ for all $1 \leq i \leq n$,
  then $(t_1, \ldots, t_n) \in T_{(s_1, \ldots, s_n)}(\Sigma, X)$.   (tuple)

- If $t_f \in T_{(s_1 \to s_2)}(\Sigma, X)$ and $t_a \in T_{s_1}(\Sigma, X)$,
  then $(t_f(t_a)) \in T_{s_2}(\Sigma, X)$.   (application)
- If $t_f \in T_{(s_1 \to s_2)}(\Sigma, X)$ and $t_g \in T_{(s_2 \to s_3)}(\Sigma, X)$,
  then $(t_g \circ t_f) \in T_{(s_1 \to s_3)}(\Sigma, X)$.   (composition)
- If $t_b \in T_{bool}(\Sigma, X)$ and $t_t, t_e \in T_s(\Sigma, X)$,
  then $(\underline{\text{if }} t_b \underline{\text{ then }} t_t \underline{\text{ else }} t_e) \in T_s(\Sigma, X)$.   (conditional)

The $S^{\times, \to}$-indexed family of *constructor terms* is defined similarly by confining the function symbols to C_s for all $s \in S^{\times, \to}$. For a term $t \in T_s(\Sigma, X)$, $VAR_{s_1}(t)$ denotes the set of all variables of sort s_1 occuring in t.

A Σ-*equation* is a universally quantified formula $\forall x_1 : s_1, \ldots, \forall x_n : s_n \ (l = r)$ relating two terms $l, r \in T_s(\Sigma, X)$ of the same sort $s \in S^{\times, \to}$.

In writing sorts, terms and equations, we drop brackets with usual conventions. We also omit the universal quantification ranging over all variables in an equation.

A *substitution* $\sigma : X \to T(\Sigma, X)$ is an $S^{\times, \to}$-indexed family of functions $\sigma_s : X_s \to T_s(\Sigma, X)$. The inductively defined application $t[\sigma]$ of a substitution σ to a term t replaces all variables x in t by the term $\sigma(x)$.

An *algebraic specification* (Σ, E) is composed of a signature Σ and a set E of Σ-equations.

The axioms of specifications are equations between higher order terms built from the function symbols of the signature and variables of derived sorts. As a specific feature, the specification language of LTS supports curried function symbols. The curry operator transforms an n-ary function symbol f of type $(s_1, \ldots, s_n)s$ into a term of type $(s_1 \to \ldots \to (s_n \to s) \ldots)$. Higher order terms can be built by applying a curried function symbol to argument terms and by function composition. There is no λ-abstraction in the specification language, since it causes problems in the mechanical treatment of terms and specifications. If a function is not denotable with currying and composition in LTS , the user may insert new function symbols and axioms characterizing that function.

In summary, the specification language offers higher order constructs to support advanced concepts from functional programming. At the same time, the specification language is confined to higher order equations which can effectively be manipulated by the transformation system.

The conditional term supports clear function definitions. Its condition is not an equation, but a term of the primitive sort `bool` . This avoids a theory with conditional equations; it also results in a strict distinction between the specification `Bool` and the truth values used in equational logic.

3.3 Semantics

Σ-algebras associate a fixed interpretation with the names of a signature.

A Σ-*algebra* $A = ((s^A)_{s \in S^{\times, \to}}, (f^A)_{f \in F})$ provides a non-empty *carrier set* s^A for every basic sort $s \in S$ and an element $f^A \in s^A$ for every function symbol $f \in F_s$. In the inductively defined interpretation of derived sorts, $(s_1 \to s_2)^A = [s_1^A \to s_2^A]$ denotes the set of all total functions, and $(s_1, \ldots, s_n)^A = s_1^A \times \cdots \times s_n^A$ the Cartesian product.

A Σ-algebra satisfying all equations in E is called a *model* of the specification. The *semantics CGEN(Σ, E)* of an algebraic specification (Σ, E) comprises the class of all constructor generated models.

The loose semantics based on constructor generated models is well-suited for the transformational approach. The class of models reflects the variety of all possible implementations; this supports a wide range of refinement steps. The generation principle enables the transformation system to derive properties on constructor sorts by term induction.

In general, the function spaces being interpretations of function sorts, are not term generated. As a consequence, the generation principle does not guarantee that every element of a carrier of a constructor sort can be denoted by a ground term.

3.4   Structuring Specifications

Large specifications can be structured into subspecifications; vice versa, existing subspecifications can be combined into larger specifications. The combining operators for specifications structure large specifications in a syntactic way. For the time being, they do not impose an additional structure on the class of models. The semantics of structured specifications is determined by reducing the structure to the form (Σ, E).

Algebraic specifications can be structured by extension, union and signature morphisms with aliasing. Specifications may be parameterized by sorts; a parameter sort is interpreted by a carrier without a generation constraint. In LTS there is no explicit method of parameter instantiation like in CASL . Parameterisation can be expressed by union and a mapping describing a signature morphism following the pushout construction.

4   Analysing Specifications

The Lübeck Transformation System assists the user by checking various properties of the specification to be transformed. The analysis of a specification might reveal its inconsistency or discover an underspecification. As a major benefit, the analysis should reveal useful properties that guide the user selecting the next transformation step. The properties refer to the different constituents of specifications, viz. single equations, single operations or entire subspecifications.

During its life cycle, LTS analyses a specification completely when it is loaded into the system, and incrementally, after it has been transformed.

4.1   Properties

LTS checks various properties of the specification that are important for its further development or its automatic compilation into SML code. Fig. 2 lists the properties to be analysed along with their abbreviations. In the sequel, we shortly introduce the relevant notions.

A	=	algorithmic		OLF	=	non-overlapping
C	=	constructive		CMPL	=	complete
L	=	left-linear		CATA	=	catamorphism
T	=	terminating		CNSTR	=	data type constraint

Fig. 2. Properties analysed by LTS

In the following definitions $\widehat{=}$ denotes the syntactic equality. Let $f \in F$ be a function symbol with functionality $(s_1 \to s_2)$. An equation $f(tc) = r$ is called *constructive*, if the argument $tc \in T_{s_1}(\Sigma, X)$ is a constructor term and $VAR_s(r) \subseteq VAR_s(f(tc))$ holds for all $s \in S^{\times, \to}$.

A constructive equation $f(tc) = r$ is called *left-linear*, if each variable occurs at most once on the left-hand side.

Two constructive equations $f(tc_1) = r_1$ and $g(tc_2) = r_2$ are called *overlapping*, if $f \widehat{=} g$ holds and there are substitutions σ_1, σ_2 with $f(tc_1)[\sigma_1] \widehat{=} g(tc_2)[\sigma_2]$. A set of equations is called *non-overlapping*, if each two distinct equations are not overlapping.

Let $E_f \subseteq E$ denote the set of constructive equations with outermost operation f on the left-hand side. The operation f is called *complete*, if for all constructor terms tc there is a substitution σ and an equation $f(tc_1) = r \in E_f$ such that $tc_1[\sigma] \widehat{=} tc[\sigma]$.

The operation $f \in F_{(s_1 \to s_2)}$ is called a *catamorphism*, if $s_1 \in SC$ is a constructor sort and for all constructors $c_1, \ldots, c_n$ with result sort s_1 there is a constructive, left-linear equation $f(c_i(x_{i,1}, \ldots, x_{i,n_i})) = r_i$ in E such that all occurrences of a variable $y \in VAR_{s_1}(r_i)$ in the term r_i have the form $f(y)$.

An operation f is specified *algorithmically*, if the set E_f is non-overlapping and each equation in E_f is left-linear.

4.2 Hierarchical Property Checking

The properties may refer to single equations, single operations or entire specifications. For an equation we analyse whether its structure meets the patterns used in SML function definitions. The properties of an operation are derived from the properties of all constructive equations with this operation as outermost function symbol. Finally the properties of a specification are synthesized from the properties of the equations and the operations.

Properties of Single Equations. LTS checks every single equation whether it is constructive. In this case, the equation contributes to the definition of the outermost operation on the left-hand side. The equation is then directed and associated with the definition of that operation.

If an equation proves to be constructive, LTS additionally checks whether the equation is left-linear.

Next the system examines the terms on the left-hand and the right-hand side whether they are both constructor terms of a sort $sc \in SC$. If this applies to the equation, it is marked to constrain the constructor sort sc. Such equations are called data type constraints.

Finally LTS tests whether the equation is terminating. The termination test implements the lexicographic path order extended to higher order terms without λ-abstraction [11].

Revisiting the specification Sumsq, we identify the constructor sort list constrained by the associativity of concatenation:

```
ass:   conc(k,conc(l,m)) = conc(conc(k,l),m) PROP: RF(conc) C L T CNSTR
```

LTS also detects the termination of this equation. The equation

```
sum3:   sum (conc(k,l)) = add(sum k,sum l) PROP: RF(sum) C L T
```

is constructive, leftlinear and terminating. The equation contributes to the definition of sum indicated by the cross reference RF.

In fine tuning mode, see Section 7, the user is endowed with additional checks which can be invoked manually. For example, terms and equations can be tested whether they are in normal form or algorithmic. These properties guide the selection of further refinement steps in fine tuning mode.

Properties of Operations. The properties of an operation are partly synthesized from the properties of the equations describing the operation. An operation is specified algorithmically if all corresponding equations are left-linear and the set of all these equations is non-overlapping. Then this set of equations can be compiled into a function definition in SML. After testing the completeness, LTS checks the catamorphism property.

Again we illustrate the analysis using the motivating example Sumsq. The three constructive equations

```
sum1:   sum nil        = zero,
sum2:   sum (sglt(x))  = x,
sum3:   sum (conc(k,l)) = add(sum k,sum l)
```

contribute to the definition of the operation sum. This operation is algorithmic, since the equations are left-linear and non-overlapping. The set {sum1,sum2,sum3} of equations covers all constructor patterns, therefore the operation sum is complete. The definition of the operation sum follows the inductive structure of the data type list and forms a catamorphism.

```
sum : (nL.list)nat > PROP: A OLF C CATA CMPL
```

The catamorphism property is a prerequisite for the fusion step, compare 6.2.

Properties of Specifications. The analysis of an entire specification is based on the properties of its function symbols and equations.

If all equations are terminating, the specification is terminating as well. The non-constraint equations are then checked whether they are constructive and non-overlapping. If this is the case and all function symbols are algorithmic, the specification is algorithmic as well. If the specification is terminating, LTS additionally tests whether the set of directed equations forms a confluent rewrite system. For this purpose LTS generates the set of critical pairs of terms and tries to join them by the rewrite system. If all critical pairs are joinable, the rewrite system is confluent [1]. An unjoinable critical pair may indicate an inconsistency of the specification. The set of unjoinable critical pairs of a specification Spec is displayed with the command show_cps("Spec").

The accompanying specification Sumsq is algorithmic, since all equations are constructive and non-overlapping, and all operations are algorithmic. LTS does not detect any critical pairs, all equations are terminating — henceforth the rewrite system of Sumsq is confluent. After loading the specification Sumsq, the properties analysed by LTS are shown by displaying the specification with the command print_spec:

```
SPEC Sumsq = Nat + nL as Clist[T/nat] ;
    . . .
END
PROPERTIES:
      ALGORITHMIC
      CONFLUENT
      NONOVERLAPPING
      CONSTRUCTIVE
      TERMINATING
```

5   Refinement by Transformation

The transformation of a specification effects a refinement of its behavioural semantics. In this section, we survey the transformation rules and strategies offered by the Lübeck Transformation System. Each transformation rule ensures a sound refinement step for the entire specification. We distinguish between the transformation of specifications and the manipulation of single equations, compare Sections 6 and 7. This gives rise to the different working modes of LTS.

5.1   Refinement Relation

A refinement step enriches the signature of a specification while retaining the properties of its models.

A signature $\Sigma_1 = (S_1, F_1, C_1)$ is called a *subsignature* of the signature $\Sigma_2 = (S_2, F_2, C_2)$, if $S_1 \subseteq S_2$ and $F_1 \subseteq F_2$ holds. The Σ_1-reduct denoted by $A|_{\Sigma_1} = ((s^{A|_{\Sigma_1}})_{s \in S_1^{\times,\to}}, (f^{A|_{\Sigma_1}})_{f \in F_1})$ of a Σ_2-algebra $A = ((s^A)_{s \in S_2^{\times,\to}}, (f^A)_{f \in F_2})$ keeps all carriers and functions of the subsignature, that is $s^{A|_{\Sigma_1}} = s^A$ for all $s \in S_1^{\times,\to}$ and $f^{A|_{\Sigma_1}} = f^A$ for all $f \in F_1$. A specification (Σ_1, E_1) is *refined to* a specification (Σ_2, E_2), if Σ_1 is a subsignature of Σ_2, and $CGEN(\Sigma_1, E_1) \supseteq CGEN(\Sigma_2, E_2)|_{\Sigma_1}$ holds.

This notion of refinement is general enough for the diversity of applications faced in a transformation system. The refinement relation extends the signature of a specification and narrows the set of possible implementations.

5.2   Working Modes

The user of LTS refines a specification by invoking commands that manipulate the specification in an interactive way. The commands implement sound refinement steps of different granularity for (the constituents of) specifications.

The Lübeck Transformation System supports two distinct working modes. The *specification mode* provides elementary and complex refinement steps affecting the entire specification, whereas the *fine tuning mode* allows the transformation of single equations using different logic and algebraic rules.

In the specification mode, the user can invoke various elementary refinement steps, for example extending the signature, adding new axioms or dropping proven axioms. He can also invoke complex transformations like the fold/unfold paradigm or a fusion step for catamorphisms.

If the system is in the specification mode, the fine tuning process on a single equation can be started by the command start_task.

5.3   Administrating Transformation Nodes

When starting a fine tuning process, a transfomation node is generated containing the selected equation along with its status, a set of induction hypotheses and a substitution. In the beginning the latter two are empty. The new transformation node is pushed onto the top of the initially empty stack of transformation nodes. The transformation node on top of the stack can now be refined by invoking particular commands like rewriting (application of other axioms), complete case analysis (induction), application of induction hypotheses, generalization and decomposition.

Alternatively, the transformation node on top of the stack can be postponed. Then it is moved to the bottom of the stack, and the next transformation node becomes active.

Each command may generate several new transformation nodes, specialize the current substitution, and add induction hypotheses accompanied by an induction order. After several refinement steps, the user can replace the start equation by the derived equations residing on the transformation stack. Checking in the derived equations, concludes the fine tuning process of the start equation.

The entire fine tuning process can be seen as one large refinement step in the specification mode. In fact, complex transformation steps of the specification mode like fusion and fold/unfold use automatically performed sequences of fine tuning steps. In contrast to (semi-)automated steps, the fine tuning process entirely relies on user interaction.

6   Refinements in Specification Mode

The specification mode offers coarse refinement steps enabling major design decisions on the development process. These transformations affect the entire specification and comprise various elementary steps and meta strategies.

6.1   Elementary Steps

A specification can be enriched by adding new function symbols, sorts, and equations. The transformation commands add_function(es), add_sort(es), and add_axiom(es) expect a string es as argument which is parsed into the declaration of a function symbol, a sort or an axiom, respectively. The required syntax follows the syntax of the specification language. The parsed argument is context checked and then inserted into the specification.

Axioms can also be deleted using the command drop_axiom(ax) which causes a proof obligation.

6.2   Meta Strategies

Meta strategies represent wide-spanning refinement steps composed from elementary steps and fine tuning steps. If a meta strategy does not succeed automatically, the user can complete the step also manually.

Fine Tuning Strategies. The command applystrat(s:strategy,ax:string) applies a strategy s for the fine tuning mode to an axiom ax. If the strategy, compare Subsection 7.3 fails, the system stops for the user to proceed manually.

Fold/Unfold Step. The fold/unfold step of LTS is based on a fine tuning strategy which is described in Subsection 7.3 in more detail. The command foldunfold(ax:string,f:string) refines the equation ax by fold and unfold steps, until the operation f does not occur anymore on its right-hand side. If the strategy fails, the system stops in the fine tuning mode, and the user can finish the refinement manually.

Fusion Step. Let the operation φ be a catamorphism, i.e. an operation compatible with all constructors of its argument sort. The fusion strategy aims at synthesizing a new axiomatization of the function

$$h = f \circ \varphi$$

using the fusion theorem for catamorphisms [7].

We demonstrate the fusion strategy for lists revisiting the specification Sumsq from Section 2. The definition of the operation sumsq

```
sumsq:   sumsq l = (sum ~ (@map sq)) l
```

can be used to synthesize a new axiomatization which avoids the construction of the intermediate data structure by the operation map. The fusion strategy generates a new combine operation cop1_conc which describes the result of the operation sumsq in terms of the results of its recursive application to the concatenated sublists. The combine operation cop1_conc has to distribute over the operation sum. The strategy uses this property to derive an algorithmic version of the combine operation. This equation can then be used to synthesize the new axiomatization of the operation sumsq:

```
sumsq.nil:    sumsq nL.nil = zero
sumsq.sglt:   sumsq (nL.sglt x_9) = mult(x_9,x_9)
sumsq.conc:   sumsq (nL.conc(k_10,k_11)) = add(sumsq k_10,sumsq k_11)
```

The fusion step may introduce new properties on the operation add stemming from the data type constraints of lists. These properties result in proof obligations which are explicitly inserted into the specification.

```
sumsq._ass:   cop1_conc(cop1_conc(x_1,x_2),x_3) = add(x_1,add(x_2,x_3))
sumsq._ntl:   cop1_conc(sum nil,x_4) = x_4
sumsq._ntr:   cop1_conc(x_5,sum nil) = x_5
```

If the automatic derivation of the combine operations fails, the fusion strategy stops in the fine tuning mode for further interactive steps.

7   Refinements in the Fine Tuning Mode

The Lübeck Transformation System provides several functions to perform fine tuning steps. These functions can be combined into strategies; single step strategies are predefined. In the implementation, the commands map a transformation node into a list of transformation nodes. In this section, we survey the most important commands and explain their effects.

7.1   Single Step Commands

The fine tuning commands reside in different modules which are mutually independent and can be loaded as needed. Each command is implemented by a transformation function which applies the single step strategy associated with it. The transformation function uses the transformation node on top of the stack as an implicit argument. Further arguments are supplied as parameters when invoking the transformation function. The command dopostpone postpones the transformation node on top of the transformation stack to proceed with the next node.

Rewriting Steps. The standard fine tuning steps for rewriting are summarized in Fig. 3. Further rewrite steps comprise, among others, the simplification of terms, the treatment of higher order terms, the manipulation of conditions and of conditional terms. Miscellaneous fine tuning commands are illustrated in Fig. 5.

Induction. The induction principle in LTS is based on *term induction* [9].

The induction process can be started with three commands. The command induct(va:string,or:order) orients the case analysis on the construction principle of the sort of the variable va. The command inductf(f:string,or:order) introduces a structural induction by a case analysis on the arguments of the function symbol f — provided this operation is completely defined. Both commands must be supplied with an induction order or.

onerew	Apply one rewrite step with all terminating equations.
onerewall	Apply one rewrite step with all equations
rew(rsl)	Apply one left-to-right rewrite step with equations rsl.
rewrl(rsl)	Apply one right-to-left-rewrite step with equations rsl.
unfold(f)	Unfold the function symbol f. Assumption: f is constructive.
fold(f)	Fold the function symbol f. Assumption: f is constructive.
nfl(rsl)	Form normal form with equations rsl.
nf	Form normal form with all terminating equations.

Fig. 3. Fine tuning commands for rewriting

The third command $\mathtt{minduct(va:string,os:string)}$ resembles the command $\mathtt{induct}$, but introduces a measure function $\mathtt{os}$ as a string which is parsed and type checked.

Inductions in LTS are complete case analyses which also cause the generation of induction hypotheses and the specialization of the substitutions for the resulting transformation nodes. The hypotheses can be used in the fold step.

With every hypothesis, we attach an induction order. Frequently used syntactic orders like the lexicographic path order or the multiset order are built in. Semantical orders can be user-defined by a measure function mapping the elements of the induction sort into the standard sort $\mathtt{nat}$. The application of the induction hypothesis forms a conditional transformation; before using the induction hypothesis, the application condition must be verified.

Assume that the current transformation node has a substitution σ and an induction hypothesis $l = r$ with the syntactic order $\sqsubset$. The substitution reflects the history of the case analyses introduced on the current transformation node sofar; the induction hypothesis was generated by a former induction step. If the induction hypothesis could be applied to a subterm t of the current transformation node with matching $l[\tau] \cong t$, then LTS attempts to prove automatically the application condition $l[\tau] \sqsubset l[\sigma]$ of the induction hypothesis. If $\sqsubset$ is a semantic order with measure operation $\mathtt{mes}$, and the induction variable is $\mathtt{v}$, then the application condition $\mathtt{lt(mes(\tau(v)),mes(\sigma(v)))=true}$ has to be proved.

The commands $\mathtt{applhyp}$ resp. $\mathtt{applhyprl}$ apply the induction hypothesis in left-to-right-direction resp. right-to-left direction. Variants provide a location-oriented search of applicable redexes in the equation. Assume the induction hypothesis matches a subterm. In case of a syntactic order, the system automatically compares the two terms. In case of a semantical order, the system applies the induction hypothesis in any case, but generates a proof obligation.

7.2   Example Session with Fine Tuning Steps

We present several fine tuning steps with the specification $\mathtt{Sumsq}$ from Section 2. We aim at deriving a directly recursive version of the operation $\mathtt{sumsq}$ by eliminating the auxiliary operations $\mathtt{map}$ and $\mathtt{sum}$ in its definition.

induct(va,or)	Introduce case analysis on constructor variable va with induction order or .
inductf(f,or)	Introduce case analysis based on the completely defined operation f with induction order or .
minduct(va,os)	Introduce case analysis on constructor variable va with measure function os .
applhyp(i)	Apply hypothesis number i in left-to-right direction .
applhyprl(i)	Apply hypothesis number i in right-to-left direction .

Fig. 4. Fine tuning commands for induction

To this end, we refine the axiom sumsq in fine tuning mode. In the system output, a directed equation $l = r$ is denoted by $l := r$ or $r := l$, respectively. When a specification refinement has been started on the specification Sumsq, the fine tuning mode can be entered by the command start_task .

```
- start_task("sumsq");
```

```
TASK sumsq
    sumsq l := (sum ~ (@map sq)) l
```

The term on the right-hand side consists of a functional composition applied to the variable l. In order to rewrite it with the defining axioms of the operations sum and map, we transform it into a nested application using the command docomp2appl :

```
- docomp2appl();
```

```
TASK sumsq
    sumsq l := sum ((@map sq) l)
```

We proceed with an induction on the variable l performing a case analysis following the constructors of sort list . As induction order we use the lexicographic path order:

```
- doinduct("l",lpo);
```

```
TASK sumsq.1
    sumsq nL.nil := sum (@map sq nL.nil)
```

The system generates three new transformation nodes sumsq.1, sumsq.2 and sumsq.3 reflecting the cases for the constructors nil, sglt and conc. The elementary transformation nodes sumsq.1 and sumsq.2 can be finished by a few rewriting steps. The transformation node sumsq.3 can be simplified by rewriting to the following transformation node:

decomp	Apply a congruence step.
	Assumption: Outermost function symbols are equal.
caseb(b)	Introduce a conditional term with condition **b**.
iflift	Lift conditional in an application.
ifsimp	Simplify term by serveral simplification patterns.
gener(t)	Generalize term **t** by a fresh variable.
gener_heur	Find a generalizable term and generalize it.
catafusion	Apply the fusion theorem for catamorphisms.
swap	Swap the two sides of an equation.
applext	Apply the extensionality rule on higher order equations.
comp2appl	Transform a function composition into application.

Fig. 5. Miscellaneous fine tuning commands

```
TASK sumsq.3

    sumsq (nL.conc(k_3,k_4)) := add(sum (map(sq,k_3)),sum (map(sq,k_4)))

HYPS:
    (1) sumsq l = sum (@map sq l) ORDER: lpo
WITH
    l <-- nL.conc(k_3,k_4)
```

The long form of the transformation node above shows the induction hypothesis generated by the former induction step. The induction hypothesis can now be applied twice using the command `doapplhyprl(1)`. The equation of the resulting transformation node

```
- doapplhyprl(1);
TASK sumsq.3
    sumsq (nL.conc(k_3,k_4)) := add(sumsq k_3,sumsq k_4)
```

can now be inserted into the original specification `Sumsq` using the command `checkin_task`. Together with the results of the elementary tasks `sumsq.1` and `sumsq.2`, it replaces the axiom `sumsq`.

7.3   Transformation Strategies

The single-step commands can be combined into complex strategies using a fixed collection of strategy constructors. This idea was originally proposed in the LCF system [16]. Pre-defined and user-defined strategies can be executed by the command `try`. The transformation system also provides various predicates for formulating conditions in user-defined strategies.

Internally, a single-step command maps a transformation node into a list of transformation nodes. The application of a strategy transforms a stack of transformation nodes into a new stack of transformation nodes.

```
skip
postpone
setStatus of string
step of (transfnode -> transfnode list)
scomp of strategy list
repeat of strategy
ifSuc of strategy*strategy*strategy
ifCond of (transfnode list -> bool)*strategy*strategy
whileLoop of (transfnode list -> bool)* strategy
```

Fig. 6. Constructors for strategies

isStatus(st)	Is the status of the actual task st ?
isGeneralisable	Is the actual task generalizable?
isSpecialisable	Should the actual task be specialised?
isNormalform	Is the actual task in normal form?
NormalformAll	Are all tasks in normal form?
existsInRhs(f)	Does the function f occur on the right-hand side?
isConstr	Is the actual task constructive?

Fig. 7. Predicates on the transformation stack

Constructing Strategies. The constructors for building strategies are summarized in Fig. 6.

The empty strategy skip has no effect. The strategy postpone moves the top of the transformation stack to the bottom. As a result the second transformation node becomes the focus of the next step. The strategy step(f) attempts to apply the command f to the top of the transformation stack. The constructor scomp([s1...,sn]) combines a list of strategies si expressing the sequential execution of the associated refinement steps. The strategy repeat(s) repeatedly applies the strategy s until its application has no further effect on the transformation stack. The command ifSuc(c,s,t) denotes a conditional strategy. If the first strategy c succeeds, i.e. it has an effect on the top transformation node, then the system proceeds with the second strategy s, otherwise with the third strategy t. The conditional strategy ifCond(cond,s,t) and the while strategy whileLoop(cond,s) both require a condition on the transformation stack. The conditional strategy executes s, if the condition cond evaluates to true and t otherwise. The while strategy repeats the strategy s as long as the condition cond holds. Note that the repeat and the while strategies may cause nontermination.

Predicates on the Transformation Stack. For expressing conditions in writing strategies, LTS provides a collection of useful predicates for referencing the transformation stack, compare Fig. 7.

```
step(inductf fname lpo)
whileLoop (isStatus ''active")
    repeat
        ifSuc (step(applhyprl 1))
            skip
        else
            ifSuc (step(onerew))
                skip
            else step(onerewall)
    ifCond (existsInRhs fname)
        setStatus ("fail")
        else setStatus("ready")
    postpone
endwhile
```

Fig. 8. Fold/Unfold strategy (in sugared syntax)

The Fold/Unfold Strategy. As an example, we construct a strategy implementing the fold/unfold paradigm. The strategy aims at eliminating the operation from the right-hand side of an equation.

In LTS a strategy has to be built by using the constructors of Fig. 6. The strategy is presented in Fig. 8 in a syntactically sugared form. The strategy first invokes a case analysis based on the definition of the function. Each resulting transformation node is treated by repeatedly trying to apply the induction hypothesis and applying one rewrite step until there is no further possibility for rewriting. In the status field of the transformation nodes we record whether the node has already been treated and whether it succeeded.

8   Architecture and Implementation

The Lübeck Transformation System was designed with emphasis on flexible modification and extension. Therefore LTS is completely implemented in Standard ML 97 and uses the top-level environment as the basic user interface. The approximately 8 000 lines of code are structured into 21 modules and 10 layers. LTS is developed with Moscow ML, but also runs under SML/NJ.

In the sequel, we describe the architecture of the system, survey the status and explain some implementation decisions.

The transformation system is roughly composed of ten software layers, compare Fig. 9. The bottom layers are completely implemented and will not significantly change in the future, whereas the upper layers are still under construction.

The abstract syntax layer comprises the data structures used for the specification language and the transformations. Some node types of the abstract syntax tree like axioms, function symbols and specifications contain a list of at-

10	graphical user interface	in design
9	code generation	almost completed
8	strategies	to be extended
7	manipulation of specifications	to be extended
6	analysis of specifications	to be extended
5	manipulation of equations	to be extended
4	manipulation of terms	complete
3	semantical analysis	complete
2	syntactical analysis	complete
1	abstract syntax	complete

Fig. 9. Layers of LTS with current status

tributes recording properties computed in the analysis phase. Further attributes can easily be implemented.

The syntactical analysis layer consists of two parsers: a bottom-up-parser built by the lex and yacc tools which is used for loading new specifications, and a combinator parser [18] providing the possibility to parse new specification elements like new axioms manually invoked by the user.

The semantics layer contains all functions for context checking, in particular polymorphic type checking.

The term manipulation layer provides functions for simplifying, currying und uncurrying terms, matching, rewriting and unification modulo curry/uncurry. Most algorithms are variants of the algorithms in the text book [1]. The system automatically curries resp. uncurries a term during matching and unification. The term manipulation functions form the foundation for the next level.

The layer for manipulating equations provides the functions mainly explained in Section 7. These functions are separated into mutually independent modules such that the collection of needed modules can be loaded. This enables an application dependent usage of the manipulation functions by choosing the relevant modules. In this way, the system can be specialized to particular application fields, for example stream processing functions or high level parallel programming.

The analysis layer provides all necessary functions needed for the analysis of specifications. A specification is analysed whenever it is loaded or refined. In this way, the user is continuously informed about its properties making profound design decisions. The properties of a specification found during analysis are also used for verifying the applicability conditions in various refinement steps.

The layer for manipulating specifications encapsulates all the functions and data structures used for the refinement of specifications, compare Section 6.

The strategy layer is completely based on the layers 4 to 7; it performs all manipulations for specifications and equations using these layers. The bottom layers 1 to 7 can therefore be seen as kernel architecture of the system. All future developments will be set on top of this kernel architecture.

The almost completed implementation of the code generation layer compiles algorithmic specifications into Standard ML code.

For the graphical user interface we currently develop in parallel two solutions. The first approach uses the graphics library mGTK which is connected to Moscow ML. In the second approach the graphical user interface is designed and implemented in Java. The connection between LTS and the Java application is based on sockets.

9   Conclusion and Related Work

Computer support for transformational programming has a long tradition in the meantime. In the last decades, several transformation systems were proposed differing, among others, in the language, the area of support, and the degree of mechanization.

The early transformation system from Burstall and Darlington [5] supports fold and unfold steps of recursive first order equations. The system has a low degree of interaction; the transformations aim at improving the efficiency of functional programs.

The CIP system [3] deals with more powerful, language dependent specifications and programs. The system supports a large variety of transformation steps; it was primarily used for program derivations in the wide-spectrum language CIP-L [2].

Extended ML [10], for short EML, supports the specification of signatures and modules. It extends Standard ML by specification constructs like axioms, but provides no mechanization of transformation steps. Formulas in EML are expressions of data type bool. This integrates them into SML, but makes the underlying semantics more complex.

A generic approach is provided by the TAS system [12]. It is based on the Isabelle theorem prover [17] and offers a sophisticated user interface following the principle of direct manipulation. The user can define new transformation rules whose correctness can be proved by Isabelle itself. The system is flexible, since the user can widen the range of possible transformation steps. The TAS system neither supports the transformation of entire specifications nor the analysis of specifications.

The KIDS System [20] synthesizes software using a knowledge base of algorithmic design principles. The user applies such principles to problem specifications by instantiating them. The system also supports transformational derivation techniques like partial evaluation and finite differencing. A formal specification can be transformed into a program by interactively applying a sequence of high level transformations followed by several optimization steps.

The process of transformational programming can also be supported by proof systems [15,17] handling proof obligations during the derivation. This kind of mechanical assistance only covers part of the development process and does in general not support genuine transformation steps. As an exception, [6] performs an application of algorithm design patterns by PVS. Here the user has to instantiate the pattern for application manually; there is no automatic utility.

Our prototype system LTS can deal with higher order algebraic specifications. Thus it also supports advanced concepts of functional programming. Yet, the specification language allowing arbitrary equations is more expressive than pure functional programs. The chosen compromise between algebraic specifications and functional programming concepts offers a coherent framework for both property-oriented specifications and high level algorithms. LTS combines the semantic analysis of higher order theories with a transformation engine to assist the user with her design decisions. This combination leads to a better integration of computer-inferred information and user guidance. The distinction between coarse and fine tuning steps clearly separates the concerns in a hierarchical way. The transformation of entire specifications affecting several axioms and operations renders wide spanning transformation steps possible.

Future work will not only complete the prototype implementation. Rather we also continually strive for improving the system environment, the user interaction, the rule basis and the degree of mechanization. Currently we extend the standard prelude of ready-made specifications by theories for specific application areas, for example stream processing functions for distributed systems and skeletons for high level parallel programming. For the user interaction we experiment with different manipulation models, for example direct manipulations vs. meta programming. For a better acceptance, the degree of mechanization must be increased to a level where the user contributes the essential design decisions only.

So far we have carried out several case studies for derivations of higher order functional programs, data structure refinements and stream processing functions with LTS. LTS increases our understanding of the mechanization of transformational programming in a variety of areas. Moreover, LTS seems to provide a good platform for studying new approaches in this field.

Acknowledgements

The authors thank the organizers for granting an extension of the deadline, the three anonymous referees for their knowledgable amendments and A. STÜMPEL for carefully reading the manuscript. LTS was mostly implemented by the second author. J. BALSTER contributed parts of the code generation. C. KÜHNEL and K. TROJAHNER are currently implementing the graphical user interface. M. KRÜGER, E. LANGE and J. TEXTOR contributed their experiences with the system when working through various case studies.

References

1. F. Baader and T. Nipkow. *Term Rewriting and All That*. Cambridge Press, 1998.
2. F.L. Bauer, M. Broy, W. Dosch, F. Geiselbrechtinger, W. Hesse, R. Gnatz, B. Krieg-Bückner, A. Laut, T. Matzner, B. Möller, F. Nickl, H. Partsch, P. Pepper,

K. Samelson, M. Wirsing, and H. Wössner. *The Munich Project CIP: The Wide Spectrum Language CIP-L*, volume 183 of *LNCS*. Springer, 1985.

3. F.L. Bauer, H. Ehler, A. Horsch, B. Möller, H. Partsch, O. Paukner, and P. Pepper. *The Munich Project CIP: The Program Transformation System CIP-S*, volume 292 of *LNCS*. Springer, 1987.

4. F.L. Bauer, B. Möller, H. Partsch, and P. Pepper. Formal program construction by transformation – computer-aided, intuition-guided programming. *IEEE Transactions on Software Engineering*, 15:165–180, 1989.

5. M. Burstall and J. Darlington. A transformation system for developing recursive programs. *Journal of the ACM*, 1(24):44–67, 1977.

6. A. Dold. Representing, verifying and applying software development steps using the PVS system. In V.S. Alagar and M. Nivat, editors, *Proceedings of the Fourth International Conference on Algebraic Methodology and Software Technology, AMAST'95, Montreal, 1995*, volume 936 of *LNCS*, pages 431–435. Springer, 1995.

7. W. Dosch and S. Magnussen. Computer aided fusion for algebraic program derivation. *Nordic Journal of Computing*, 9, 2001. (to appear).

8. M. Feather. A survey and classification of some program transformation approaches and techniques. In L.G.L.T. Meertens, editor, *Proceedings TC2 Working Conference on Program Specification and Transformation*, pages 165–195. North Holland, 1987.

9. U. Fraus and H. Hussmann. Term induction proofs by a generalisation of narrowing. In C. Rattray and R.G. Clark, editors, *The Unified Computation Laboratory*, pages 43–55. Clarendon Press, 1992.

10. S. Kahrs, D. Sannella, and A. Tarlecki. The definition of Extended ML: a gentle introduction. *Theoretical Computer Science*, 173:445–484, 1997.

11. M. Lifantsev and L. Bachmair. An LPO-based termination ordering for higher-order terms without λ-abstraction. In J. Grundy and M. Newey, editors, *11th International Conference, TPHOLs'98, Canberra, Australia, 1998*, volume 1479 of *LNCS*, pages 277–293. Springer, 1998.

12. C. Lüth, H. Tej, Kolyang, and B. Krieg-Brückner. TAS and IsaWin: Tools for transformational program development and theorem proving. In J.-P. Finance, editor, *Fundamental Approaches to Software Engineering FASE'99. Joint European Conferences on Theory and Practice of Software ETAPS'99*, volume 1577 of *LNCS*, pages 239– 243. Springer, 1999.

13. K. Meinke. Universal algebra in higher types. *Theoretical Computer Science*, 100:385–417, 1992.

14. P. D. Mosses. CASL: a guided tour of its design. In J. L. Fiadeiro, editor, *Recent Trends in Algebraic Development Techniques. 13th International Workshop, WADT'98 Lisbon, Portugal, 1998*, volume 1589 of *LNCS*, pages 216–240. Springer, 1999.

15. S. Owre, N. Shankar, J. M. Rushby, and D.W.J. Stringer-Calvert. PVS system guide. Technical report, Computer Science Laboratory, SRI International, Menlo Park, CA, 1999.

16. L.C. Paulson. *Logic and Computation: Interactive Proof with Cambridge LCF*. Cambridge University Press, 1990.

17. L.C. Paulson. *Isabelle: A Generic Theorem Prover*, volume 828 of *LNCS*. Springer, 1994.

18. L.C. Paulson. *ML for the Working Programmer*. Cambridge University Press, 1996.

19. A. Pettorossi and M. Proietti. Rules and strategies for transforming functional and logic programs. *ACM Computing Surveys*, 28(2):360–414, 1996.
20. D.R. Smith. Automating the design of algorithms. In B. Möller, editor, *Formal Program Development (IFIP TC2/WG 2.1)*, volume 755 of *LNCS*, pages 324–354. Springer, 1993.
21. M. Wirsing. Algebraic specification. In J. van Leeuwen, editor, *Handbook of Theoretical Computer Science*, volume B, pages 675–788. Elsevier Science Publishers, 1990.
22. M. Wirsing, H. Partsch, P. Pepper, W. Dosch, and M. Broy. On hierarchies of abstract data types. *Acta Informatica*, 20:1–33, 1983.

On the Compatibility
of Model and Model-Class Transformations*

Maike Gajewsky[1] and Francesco Parisi-Presicce[2]

[1] Technical University Berlin, Germany
Institute for Communication and Software Technology
gajewsky@cs.tu-berlin.de
[2] Universitá di Roma "La Sapienza", Italy
Dipartimento di Scienze dell'Informazione
parisi@dsi.uniroma1.it

Abstract. The incremental approach to stepwise development of models requires the ability to transform models (by replacing a submodel with another one) and to change the class of models (by adding features previously ignored). In the context of Petri Nets, this paper provides rigorous foundations for model transformations and net class transformations in the framework of high level replacement systems. We define net class transformations between elementary nets, place/transition nets and a new notion of algebraic high level nets and prove, among other results, that the two kinds of transformations are compatible. This is an important results for stepwise software development with Petri nets. The high abstraction level of the results allows an easy adaption to other specification techniques, such as e. g. Algebraic Specifications.

1 Introduction

The incremental approach to stepwise development of models requires the ability to transform models (by replacing a submodel with another one) and to change the class of models (by adding features previously ignored). These two different levels of transformations, called *model transformation* and *model-class transformation*, respectively, allow arbitrary modifications of a model and are thus adequate for incremental development. For a high flexibility of the development process itself, including concurrent transformations, compatibility of transformations is of central importance. Compatibility here means that two transformations can be performed in an arbitrary order leading to the same result. It refers to transformations within one level as well as transformations across two different levels.

In this paper we present formal transformations of Petri nets which support Petri net based system development. Both kinds of transformations (*net*

---

* This work is part of the joint research project "DFG-Forschergruppe PETRINETZ-TECHNOLOGIE" supported by the German Research Council (DFG). This research was also supported in part by the TMR network "General Theory of Graph Transformation Systems" (GETGRATS) and by the Esprit Working Group APPLIGRAPH.

M. Cerioli and G. Reggio (Eds.): WADT/CoFI 2001, LNCS 2267, pp. 109–127, 2002.

class transformation and *model transformation*) are formalized on a rigorous mathematical foundation, We consider model transformations given by rule-based modifications in the context of High-Level Replacement (HLR) Systems [EHKP91,EGP99] and model-class transformations by categorical functors [AHS90]. HLR systems have been originally introduced [EHKP91] as a generalization of graph grammars which can also be applied to Petri nets, see e. g. [EGP99]. Our motivation for considering rule-based modifications is that the corresponding theory already comprises compatibility results for rule-based transformations.

Net class transformations perform a syntactical enrichment or abstraction of models in order to gain different abstraction layers within the development process. Thus, it is possible to start with a simple model on a high abstraction layer and to gradually enhance it by adding system aspects such as data, multiplicities, and time. There are already numerous Petri net variants comprising (some of) these aspects, but no constructive techniques to employ them systematically for a stepwise enhancement of the model. Support in this sense is provided by net class transformations which may add or abstract from system aspects.

In classical Petri net literature, there are already transformations between net classes. These are either used to define the semantics of a newly introduced Petri net variant or to represent a model in a more compact form, see e. g. [Lil95,Jen87]. A common property of these transformations is that they preserve (or define) the behaviour of the net, i. e. they yield an exact translation of nets. These transformations do not allow abstraction or enhancement of the model, which are essential for employing transformation in an incremental system development process. For other approaches to system refinement within one Petri net class, we refer the reader to the rich Petri net literature, especially to transition refinement [Rei82,Bau90,Jen92], the Petri net calculus [BDH92,BDE93], vicinity preserving morphisms [DM90] and liveness-preserving refinement [BG92].

Model transformations allow substituting a subnet by another one. We use the rule-based approach of HLR systems, which has already been applied in several case studies [Erm96,Par90]. Any kind of change of models within one Petri net class can be realized by a rule, and thus covers all kinds of refinement known from Petri net literature.

Both kinds of transformations together allow addition, deletion and change of system aspects, so that a final model can be reached systematically from an initial one. To illustrate this approach, we define net class transformations between three important Petri net classes, namely elementary nets [RE98], place/transition nets [Rei85,Bau90], and algebraic high level nets. The classical notion of algebraic high level nets [Vau87,Rei91,PER95] has been slightly modified in order to allow net class transformations from the other Petri net classes and to gain a relaxed applicability of model transformation. In contrast to the classical notion, [PER95], the new definition allows explicit variables and their instantiation in the new definition of morphisms.

In this paper we also examine the compatibility of the different transformations, which demands that model transformations be propagated across Petri net classes. Moreover, given a net class transformation and a model transformation, the order of application does not affect the result. Our main result, compatibility of model and model-class transformations in the above sense, is formulated on a high abstraction level. It can be instantiated with various specification techniques, namely all those which yield a HLR system. In particular, numerous variants of Petri nets are captured, where both transformations play a central role in Petri net based process models for system development. Our second main result (Theorem 8) states that all of the above mentioned net class transformations are in fact compatible with model transformations.

These results support an incremental approach to system development, where enhancements can be performed concurrently on different abstraction layers. The fundamental requirement of consistency among models of different abstraction layers can be satisfied by propagating model transformations across model-classes.

The approach to the compatibility of model and model-class transformations has other instantiations, such as Algebraic Specifications. A model transformation can be viewed as the visible part of a module specification [Par90] while a model-class transformation can be induced by (certain kinds of) institution morphisms to reflect the addition of features.

The paper is structured as follows: Section 2 reviews basic definitions of high-level replacement systems and different low-level Petri nets. In Section 3 we introduce our new notion of algebraic high-level nets. Subsequently, we define net class transformations between Petri net classes and present our compatibility results in Section 4. These results are illustrated by the stepwise development of a small system with net class and model transformations. Finally, we discuss a lifting of these net class transformations and make some concluding remarks in Section 6. Due to space limitations complete proofs can be found in [GPP00].

2 Background

In this section we review the theoretical background of model transformation, which is essential for discussing the compatibility between a model and a model-class transformation in the subsequent sections. We also briefly mention the basic algebraic notions of the elementary nets and place/transition nets categories.

2.1 Categories

Basic categories used in this paper are

Alg(SPEC.)   Objects are $SPEC$-algebras and the morphisms are given by algebra homomorphisms.

CMon.   Objects are commutative monoids and morphisms are monoid-homomorphisms.

GALG$_{iso}$.   Objects are pairs $(A, SPEC)$ of an algebraic specification and a $SPEC$-algebra A. Morphisms are generalized homomorphisms, see [BPP85].

Set. Objects are sets and morphisms are functions.

SPEC. Objects are algebraic specifications and morphisms are specification morphisms in the sense of [EM85].

2.2   Rules and Rule-Based Modification

We are now going to summarize the main concepts of high-level replacement systems in the sense of [EHKP91], a categorical generalization of graph grammars. High-level replacement systems are formulated for an arbitrary category C with a distinguished class M of morphisms which is used to construct the rules.

Definition 1 (Rules and Transformations). *A rule* $p = (L \xleftarrow{l} K \xrightarrow{r} R)$ *in C consists of the objects L, K and R, called left hand side, interface (or gluing object) and right hand side, respectively, and two morphisms $K \xrightarrow{l} L$ and $K \xrightarrow{r} R$ with both morphisms $l, r \in M$, where M is a distinguished class of morphisms in C.*

Given a rule $p = (L \xleftarrow{l} K \xrightarrow{r} R)$ *a rule-based modification $G \xRightarrow{p} H$, from an object G to an object H is given by the two pushout diagrams (1) and (2) in the category C. The morphisms $L \xrightarrow{g1} G$ and $R \xrightarrow{g3} H$ are called occurrences of L in G and R in H, respectively.*

$$
\begin{array}{ccccc}
L & \xleftarrow{\ l\ } & K & \xrightarrow{\ r\ } & R \\
{\scriptstyle g1}\downarrow & (1) & {\scriptstyle g2}\downarrow & (2) & \downarrow{\scriptstyle g3} \\
G & \xleftarrow[c1]{} & C & \xrightarrow[c2]{} & H
\end{array}
$$

By an occurrence of a rule $p = (L \xleftarrow{l} K \xrightarrow{r} R)$ *in a structure G we mean an occurrence of the left hand side L in G.*

Definition 2 (High-Level Replacement System). *Given a category C together with a distinguished class of morphisms M then (C, M) is called a HLR-category if (C, M) satisfies the following HLR-Conditions:*

1. *Existence of semi-M pushouts.*
   For all objects A, B, C and morphisms $C \leftarrow A \rightarrow B$, where at least one is in M there exists a pushout $C \rightarrow D \leftarrow B$:

$$
\begin{array}{ccc}
A & \longrightarrow & B \\
\downarrow & (1) & \downarrow \\
C & \longrightarrow & D
\end{array}
$$

2. *Existence of M pullbacks.*
   For all objects B, C, D and morphisms $B \rightarrow D \leftarrow C$, where both morphisms are in M there exists a pullback $B \leftarrow A \rightarrow C$.
3. *Inheritance of M.*
   (a) under pushouts:
       For each pushout square (1) as above $A \rightarrow B \in M$ implies $C \rightarrow D \in M$.

(b) *under pullbacks:*
   For each pullback square as above: $B \to D \in \mathcal{M}$ and $C \to D \in \mathcal{M}$ implies $A \to B \in \mathcal{M}$ and $A \to C \in \mathcal{M}$.

4. *Existence of binary coproducts and compatibility with $\mathcal{M}$.*

   (a) *For each pair of object A, B there is a coproduct $A + B$ with the universal morphisms $A \to A + B$ and $B \to A + B$.*

   (b) *For each pair of morphisms $A \xrightarrow{f} A'$ and $B \xrightarrow{f'} B'$ in $\mathcal{M}$ the coproduct morphism $A + B \xrightarrow{f+f'} B + B'$ is also in $\mathcal{M}$.*

5. *$\mathcal{M}$-pushouts are pullbacks.*
   Pushout squares of $\mathcal{M}$-morphisms are pullback squares.

6. *$\mathcal{M}$-pushout-pullback-decomposition.*
   For each diagram of the form:

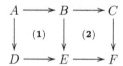

   we have that if $(1 + 2)$ is a pushout square, (2) is a pullback square, and $A \to D$, $B \to E$, $C \to F$, $B \to C$ and $E \to F$ are $\mathcal{M}$-morphisms, then also (1) is a pushout square.

2.3   Petri Net Categories

In this section we review the Petri net categories **PT** (category of place/transition nets) and **EN** (category of elementary nets), known from literature, and their properties with respect to cocompleteness and high-level replacement system.

Definition 3 (Categories PT and EN).

Category PT of Place/Transition Nets

- *A place/transition net is given by $N = (P, T, pre, post)$ with P the set of places, T the set of transitions, $pre, post : T \to P^{\oplus}$ the pre- and post-domain of transitions, where $(\_)^{\oplus}$ designates the free commutative monoid[1].*

- *Given two place/transition nets $N_i = (P_i, T_i, pre_i, post_i), i \in \{1, 2\}$, a morphism $f = (f_P, f_T) : N_1 \to N_2$ with functions $f_P : P_1 \to P_2$, and $f_T : T_1 \to T_2$ is called*
   - *transition preserving if $f_P^{\oplus} \circ pre_1 = pre_2 \circ f_T$ and $f_P^{\oplus} \circ post_1 = post_2 \circ f_T$.*
   - *strict if it is transition preserving and f_P and f_T are injective.*

Category EN of Elementary Nets

---

[1] Elements w of the free commutative monoid $X^{\oplus}$ for some set X can be represented as a linear sum $w = \sum_{x \in X} \lambda_x x$ with coefficients $\lambda_x \in \mathbb{N}$. They can be considered as multisets. Free commutative monoids induce the operations $\oplus, \ominus$ and the relation $\leq$ on linear sums.

- *An elementary net is given by $N = (P, T, pre, post)$ with P the set of places, T the set of transitions, and $pre, post : T \to \mathcal{P}(P)$ the pre- and post-domain of transitions.*
- *Given two elementary nets $N_i = (P_i, T_i, pre_i, post_i), i \in \{1, 2\}$, a morphism $f = (f_P, f_T) : N_1 \to N_2$ with functions $f_P : P_1 \to P_2$, and $f_T : T_1 \to T_2$ is called*
  - transition preserving *if $\mathcal{P}(f_P) \circ pre_1 = pre_2 \circ f_T$ and $\mathcal{P}(f_P) \circ post_1 = post_2 \circ f_T$*
  - strict *if it is transition preserving and f_P and f_T are injective.*

The following properties of these two categories will be used in the sequel.

Fact 1 (Properties of PT and EN).

- *The category **PT** of place/transition net and morphisms is finitely cocomplete.*
- *The category **PT** together with strict morphisms forms a high-level replacement system, i. e. the HLR conditions (see [EGP99], Definition 6.2.4) are satisfied.*
- *The category **EN** of elementary net and morphisms is finitely cocomplete.*
- *The category **EN** together with strict morphisms forms a high-level replacement system.*

3   New Category A*HL* of Algebraic High Level Nets

We are now going to introduce our new notion of algebraic high level nets. Unlike the classical notion [PER95], the new definition comprises explicit variables. These are essential for defining (see Fact 5) a net class transformation from place/transition nets. Moreover, variables may be substituted by terms in our new more flexible notion of morphisms between algebraic high level nets. We show that these new notions of algebraic high level net and algebraic high level net morphism form a high-level replacement system. Consequently, these can be employed in the context of rule-based modification. The laxer notion of morphisms allows an occurrence from an algebraic high level net inscribed with variables to a more specific one inscribed by terms and thus a relaxed applicability of rules.

Definition 4 (Algebraic High Level Net). *An* algebraic high level net *is given by*

$$N = (X, A, SPEC, P, T, pre, post, cond)$$

with

- *X : a set of variables,*
- *$A \in |\mathbf{Alg(SPEC)}|$ a SPEC algebra,*
- *$SPEC \in |\mathbf{SPEC}|$: an algebraic specification with $SPEC = (\Sigma, E)$ (see [EM85]),*
- *P : a set of places,*

- T : a set of transitions,
- $pre, post : T \to (T_{OP}(X) \times P)^{\oplus}$
  the pre- and postcondition functions of T (also see footnote 1 on page113),
  defining for each transition with adjacent arcs the arc inscriptions and the
  weight. $T_{OP}(X)$ denotes the term algebra over Σ and X, and
- $cond : T \to \mathcal{P}_{fin}(EQNS(\Sigma))$
  the function that maps each transition to a finite set of equations over the
  signature with variables representing the firing conditions.

Definition 5 (Algebraic High Level Net Morphisms). *Given two algebraic
high level nets* $N_i = (X_i, A_i, SPEC_i, P_i, T_i, pre_i, post_i, cond_i)$, *for* $i = 1, 2$. *An
algebraic high level net morphism* $f : N_1 \to N_2$ *is given by* $f = (f_X, f_{SPEC}, f_A, f_P, f_T)$ *where*

- $f_X : X_1 \to T_{OP_2}(X_2)$ *maps variables to terms,*
- $f_{SPEC} : SPEC_1 \to SPEC_2$ *is a specification morphism,*
- $f_A : A_1 \to V_{f_\Sigma}(A_2)$ *is a homomorphism in* **Alg(SPEC$_1$)**, *where* $V_{f_\Sigma} :$
  Alg(SPEC$_2$) $\to$ **Alg(SPEC$_1$)** *is the forgetful functor with respect to the
  signature morphism* $f_\Sigma : \Sigma_1 \to \Sigma_2$,
- $f_P : P_1 \to P_2$ *maps places to places in* **Set**, *and*
- $f_T : T_1 \to T_2$ *maps transitions to transitions in* **Set**,

such that the following diagram commutes componentwise

$$
\begin{array}{ccccc}
\mathcal{P}_{fin}(EQNS(SIG_1)) & \xleftarrow{\;cond_1\;} & T_1 & \xrightarrow[post_1]{\;pre_1\;} & (T_{OP_1}(X_1) \times P_1)^{\oplus} \\
\mathcal{P}_{fin}(f_X^{\sharp}) \downarrow & (1) & f_T \downarrow & (2) & \downarrow (f_{SPEC}^{\sharp} \times f_P)^{\oplus} \\
\mathcal{P}_{fin}(EQNS(SIG_2)) & \xleftarrow{\;cond_2\;} & T_2 & \xrightarrow[post_2]{\;pre_2\;} & (T_{OP_2}(X_2) \times P_2)^{\oplus}
\end{array}
$$

These morphisms give rise to the category **AHL** *of algebraic high level nets.*
A strict algebraic high level net morphism *is an algebraic high level net morphism which has the following additional properties:*

- $f_X : X_1 \to X_2$ *is an injective mapping of variables,*
- $f_{SPEC} : SPEC_1 \to SPEC_2$ *is a strict specification morphism (see [EGP99]
  for definition of strictness),*
- $f_A : A_1 \to V_{f_\Sigma}(A_2)$ *is an isomorphism in* **Alg(SPEC$_1$)**,
- f_P *and* f_T *are injective.*

An example of strict algebraic high level net morphisms between algebraic
high level nets is depicted in Figure 5.

Remark 1 (**AHL** *is not finitely cocomplete*). The category **AHL** is not finitely
cocomplete in general. Pushouts of different term substitutions do not exist.
However, pushouts exists, if one of the morphisms is strict, shown in the first
part of the proof of Fact 2.

Remark 2 (Induced algebraic high level net morphism). Given an algebraic high level net

$$N = (X, A, SPEC, P, T, pre, post, cond)$$

a specification morphism $f_{SPEC} : SPEC \to SPEC'$ induces an algebraic high level net morphism $f : N \to (X, A', SPEC', P, T, pre', post', cond')$ where

- f_X is the identity on the set of variables X,
- f_{SPEC} is the given specification morphism,
- $f_A : A \to A' = V_{f_\Sigma}(Free_{f_\Sigma}(A))$ is the usual universal morphism in **Alg(SPEC)**,
- $f_P : P \to P$ is the identity on places, and
- $f_T : T \to T$ is the identity on transitions

and the functions $pre, post, cond$ are adjusted to reflect the "translation" operated by f_{SPEC}.

If $f_{SPEC} : SPEC \to SPEC'$ is strict and the corresponding free functor $Free_{f_\Sigma}$ is conservative, then the induced algebraic high level net morphism is also strict.

Fact 2 (AHL is High-Level Replacement System). *The category* **AHL** *as defined in Def. 5 together with strict morphisms forms a high-level replacement system, denoted as* (**AHL**, $\mathcal{M}_{strict}$).

Proof.   1. Existence of semi-$\mathcal{M}$ pushouts
Given $N_i = (X_i, A_i, SPEC_i, P_i, T_i, pre_i, post_i, cond_i)$ for $i = 0, 1, 2$ and algebraic high level net morphisms $f : N_0 \to N_1$ and $f' : N_0 \to N_2$.
We construct the pushouts

$$SPEC_2 \xrightarrow{g'_{SPEC}} SPEC_3 \xleftarrow{g_{SPEC}} SPEC_1,$$

$$V_{g'_{SPEC}}(A_2) \xrightarrow{g'_A} A_3 \xleftarrow{g_A} V_{g_{SPEC}}(A_1),$$

$$P_2 \xrightarrow{g'_P} P_3 \xleftarrow{g_P} P_1,$$

$$and \qquad T_3 \xrightarrow{g'_T} T_3 \xleftarrow{g_T} T_2$$

with respect to the underlying morphisms in the corresponding categories. On the X-component we have the following diagram:

$$
\begin{array}{ccc}
X_0 & \xrightarrow{f_X} & X_1 \\
\downarrow{\scriptstyle f'_X} & & \\
T_{OP_2}(X_2) & & \\
\uparrow{\scriptstyle u_{X_2}} & & \\
X_2 & &
\end{array}
$$

where u_{X_2} is the natural embedding of variables to terms. We construct the set of variables $X_3 := X_2 \uplus X_1 \setminus f_X(X_0)$ where $\uplus$ designates the disjoint union. Due to construction we get $g'_X =: inc : X_2 \to X_3$ as inclusion of sets, which can be lifted to $\overline{inc} : T_{OP_2}(X_2) \to T_{OP_3}(X_3)$. Furthermore we define $g_X : X_1 \to T_{OP_3}(X_3)$ by:

$$g_X(x) = \begin{cases} \overline{inc} \circ f'_X(x_0) & \text{if } x = f_X(x_0) \\ x & \text{otherwise} \end{cases}$$

As f_X is injective and therefore x_0 with $x = f_X(x_0)$ unique, the function g_X is well-defined. Due to the pushout property of T_3 we can construct $N_3 = (X_3, A_3, SPEC_3, P_3, T_3, pre_3, post_3, cond_3)$ where $pre_3, post_3,$ and $cond_3$ are uniquely induced.

2. *Existence of $\mathcal{M}$-pullbacks*
   *We construct the pullback componentwise for the X, $SPEC$, A, P, and T component in the corresponding categories **Set** and **GALG**$_{iso}$. As in [PER95], Thm. 5.10, we obtain a diagram (1) below as a pullback in **Set**, as*

$$\begin{array}{ccc}
(T_{OP_0}(X_0) \times P_0)^{\oplus} & \longrightarrow & (T_{OP_1}(X_1)) \times P_1)^{\oplus} \\
\downarrow & (1) & \downarrow {\scriptstyle g_X} \\
(T_{OP_2}(X_2)) \times P_2)^{\oplus} & \longrightarrow & T_{OP_3}(X_3)) \times P_3)^{\oplus}
\end{array}$$

   (a) the term algebra construction preserves pullbacks of injections,
   (b) products are compatible with pullbacks, and
   (c) the forgetful functor $V : \mathbf{CMon} \to \mathbf{Set}$, where $\mathbf{CMon}$ is the category of commutative monoids, preserves pullbacks.
   This yields functions $pre_0, post_0 : T_0 \to (T_{OP_0}(X_0) \times P_0)^{\oplus}$. For the $cond_0$ function we exploit the fact that the powerset functor preserves pullbacks if at least one morphism is injective. Thus, the function $cond_0$ is obtained, leading to a well-defined algebraic high level net morphism
3. *Inheritance of $\mathcal{M}$*
   (a) under pushouts
       *The morphism g' (see item 1) is strict: g'_X is injective, g'_{SPEC} is a strict specification morphism as these are inherited in **SPEC**, g'_A is also due to the construction of pushouts in **GALG**$_{iso}$, and g'_P and g'_T are injective, as injective morphisms are inherited under pushouts in **Set**.*
   (b) under pullbacks:
       In the case of pullbacks, we have for strict f' and g' :
       *Injective morphisms are preserved under $\mathcal{M}$-pullbacks in **Set** and strict injective morphisms are preserved under $\mathcal{M}$-pullbacks in **SPEC**.*
4. *Existence of binary coproducts and compatibility*
   (a) Coproducts are constructed componentwise. Due to the coproduct property of $T_1 + T_2$ we obtain $pre_C, post_C : T_1 + T_2 \to (T_{OP_C}(X_1 + X_2) \times (P_1 + P_2))^{\oplus}$ and similarly for $cond_C$, leading to a well-defined algebraic high level net morphism.

*(b) Moreover, injective morphisms in **Set** and strict injective morphisms in **SPEC** are compatible with coproducts. Hence the coproduct morphism is strict.*

5. *$\mathcal{M}$-pushouts are pullbacks*
   This property is due to the componentwise construction of pushouts and pullbacks and the corresponding property in **Set**, **SPEC**, *and* **GALG**$_{iso}$.

6. *$\mathcal{M}$-pushout-pullback-decomposition*
   For algebraic high level nets without explicit variables, this has been shown in [PER95]. We consider the X component:

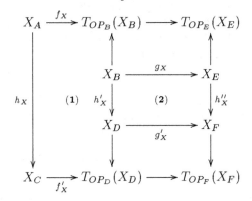

Let there be an algebraic high level net N_Y with X_Y, and $TOP_Y(X_Y)$ and morphisms $k_X : X_B \rightarrow TOP_Y(X_Y)$ and $k'_X : X_C \rightarrow TOP_Y(X_Y)$ such that $\overline{k_X} \circ \overline{f_X} = k'_X \circ h_X$.

Let $\alpha_D \in X_D$. We have $\alpha_F = g'_X(\alpha_D) \in X_F$ and due to pushout property of X_F we have

(a) there is $\alpha_E \in X_E$ with $h''_X(\alpha_E) = \alpha_F$. Thus, we have $\alpha_D \in X_D$, $\alpha_E \in X_E$, and $\alpha_F \in X_F$ with $\alpha_D \xmapsto{g'_X} \alpha_F \xleftarrow{h''_X} \alpha_E$. Due to pullback property of (2) we have
   $$\exists \alpha_B \in X_B : \quad h'_X(\alpha_B) = \alpha_D \text{ and } g_X(\alpha_B) = \alpha_E \text{ or}$$

(b) there exists $\alpha_C \in X_C$ with $\overline{g'_X \circ f'_X}(\alpha_C) = g'_X \circ f'_X(\alpha_C) = \alpha_F$. As g'_X is injective we have $f'_X(\alpha_C) = \alpha_D$.

We define $l_X : X_D \rightarrow TOP_Y(X_Y)$ by

$$l_X(\alpha_D) = \begin{cases} k_X(\alpha_B) & \text{in case of 6a} \\ k'_X(\alpha_C) & \text{in case of 6b} \end{cases}$$

We have to show well-definedness for the case where both cases 6a and 6b are true, i.e. we have to show that $k_X(\alpha_B) = k'_X(\alpha_C)$ for $h'_X(\alpha_B) = \alpha_D = f'_X(\alpha_C)$. In this case

$$\begin{aligned} \alpha_F &= h''_X \circ g_X(\alpha_B) \\ &= g'_X \circ h'_X(\alpha_B) \\ &= g'_X(\alpha_D) \\ &= \overline{g'_X \circ f'_X}(\alpha_C) \end{aligned}$$

As $(1+2)$ is a pushout it follows that there exists an $\alpha_A \in X_A$ with $h_X(\alpha_A) = \alpha_B$ and $g_X \circ f_X(\alpha_A) = \alpha_E$. Since g_X is injective, we have $f_X(\alpha_A) = \alpha_B$
Summarizing, we have $k_X(\alpha_B) = k_X \circ h_X(\alpha_A) = \overline{k'_X} \circ f_X(\alpha_A) = k'_X \circ \underline{f_X}(\alpha_A) = k'_X(\alpha_B)$ and thus well-definedness of l_X. Furthermore, we have $\overline{l_X} \circ h'_X = k_X$ and $\overline{l_X} \circ f'_X = k'_X$ due to definition of l_X. Uniqueness follows from totality of l_X and commutativity property. Thus (1) is a pushout.

For the remaining components this has already be shown and follows from the corresponding properties in **Set**, **SPEC**, respectively **GALG**$_{\mathrm{iso}}$.

4   Compatibility of Net Class and Model Transformations

In this section we define and examine net class transformations between the standard Petri net categories given in Section 2.3 (Def. 3) and the new category of algebraic high level nets, introduced in Section 3. In Theorem 7 we state sufficient conditions for the preservation of model transformation by net class transformations in the context of high-level replacement systems and categorical functors. Consequently, this result does not only apply to (different kinds of) Petri nets, but to any specification technique which yields a high-level replacement system, e. g. algebraic specifications, and corresponding functors. At the end of this section we prove that all the previously defined net class transformations satisfy these conditions, leading to the conclusion (Theorem 8) that all of them are compatible with model transformations.

We start by formalizing net class and model transformations.

Concept 1 (Formalization of Transformations). *We formalize net class transformations as categorical functors between Petri net categories.*
A model transformation is formalized as rule-based modification in the sense of Def. 1 in Section 2.2

In the following, we use functors and net class transformations, respectively rule-based modification and model transformation, as synonyms.

4.1   Functors between Petri Net Categories

Fact 3 (Functor *Causality*). *The functor Causality* : **PT** $\rightarrow$ **EN** *is defined as follows:*

On objects. *For a place/transition net $N = (P_N, T_N, pre_N, post_N)$, we define Causality(N) is the elementary net $(P_{N'}, T_{N'}, pre_{N'}, post_{N'})$ with $P_{N'} = P_N$, $T_{N'} = T_N$, $pre_{N'}(t) = \{p \mid p \leq pre_N(t) \text{ and } p \in P_{N'}\}$, and $post_{N'}(t) = \{p \mid p \leq post_N(t) \text{ and } p \in P_{N'}\}$, where $\leq$ denotes the multiset inclusion.*
On morphisms. *For a place/transition net morphism $f = (f_P, f_T) : N_1 \rightarrow N_2$, Causality(f) $= f : Causality(N_1) \rightarrow Causality(N_2)$ is an elementary net*

morphism since for $t_1 \in T_1$:

$$p_1 \in pre_{N_1'}(t_1) \Longleftrightarrow p_1 \leq pre_{N_1}(t_1) \quad \text{due to the definition of } pre_{N'}$$
$$\Longleftrightarrow f_P^{\oplus}(p_1) \leq f_P^{\oplus}(pre_{N_1}(t_1)) \quad \oplus \text{ is linear}$$
$$\Longleftrightarrow f_P(p_1) \leq pre_{N_2}(f_T(t_1)) \quad \text{property of } f \text{ as morphism}$$
$$\Longleftrightarrow f_P(p_1) \in pre_{N_2'}(f_T(t_1)) \quad \text{due to the def. of } pre_{N'}$$

Therefore, we have $\mathcal{P}(f_P) \circ pre_{N_1'} = pre_{N_2'} \circ f_T$ for each $t_1 \in T_1$.

Fact 4 (Functor $Weight$). *The functor $Weight : \mathbf{EN} \to \mathbf{PT}$ is defined as follows:*

On objects. *For an elementary net $N = (P_N, T_N, pre_N, post_N)$, $Weight(N)$ is the place/transition net $(P_{N'}, T_{N'}, pre_{N'}, post_{N'})$ with $P_{N'} = P_N$, $T_{N'} = T_N$, $pre_{N'}(t) = \oplus_{p \in pre_N(t)} p$, and $post_{N'}(t) = \oplus_{p \in post_N(t)} p$.*

On morphisms. *For an elementary net morphism $f = (f_P, f_T) : N_1 \to N_2$, $Weight(f) = f : Weight(N_1) \to Weight(N_2)$ is a place/transition net morphism since for $t_1 \in T_1$:*

$$p_1 \leq pre_{N_1'}(t_1) \Longleftrightarrow p_1 \in pre_{N_1}(t_1) \quad \text{due to def. of } pre_{N'}$$
$$\Longleftrightarrow \mathcal{P}(f_P)(p_1) \in \mathcal{P}(f_P)(pre_{N_1}(t_1)) \quad \text{linear}$$
$$\Longleftrightarrow f_P(p_1) \in pre_{N_2}(f_T(t_1)) \quad \text{property of } f \text{ as morphism}$$
$$\Longleftrightarrow f_P(p_1) \leq pre_{N_2'}(f_T(t_1)) \quad \text{due to def. of } pre_{N'}$$

As all arc weights are equal to 1, $f_P^{\oplus} \circ pre_{N_1'} = pre_{N_2'} \circ f_T$.

The net class transformation *Data* constructs the set X of variables which are used to build arcs in the algebraic high level net.

Fact 5 (Functor $Data$.). *The functor $Data : \mathbf{PT} \to \mathbf{AHL}$ is defined as follows:*

On objects. *For a place/transition net $N = (P_N, T_N, pre_N, post_N)$, define $Data(N) = (X_{N'}, A_{N'}, SPEC_{N'}, P_{N'}, T_{N'}, pre_{N'}, post_{N'}, cond_{N'})$ where:*
 – *$X_{N'}$: The set of variables (of the form (t, p)) is the set $X_{pre} \uplus X_{post}$, where X_{pre} is given by the set $inc \circ (id_T, pre_N)(T)$ in the following diagram (and in a similar one for X_{post}:*

$$
\begin{array}{ccc}
& T \times P^{\oplus} & \xrightarrow{\quad inc \quad} \uplus(T \times P) \\
\nearrow^{(id_T, pre_N)} & \searrow^{pr_{P^{\oplus}}} & \\
T \xleftarrow{\quad pr_T \quad} & \searrow & \\
& \xrightarrow{pre} P^{\oplus} &
\end{array}
$$

where pr_T and $pr_{P^{\oplus}}$ are the projections to T, resp. $P^{\oplus}$ and $inc : T \times P^{\oplus} \to \uplus(T \times P)$ is defined by

$$
inc(t, w) == \begin{cases} \underbrace{(t, p_1) \uplus \cdots \uplus (t, p_1)}_{\lambda_1 \, times} \uplus \cdots \uplus \underbrace{(t, p_n) \uplus \cdots \uplus (t, p_n)}_{\lambda_n \, times} \\ \quad \quad if \ w = \lambda_1 \cdot p_1 \oplus \cdots \oplus \lambda_n \cdot p_n \in P^{\oplus} \\ \emptyset \quad \quad if \ w = \epsilon \end{cases}
$$

The morphism inc is the universal morphism of the free construction
$inc : T \times P^{\oplus} \to \uplus(T \times P) \cong (T \times P^{\oplus})^{\oplus}$

- $SPEC_{N'} = \emptyset$ *is the empty algebraic specification,*
- $A_{N'} = \emptyset$ *is the empty algebra,*
- $P_{N'} = P_N$,
- $T_{N'} = T_N$,
- $pre_{N'}(t) = \bigoplus_{(t,p)\in inc\circ(id_T,pre)(t)}((t,p),p)$
- $post_{N'}(t) = \bigoplus_{(t,p)\in inc\circ(id_T,post)(t)}((t,p),p)$
- $cond_{N'}(t) = \emptyset$ *for each* $t \in T_{N'}$

On morphisms. *For a place/transition net morphism* $f = (f_P, f_T) : N_1 \to N_2$,
define $Data(f) = (f_X, f_{SPEC}, f_A, f_P, f_T) : Data(N_1) \to Data(N_2)$ *with*

- f_X *given by* $(f_T, f_P)^{\oplus}$. *As in the following diagram:*

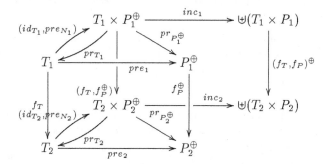

It is well defined due to the commutativity of the above diagram.

- $f_{SPEC} = \emptyset$ *is the initial SPEC morphism*
- $f_A = \emptyset$ *is the initial algebra morphism*

The following diagram commutes due to the definition of cond and f_X:

$$
\begin{array}{ccccc}
\emptyset & \xleftarrow{cond_1} & T_1 & \xrightarrow{pre_{N_1}} & (X_1 \times P_1)^{\oplus} \\
{\scriptstyle\emptyset}\downarrow & & {\scriptstyle f_T}\downarrow & & \downarrow{\scriptstyle(f_X,f_P)^{\oplus}} \\
\emptyset & \xleftarrow{cond_2} & T_2 & \xrightarrow{pre_{N_2}} & (X_2 \times P_2)^{\oplus}
\end{array}
$$

Thus, $Data(f)$ *is a well-defined algebraic high level net morphism.*

Fact 6 (Functor Skeleton). *The functor Skeleton :* **AHL** $\to$ **PT** *is defined as
follows:*

On objects. *Given* $N = (X_N, A_N, SPEC_N, P_N, T_N, pre_N, post_N, cond_N)$ *define* $Skeleton(N) = (P_{N'}, T_{N'}, pre_{N'}, post_{N'})$ *where*

- $P_{N'} = P_N$,
- $T_{N'} = T_N$,
- $pre_{N'}(t) = pr_{P^{\oplus}} \circ pre_N(t)$, *and* $post_{N'}(t) = pr_{P^{\oplus}} \circ post_N(t)$ *where* $pr_{P^{\oplus}} :$
  $(T_{OP}(X_N) \times P_N)^{\oplus} \to P_N^{\oplus}$ *is a projection.*

On morphisms. *An algebraic high level net morphism* $f = (f_X, f_{SPEC}, f_A, f_P,$
$f_T) : N_1 \to N_2$ *due to the corresponding commutativity of* f *with pre and
post functions, is a well-defined place/transition net morphism.*

4.2   Compatibilty Results

We are now going to state our main results concerning the preservation of rules and rule-based modification by functors. This result is formulated in the context of high-level replacement systems and categorical functors only. Consequently, it can be instantiated by any high-level replacement system and can be easily transferred to e. g. Algebraic Specifications or Graph Grammars by just verifying the relatively simple preconditions of Theorem 7.

At the end of this section we prove these preconditions for the functors of the previous section. All these functors satisfy the preconditions and therefore preserve rule-based modification.

Definition 6 (Derivation-Compatible Functors). *Given high-level replacement systems* $(\mathcal{C}_1, \mathcal{M}_1)$ *and* $(\mathcal{C}_2, \mathcal{M}_2)$, *a functor* $F : \mathcal{C}_1 \to \mathcal{C}_2$ *is derivation-compatible if the following conditions are satisfied:*

1. *($\mathcal{M}$-compatibility): For each morphism* $m_1 \in \mathcal{M}_1$ *we have* $F(m_1) \in \mathcal{M}_2$.
2. *(PO-Preservation): F preserves pushouts, i. e. given a pushout* (N_3, g, g') *of* (N_0, f, f') *in* $\mathcal{C}_1$ *then* $(F(N_3), F(g), F(g'))$ *is the pushout of* $(F(N_0), F(f), F(f'))$ *in* $\mathcal{C}_2$.

Theorem 7 (Derivation-compatible Functors Preserve Rule-Based Modification). *Given two high-level replacement systems* $(\mathcal{C}_1, \mathcal{M}_1)$ *and* $(\mathcal{C}_2, \mathcal{M}_2)$, *a rule p and a rule-based modification* $G \overset{p}{\Longrightarrow} H$ *in* $\mathcal{C}_1$, *any derivation-compatible functor* $F : \mathcal{C}_1 \to \mathcal{C}_2$. *preserves rule-based modification, i. e. the diagram commutes.*

$$\begin{array}{ccc} G & \overset{p}{\Longrightarrow} & H \\ F \Big\Vert & & \Big\Vert F \\ F(G) & \overset{F(p)}{\Longrightarrow} & F(H) \end{array}$$

Proof. As F is a functor, the rule $p : L \overset{l}{\longleftarrow} K \overset{r}{\longrightarrow} R$ in $\mathcal{C}_1$ is translated into the rule $F(p) : F(L) \overset{F(l)}{\longleftarrow} F(K) \overset{F(r)}{\longrightarrow} F(R)$ in $\mathcal{C}_2$. The rule is well-defined as F is $\mathcal{M}$-compatible. The rule-based modification $G \overset{r}{\Longrightarrow} H$ in $\mathcal{C}_1$ is translated by the functor F into the diagram below in $\mathcal{C}_2$:

$$\begin{array}{ccccc} F(L) & \overset{F(l)}{\longleftarrow} & F(K) & \overset{F(r)}{\longrightarrow} & F(R) \\ F(g_1) \Big\downarrow & \textbf{(PO)} & \Big\downarrow F(g_2) & \textbf{(PO)} & \Big\downarrow F(g_3) \\ F(G) & \underset{F(c1)}{\longleftarrow} & F(C) & \underset{F(c2)}{\longrightarrow} & F(H) \end{array}$$

Both squares are pushout squares in $\mathcal{C}_2$ due to the preservation of pushouts by F. Moreover, as $F(l), F(r) \in \mathcal{M}_2$, the rule $F(p)$ is well defined and the pushout complement $F(C)$ is unique. Consequently, the result of the rule-based modification $F(H)$ is unique.

Theorem 8 (Petri Net Functors Preserving Rule-Based Transformation). *The following functors preserve Rule-Based Transformations:*

- *Functor Weight* : **EN** → **PT**
- *Functor Causality* : **PT** → **EN**
- *Functor Data* : **PT** → **AHL**
- *Functor Skeleton* : **AHL** → **PT**
- *Functor Data* ∘ *Weight* : **EN** → **AHL**
- *Functor Causality* ∘ *Skeleton* : **AHL** → **EN**

induced by composition

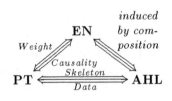

Proof (sketch).

We have to show that all of the above functors are derivation-compatible, in the sense of Def. 6: $\mathcal{M}$-compatibility follows immediately from the Definition of $\mathcal{M}$-morphisms and the Definitions of the functors. PO-Preservation of Weight, Causality and Skeleton follows from the componentwise construction of pushouts in the source and target categories. For PO-Preservation of Data we have to show that the X component of the pushout-object is a pushout in **Set**. This can be shown by contradiction using the fact that the X-component corresponds to the arcs in the place/transition net.

5 Example: Development of a Producer/Consumer System

In this section we illustrate Petri net transformations by an example of the development of a producer/consumer system. We employ net class transformations and model transformations in order to enhance the abstract and simple model given in Figure 1. The elementary net $N_{0_{EN}}$ models two separate systems of a producer (on the left-hand side), cycling between an idle state and a producing one, and a consumer with a similar behaviour.

In order to synchronize the two isolated systems of the producer and consumer, we modify the initial model by applying the rule $buffer_{EN}$ in Figure 2.

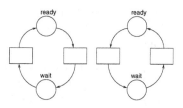

Fig. 1. Initial model $N_{0_{EN}}$ of producer and consumer

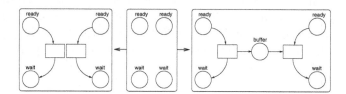

Fig. 2. Rule $buffer_{EN}$ for cooperation between producer and consumer

The rule in Figure 2 adds a buffer between the systems, where the producer puts a produced item and the consumer takes it. The rule-based modification via the rule $buffer_{EN}$ to the initial net, depicted as $N_{0_{EN}} \overset{buffer_{EN}}{\Longrightarrow} N_{1_{EN}}$, leads to the net depicted in Figure 3.

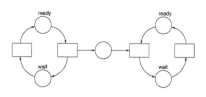

Fig. 3. Producer and consumer system $N_{1_{EN}}$

We are now going to perform a net class transformation from elementary nets to place/transition nets. The initial model $N_{0_{EN}}$ is syntactically transformed into a place/transition net, by weighting all arcs with 1, by the functor *Weight* defined in Definition 4. The graphical illustration of the resulting net $N_{0_{PT}}$ looks like $N_{0_{EN}}$ in Figure 1, because an arc weight of 1 is omitted by convention. Moreover, we transform the rule $buffer_{EN}$ into the rule $buffer_{PT}$ operating on place/transition nets. Its application yields a place/transition $N_{1_{PT}}$ net corresponding to the net $N_{1_{EN}}$. A further rule-based modification by rule $quick_{PT}$ changes the pace of the producer; it produces two items instead of one in a single production step.

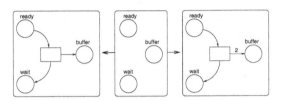

Fig. 4. Rule $quick_{PT}$ for changing arc weight

Place/Transition Nets. Causality translates this new rule $quick_{PT}$ back to **EN**, the category of elementary nets. The resulting rule $quick_{EN}$ is the identical rule, where exactly those parts are added which are deleted. The transformations are depicted in Figure 6.

Algebraic High-Level Nets We transform the place/transistion net $N_{0_{PT}}$ into an algebraic high level net $N_{0_{AHL}}$ via the functor *Data* defined in Definition 5. The empty algebraic specification and algebra are added. Moreover, arcs are inscribed by unique variables, where the weight of the arc is also taken into account. For each arc with arc of weight n we create n unique variables attached to the arc in the algebraic high level net. Firing conditions are left empty by this transformation. Correspondingly the rule $buffer_{PT}$ is translated into **AHL**, the category of algebraic high level nets. This leads to the rule $buffer_{AHL}$. A subsequent model transformation via $buffer_{AHL}$ leads to the net $N_{1_{AHL}}$. Finally, we modify the algebraic specification and the arc inscriptions by specifying the items produced by the corresponding terms. This is done by the rule $items_{AHL}$ depicted in Figure 5.

All the transformations are summarized in Figure 6, where $\Longrightarrow$ symbolizes model transformation and $\Longleftrightarrow$ signify the net class transformations. Note, that only the rules $buffer_{EN}$, $quick_{PT}$, and $items_{AHL}$ have been given explicitly. All other rules are derived from these ones by the functors. We have developed

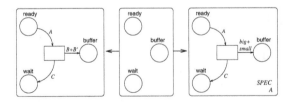

$SPEC =$
sorts: items
opns: big: $\to$ items
          small: $\to$ items
$A = (\{big - product, small - product\}, big - product, small - product)$

Fig. 5. Rule $items_{AHL}$

several models on different abstraction layers, where abstraction means a loss of information: *Skeleton* abstracts from data information and *Causality* abstracts from multiplicities. Viceversa, *Data* integrates this information on a syntactical level leading to a model with trivial data. This model is subsequently refined by specific data, and similarly for *Weight*.

Note that a refinement (by any transformation) typically does not preserve behaviour, but usually realizes a restriction, which is sensible in the context of stepwise system development. Abstraction, on the other hand, typically preserves the firing behaviour in the sense that in the abstracted model a transition may fire if its counterpart in the refined model may fire.

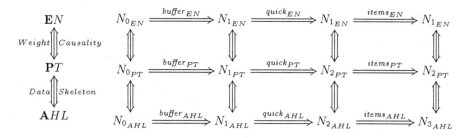

Fig. 6. Transformations used in the Example

6   Concluding Remarks

In this paper we have formalized the notion of transformation of Petri nets, both on the level of models and on the level of net classes, in a categorical framework based on high level replacement systems.

After defining a more flexible model of Algebraic High Level nets, functors have been defined that allow the lifting of any derivation (transformation of models) from one net class to another one with richer features. Compatibility results guarantee that the order of application of net class transformations and model transformations does not affect the result. The framework of high level replacement systems allows to exploit known results such as Local Confluence and Parallelism Theorems within each kind of transformation. Furthermore, the generality of the results stated in terms of Derivation-compatible functors allows the application of this framework to other specification formalisms, such as algebraic specifications.

The functors presented in this paper operate on the net structure (*Causality* and *Weight*) and on the data of a Petri net (*Data* and *Skeleton*). They add or remove one of these basically independent components. Future work will comprise a lifting of these functors, so that adding/forgetting marking and data and changing of net structure can be uniformly described independent of the net classes. The theory of parameterized net classes introduced in [Pad96] yields a suitable frame for a uniform lifting of the functors.

Still under investigation are other net classes (such as colored or timed Petri nets) and functors to relate them to the basic ones investigated here.

An important issue in the context of software development is the preservation of system properties via transformations. Preservation of properties like deadlock-freeness, safety properties in the sense that "nothing bad can happen in the system", flow structure, or lifeness might be desired when performing transformations. With respect to the preservation of safety properties by model transformation, there are already important results stated in [PGE98,PGH99,PHG00]. The extensive classification of net class transformations and model transformation is a crucial topic of future research.

References

[AHS90]   J. Adamek, H. Herrlich, and G. Strecker. *Abstract and Concrete Categories.* Series in Pure and Applied Mathematics. John Wiley and Sons, 1990.

[Bau90]   B. Baumgarten. *Petrinetze, Grundlagen und Anwendungen.* BI, 1990.

[BDE93]   E. Best, R. Devillers, and J. Esparza. General refinement and recursion operators for the Petri Box Calculus. In P. Enjalbert et al., editors, *STACS' 93*, volume 665 of *Lecture Notes in Computer Science*, pages 130–140. Springer Verlag, 1993.

[BDH92]   E. Best, R. Devillers, and J. Hall. The Box Calculus: a new causal algebra with multi-label communication. In *Advances in Petri Nets*, volume 609 of *Lecture Notes in Computer Science*, pages 21–69. Springer Verlag, 1992.

[BG92]   C. Brown and D. Gurr. Refinement and Simulation of Nets – a categorical characterization. In K. Jensen, editor, 13th *International Conference on Application and Theory of Petri Nets*, volume 616 of *Lecture Notes in Computer Science*, pages 76–92. Springer Verlag, 1992.

[BPP85]   E.K. Blum and F. Parisi-Presicce. Tha Semantics of Shared submodle Specifications. In *TAPSOFT 85*, volume 185 of *Lecture Notes in Computer Science*, pages 359–373. Springer Verlag,1985.

[DM90]   J. Desel and A. Merceron. Vicinity Respecting Net Morphisms. In *Advances in Petri Nets*, volume 483 of *Lecture Notes in Computer Science*, pages 165–185. Springer Verlag, 1990.

[EGP99]   H. Ehrig, M. Gajewsky, and F. Parisi-Presicce. *High-Level Replacement Systems with Applications to Algebraic Specifications and Petri Nets*, volume 3: Concurrency, Parallelism, and Distribution, chapter 6, pages 341–400. World Scientific, Handbook of Graph Grammars and Computing by Graph Transformations, 1999.

[EHKP91] H. Ehrig, A. Habel, H.-J. Kreowski, and F. Parisi-Presicce. From graph grammars to high level replacement systems. In *4th Intern. Workshop on Graph Grammars*, volume 532 of *Lecture Notes in Computer Science*, pages 269–291. Springer Verlag, 1991.

[EM85]   H. Ehrig and B. Mahr. *Fundamentals of Algebraic Specification 1: Equations and Initial Semantics*, volume 6 of *EATCS Monographs on Theoretical Computer Science*. Springer Verlag, Berlin, 1985.

[Erm96]   C. Ermel. Anforderungsanalyse eines medizinischen Informationssystems mit Algebraischen High-Level-Netzen. Technical Report 96-15, TU Berlin, 1996.

[GPP00] M. Gajewsky and F. Parisi-Presicce. Formal Transformations of Petri Nets. Technical Report 2000-12, Technical University Berlin, 2000.

[Jen87] K. Jensen. Computer tools for construction, modification and analysis of Petri nets. In W. Brauer, W. Reisig, and G. Rozenberg, editors, *Petri Nets: Applications and Relationships to Other Models of Concurrency*, pages 4–19. Springer Verlag, 1987.

[Jen92] K. Jensen. *Coloured Petri Nets. Basic Concepts, Analysis Methods and Practical Use*, volume 1: Basic Concepts. Springer Verlag, EATCS Monographs in Theoretical Computer Science, 1992.

[Lil95] J. Lilius. *On the Structure of High-Level Nets*. PhD thesis, Helsinki University of Technology, 1995. Digital Systems Laoratory, Research Report 33.

[Pad96] J. Padberg. *Abstract Petri Nets: A Uniform Approach and Rule-Based Refinement*. PhD thesis, Technical University Berlin, 1996. Shaker Verlag.

[Par90] F. Parisi-Presicce. A rule-based approach to modular system design. In *Proc. of 12th IEEE Internat. Conf. on Software Engineering, Nice (France)*, pages 202–211, 1990.

[PER95] J. Padberg, H. Ehrig, and L. Ribeiro. Algebraic high-level net transformation systems. *Mathematical Structures in Computer Science*, 5:217–256, 1995.

[PGE98] J. Padberg, M. Gajewsky, and C. Ermel. Rule-Based Refinement of High-Level Nets Preserving Safety Properties. In E. Astesiano, editor, *Fundamental Approaches to Software Engineering*, volume 1382 of *Lecture Notes in Computer Science*, pages 221–238. Springer Verlag, 1998.

[PGH99] J. Padberg, M. Gajewsky, and K. Hoffmann. Incremental Development of Safety Properties in Petri Net Transformations. In G. Engels and G. Rozenberg, editors, *Theory and Application of Graph Transformations(TAGT'98)*, volume 1764 of *Lecture Notes in Computer Science*, pages 410–425. Springer Verlag, 1999.

[PHG00] J. Padberg, K. Hoffmann, and M. Gajewsky. Stepwise Introduction and Preservation of Safety Properties in Algebraic High-Level Net Systems. In T. Maibaum, editor, *Fundamental Approaches to Software Engineering*, volume 1783 of *Lecture Notes in Computer Science*, pages 249–265. Springer Verlag, 2000.

[RE98] G. Rozenberg and J. Engelfriet. Elementary net systems. In W. Reisig and G. Rozenberg, editors, *Lectures on Petri Nets: Basic Models*, volume 1491 of *Lecture Notes in Computer Science*, pages 12–121. Springer Verlag, 1998.

[Rei82] Wolfgang Reisig. *Petrinetze. Eine Einführung*. Springer Verlag, 1982.

[Rei85] W. Reisig. *Petri Nets*, volume 4 of *EATCS Monographs on Theoretical Computer Science*. Springer Verlag, 1985.

[Rei91] W. Reisig. Petri Nets and Algebraic Specifications. *Theoretical Computer Science*, 80:1–34, 1991.

[Vau87] J. Vautherin. Parallel System Specification with Coloured Petri Nets. In G. Rozenberg, editor, *Advances in Petri Nets 87*, volume 266 of *Lecture Notes in Computer Science*, pages293–308. Springer Verlag, 1987.

Verifying a Simple Pipelined Microprocessor Using Maude

N.A. Harman*

Department of Computer Science, University of Wales Swansea,
Singleton Park, Swansea SA2 8PP

Abstract. We consider the verification of a simple pipelined microprocessor in *Maude*, by implementing an equational theoretical model of systems. Maude is an equationally-based language, with an efficient term
rewriting implementation, and effective meta-level tools. Microprocessors and other systems are modelled as *iterated maps* operating in time
over some state-set, and are related by means of data and abstraction
maps, and correctness is reduced to state exploration by the choice of
an appropriate *initialization function*, ensuring/enforcing consistency of
the *initial state*.

1 Introduction

This paper considers the verification of a simple pipelined microprocessor in
Maude [4], an equational, algebraic language with strong meta-language tools
and an efficient term rewriting implementation. Hardware systems, and models of
hardware correctness, are represented within a well-developed set of mathematical tools, developed by application to case studies, and based on an equational,
algebraic model. In a related paper [16], we consider the process of verification
in Maude in more detail.

Microprocessors, and related systems, are modelled as *iterated maps*

$$F : T \times A \to A,$$
$$F(0, a) = h(a),$$
$$F(t + 1, a) = f(F(t, a)),$$

where T is a clock, A is the state-set, f is a *next-state function* defining state
evolution, and h is an (optional) *initialization function*, ensuring/enforcing consistency of *initial state a*. Initialization function h is an important component of
the verification process, and careful construction is essential (Sect. 5.4 and [11,
8]) to reduce formal verification to state exploration. In this paper, we do not
consider input and output: however, they are easily accommodated [19, 8]

---

* This work is supported by UK EPSRC grants GR/N15955 and GR/M82202. The
author is grateful to the referees for their insights, and their helpful comments and
suggestions.

M. Cerioli and G. Reggio (Eds.): WADT/CoFI 2001, LNCS 2267, pp. 128–151, 2002.

Maude was chosen as the appropriate tool to implement our theoretical model because (a) it has the same mathematical basis; (b) it is fast (approximately 20K rewrites per second on a 700MHz Pentium III, when applied to hardware examples); (c) its meta-level tools allow proof strategies to be constructed quickly and flexibly; and (d) it is easy to learn. However, other tools could also be used (initial experiments were undertaken with PVS [29]).

This paper forms part of a series on theoretical models of microprocessors by the author, J V Tucker and A C J Fox. In [18, 19] mathematical models of *microprogrammed* examples are considered. In [11, 13], *correctness models* and the formal verification process are examined. In [10, 12] models of *superscalar* processors are examined by means of a substantial example. An extended account of some of this work can be found in [8]. To date, our main interest has been *theoretical* models of systems, their correctness, and their verification: mainly microprocessors, but also include programming languages and their compilers [31], including the Java Virtual Machine [32]. Work has progressed on a set of mathematical tools for modelling behaviour and correctness in a modular and software tool-independent way. However, in this paper and in [16]. we consider the implementation of these mathematical tools within Maude.

The structure of this paper is as follows. In Sect. 1.1 we consider related work. In Sect. 2 we introduce the required theoretical fundamentals. In Sect. 3 we introduce the Maude system. In Sect. 4 we introduce the architecture (specification) of a simple microprocessor SPM, in Maude. In Sect. 5 we introduce a pipelined implementation ACP. In Sect. 6 we consider the correctness and verification of ACP with respect to SPM. Finally, in Sect. 7 we summarize our techniques and their applicability.

1.1 Related Work

The main distinction between this and related work is that our main interest is building theoretical/mathematical models of system, rather than on addressing [industrial] examples within software tools, where an important theme has been efficient verification strategies. This distinction is in part simply the result of the background and interests of the author and his co-workers, J V Tucker and A C J Fox. It is also motivated by the belief that sysematic modelling is needed *in addition* to large-scale verification attempts. Such models provide the mathematical justification for correctness proofs (which is otherwise lacking); they enable the limits of particular techniques to be mapped [8]; and potentially expose new and fruitful verification strategies [8, 12]. It also enables us to consider an incorporate other aspects of the computing abstraction hierarchy (high- and low-level languages, compilers *etc.*) in a uniform way [31, 32].

Interesting work on pipelined microprocessor verification includes [28] on AAMP5, a non-trivial, industrial example, and its verification in PVS [29] (recent accounts are [7, 30]: see also [21]); [34] on UINTA, a processor of moderate complexity, and its verification in HOL [15];and [3] on a part of DLX [20]. A refinement of the approach in [3], more applicable to out-of-order systems and long pipelines is [23, 24],. In addition, work has been undertaken on the complex timing models of superscalar processors [33, 2]: [22] additionally considers exception processing in such an environment. The work in [25, 6] uses *Hawk*, a variant of the functional language *Haskell*.

Generally, the intuitive models seen are conceptually similar to our own [18, 19, 10], though significant differences exist in the approach to time. Commonly, in pipelined systems, state elements in the specification are viewed as distributed in time in the implementation. We regard specification states as [some function of] state elements in the implementation at a single point in time (see [11, 8]). In addition, time is explicitly present in our model. Although explicit time is removed from the verification process for microprocessors (Sect. 2.4), recall that our main interest has been in theoretical models of systems, of which microprocessors are only one example.

Other interesting early work on microprocessors includes that of Birtwistle and Gordon's groups using HOL: for example, [5, 14, 1]. Currently, Birtwistle and Gordon's groups are jointly working on the ARM 6 microprocessor. Current work on verification in this project is being undertaken by A C J Fox using the theoretical models described here [9], though employing HOL as a theorem proving tool.

2   Theoretical Preliminaries

Computer systems are modelled in a universal algebra framework. We do not propose to discuss universal algebra here: see (among many accounts) [35, 27]. We define functions equationally, primarily using definition by cases and primitive recursion. Time is modelled using a *clock algebra* and computer systems are modelled with [many-sorted] *algebras*.

A many-sorted algebra consists of *carrier sets* and *functions* ranging over the carrier sets: $(A_1, A_2, \ldots, A_l \mid f_1, f_2, \ldots, f_m)$, with carrier sets $A_1, A_2, \ldots, A_l$ and functions f_i of the form $f_i : A_{s_1} \times A_{s_2} \times \ldots \times A_{s_n} \to A_s$, where $1 \leq i \leq m$ and $1 \leq s, s_j \leq l$ for $1 \leq j \leq n$, and each f_i is defined by one or more equations. If $n = 0$ then f_i is called a *constant*. Many-sorted algebras straightforwardly model Maude modules.

2.1   Clocks and Iterated Maps

Synchronous systems operate in time, starting at time zero in an *initial* state $h(a)$ where h is some *initialization function* (possibly the identity function). Future system states are determined by a *next-state function* f, enumerated by a clock algebra $T = (T \mid 0, +1)$, where the carrier T is a copy of the natural numbers (that is, time is discrete).

Definition 1. *Let T be a clock and let A be any non-empty set representing a state-space. An* iterated map $F : T \times A \to A$ *is a primitive recursive function defined by*

$$F(0, a) = h(a),$$
$$F(t + 1, a) = f(F(t, a))$$

where $h : A \to A$ and $f : A \to A$ are respectively the initialization *and* next-state *functions of the state function F.*

Initialization function h (which may be the identify function) limits the number of initial states, and (if carefully chosen) acts as an invariant during verification.

Our system models consist of hierarchies of algebras. Given state set A, clock T and iterated map F, a top-level *state algebra*

$$(A, T \mid F)$$

is implemented in terms of a *next-state algebra*

$$(A, T \mid 0; t + 1, f, h),$$

where $0 \in T$ is the initial clock cycle, $+1 : T \to T$ increments clock cycles, $f : A \to A$ is the next-state function, and $h : A \to A$ is the initialization function. Next-state algebras are in turn implemented in terms of *machine algebras*, which contain the primitive sorts and operations necessary to construct f and h: typically, bit vectors of various lengths, primitive arithmetic and logical operations and so on.

2.2   Data and Timing Abstraction Mappings

Data abstraction maps are surjective functions $\psi : B \to A$ between two state-spaces. Data abstraction maps are commonly projections between two composite state-spaces, for example, a map ψ from $B = B_1 \times B_2 \times \ldots \times B_m$ to $A = B_{i_1} \times B_{i_2} \times \ldots \times B_{i_n}$ defined as follows

$$\psi(b_1, b_2, \ldots, b_m) = (b_{i_1}, b_{i_2}, \ldots, b_{i_n})$$

where $1 \le i_j \le m$, $1 \le j \le n$ and $n \le m$.

Two clocks are related using a temporal abstraction map, or *retiming* [17]) :

Definition 2. *A* retiming *λ is a surjective and monotonic map between two clocks such that $\lambda(0) = 0$. The set of all retimings from clock S to clock T is denoted by $Ret(S, T)$. The* immersion *$\bar{\lambda}$ of a retiming $\lambda \in Ret(S, T)$ is defined by*

$$\bar{\lambda}(t) = \text{least } s \in S \text{ such that } \lambda(s) = t.$$

The set of all immersions of retimings in $Ret(S, T)$ is denoted by $Imm(S, T)$.

Monotonicity ensures there is never a discrepancy, after abstraction, in the temporal ordering of events because, for all $s, s' \in S$ if $s' \ge s$, then $\lambda(s') \ge \lambda(s)$ where λ is a retiming. Surjectivity ensures that, given clocks T and S, and retiming $\lambda \in Ret(S, T)$, that there is no time $t \in T$ such that for all $s \ge \bar{\lambda}(t)$ $\lambda(s) = t$. This is necessary to ensure that our correctness models (Section 2.3) capture liveness as well as safety properties[1].

---

[1] However in considering superscalar processors [8, 12] surjectivity is relaxed in a controlled way to define *adjunct retimings*.

Given two clocks S and T related by retiming $\lambda \in Ret(S,T)$, and a clock cycle $s \in S$, we commonly wish to identify the clock cycle $s' \in S$ such that s' is the first cycle of S where $\lambda(s') = \lambda(s)$.

Definition 3. *The function* $start : Ret(S,T) \to [S \to S]$ *is defined by*

$$start(\lambda) = \bar{\lambda}\lambda.$$

Retimings and their associated formal tools are illustrated in Figure 1.

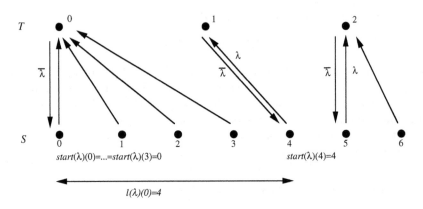

Fig. 1. A retiming and associated formal tools.

Definition 4. *A state-dependent retiming* $\lambda : A \to Ret(S,T)$ *is a map from states to retimings. The set of all state-dependent retimings from state-space A to retimings in $Ret(S,T)$ is denoted by $Ret(A,S,T)$.*

For each state of an implementation there is an associated state-dependent *uniform* retiming, defined in terms of a *duration* function over the state-space of F.

Definition 5. *Let* $F : S \times A \to A$ *be an iterated map and* $dur : A \to S^+$ *be a map from states to a positive number of clock cycles. The* uniform retiming *with respect to F and dur from a clock S to a slower clock T, is the state-dependent retiming $\lambda \in Ret(A,S,T)$ such that, for all $a \in A$ and $t \in T$*

$$\bar{\lambda}(a)(0) = 0,$$
$$\bar{\lambda}(a)(t+1) = dur(F(\bar{\lambda}(a)(t),a)) + \bar{\lambda}(a)(t)$$

where $\bar{\lambda} \in Imm(A,S,T)$ is the immersion of λ and $+$ is the addition operator. The set of all uniform retimings with respect to iterated map F is denoted by $URet_F(A,S,T)$. The singleton set containing the uniform retiming with respect to F and duration function dur is denoted by $URet_F^{dur}(A,S,T)$.

Suppose that F represents the implementation of some system over a clock S, and that T is the (slower) clock of the corresponding specification. Then specification clock cycle $t + 1 \in T$ lasts $dur(x)$ cycles of clock S, where $x = F(\bar{\lambda}(a)(t), a))$ is the state of F on clock cycle $\bar{\lambda}(a)(t) \in S$. That is, the cycle of implementation clock S corresponding with the start of the previous specification clock cycle $t \in T$. Note that dur is a function only of state, and consequently the number of cycles corresponding with any state is independent of the numerical value of $t \in T$.

2.3   Implementation Correctness

Correctness is defined in terms of the relationship between two algebras, representing implementation and specification. The state sequences specified by the implementation are mapped onto those of the specification by a data abstraction map ψ and a temporal abstraction map λ.

Definition 6. *A state function $G : S \times B \to B$ is a* correct implementation *of iterated map $F : T \times A \to A$ with respect to data abstraction map $\psi : B \to A$ and a state-dependent retiming $\lambda \in Ret(B, S, T)$ if, and only if, for all $b \in B$ and $s = start(\lambda(b))(s)$*

$$F(\lambda(b)(s), \psi(b)) = \psi(G(s, b)),$$

or, alternatively, if the following diagram commutes for all $b \in B$ and $s = start(\lambda(b))(s)$

$$
\begin{array}{ccc}
T \times A & \xrightarrow{F} & A \\
\big\uparrow{\scriptstyle (\lambda, \psi)} & & \big\uparrow{\scriptstyle \psi} \\
\bar{\lambda}(B) \times B & \xrightarrow{G} & B.
\end{array}
$$

2.4   Time-Consistency and the One-Step Theorems

Iterated map state functions are *time-consistent* if they possess the following property: for all times $s \in \bar{\lambda}(B)$, if the clock is reset to zero and the current state becomes an initial state, then is there any noticeable effect upon future state evolution? An iterated map $F : S \times A \to A$ is time-consistent if its initialization function $h : A \to A$ characterizes a state invariant. Formally, $h(a) = a$ for all states $a \in F(\bar{\lambda}(A) \times A)$ in the range of F.

Definition 7. *An iterated map $F : S \times A \to A$ is* time-consistent *with respect to uniform retiming $\lambda \in URet_F(A, S, T)$ if, and only if, for all $a \in A$ and t_1, $t_2 \in T$*

$$F(s_1 + s_2, a) = F(s_1, F(s_2, a))$$

where $s_2 = \bar{\lambda}(a)(t_2)$ and $s_1 = \bar{\lambda}(F(s_2, a))(t_1)$.

That is, $h \circ f^{s_1 + s_2} = h \circ f^{s_2} \circ h \circ f^{s_1}$.

For definition 7 to be meaningful we must establish that $s1 + s2 \in \bar{\lambda}(A)$. Ignoring the trivial case when $s2 = 0$:

$$
\begin{aligned}
s_2 &= dur(F(\bar{\lambda}(a)(t_2 - 1), a)) + \bar{\lambda}(a)(t_2 - 1), \\
&= dur(F(\bar{\lambda}(a)(t_2 - 1), a)) + \cdots + dur(F(0, a)). \\
s_1 &= dur(F(\bar{\lambda}(F(\bar{\lambda}(a)(t_2), a))(t_1 - 1), F(\bar{\lambda}(a)(t_2), a))) + \cdots \\
&\quad dur(F(0, F(\bar{\lambda}(a)(t_2), a))).
\end{aligned}
$$

That is, s_2 is the sum of $dur(x)$ for $x \in \{F(0, a), \ldots, F(\bar{\lambda}(a)(t_2 - 1), a))$ (at which time the state of F is $F(\bar{\lambda}(a)(t_2), a)))$, and s_1 is the sum of $dur(y)$ for $y \in F(0, F(\bar{\lambda}(a)(t_2), a))), \ldots, F(\bar{\lambda}(F(\bar{\lambda}(a)(t_2), a))(t_1), F(\bar{\lambda}(a)(t_2), a)))$. Observe that (trivially)

$$
F(\bar{\lambda}(a)(t_2), a)) = F(0, F(\bar{\lambda}(a)(t_2), a))),
$$

and hence from the definition of $URet_F(A, S, T)$ (definition 5), it follows that $s_1 + s_2 \in URet_F(A, S, T)$.

2.5   The Rôle of Initialisation Functions

The purpose of initialisation functions is to eliminate unwanted starting states: *not* to describe the initial behaviour of a system. For example, consider an implementation with memory m, program counter pc and instruction register ir: we may initially require $ir = m(pc)$, and hence not wish to consider starting states that do not have this property. The precise choice of initialisation function will vary according to circumstances: we could choose an initialisation function that enforced some predefined reset state; or we could choose the identity function (which could, of course, mean that the state evolution of F may not be correct). Between these alternatives is the useful class of initialisation functions that leaves initial state a unchanged *provided* a is already consistent with correct future state evolution of F: that is, those initialization functions that establish time-consistency for F (definition 7). We can regard the conjunction of the various required relationships between state components of iterated map F (for example, $ir = m(pc)$) as a consistency-checking *invariant* I that must hold, at certain times, for the correct state evolution of F: in the case where F represents the implementation of a microprocessor, those times will correspond to the start/end of machine instructions[2]. Invariant I may be checked by an initialisation function h, on initial state $a \in A$: if I holds, then $h(a) = a$. Such initialisation functions are an important part of the verification process, and are analogous to the *pipeline invariants* of [7].

In practice, h may be complex and difficult to construct. In Sect. 5.4 we describe an algorithmic method that is sometimes applicable.

The following two results are called one-step theorems. Theorem 1 states that if $\lambda \in Ret(B, S, T)$ is a uniform retiming then time-consistency with respect

---

[2] It is recognised that identifying start/end times may be problematic in superscalar examples: [8, 12].

to λ is sufficiently verified by examining the implementation at times $t = 0, 1$. Theorem 2 states that retiming uniformity and implementation time-consistency are sufficient conditions to enable correctness to be verified by examining times $t = 0, 1$.

Theorem 1. *If $F : S \times A \to A$ is an iterated map with initialization function $h : A \to A$ and if $\lambda \in URet_F^{dur}(A, S, T)$ is a uniform retiming then F is time-consistent with respect to λ if, and only if, for all $a \in A$*

$$F(0, a) = h(F(0, a)), \text{ and } F(\bar{\lambda}(a)(1), a) = h(F(\bar{\lambda}(a)(1), a)).$$

Proof. See [8, 11].

Theorem 2. *Let $F : T \times A \to A$ and $G : S \times B \to B$ be iterated maps. Let $\psi : B \to A$ be a data abstraction map and let $\lambda \in URet_G(B, S, T)$ be a uniform retiming. If F is non-initialized (that is, the map h is the identify function), and G is time-consistent with respect to λ then G is a correct implementation of F if, and only if*

$$F(0, \psi(b)) = \psi(G(0, b)), \text{ and } F(1, \psi(b)) = \psi(G(\bar{\lambda}(b)(1), b)).$$

Proof. See [8, 11].

3   Introduction to Maude

Maude is an equationally-based, algebraic language with a term rewriting implementation [4]. The following simple algebra, or module, representing a memory, illustrates the main features of interest to us[3].

```
fmod MEM is
      protecting MACHINE-WORD .
      sorts Mem .

      op _[_] : Mem MAR -> Word .          *** Memory read
      op _[_/_] : Mem Word MAR -> Mem .    *** Memory write

      var M : Mem .
      vars A B : MAR .
      var W : Word .

      eq M[W / A][A] = W .
      ceq M[W / A][B] = M[B] if A =/= B .
endfm
```

The module MEM imports a module MACHINE-WORD defining (among other things) sorts MAR and Word representing memory addresses and memory words

---

[3] Maude also includes a mechanism for defining type hierarchies, which we do not use here.

respectively. MEM introduces sort Mem, and operations _[_] and _[_/_], representing memory reading and writing, defined by a pair of equations (one of which is conditional). The operations are defined in mixfix syntax, where arguments will replace the _ characters in the operation name.

4   SPM Architecture

SPM is a simple microprocessor architecture, with five instructions (add, load, store, branch and set), separate program and data memories md and mp, a general purpose register set reg and a program counter pc. We use separate data and program memories to simplify the process of mapping to a pipelined implementation with data and instruction caches. However, this is not necessary. SPM was first used in the form described here in [12, 8]: a previous version appeared in [10]. In addition to the pipelined implementation ACP (Sect. 5), which has been verified manually [8] and using Maude (Sect. 6), a superscalar version ACS exists [12, 8]. A variant of ACS is currently being verified in Maude[4].

The SPM architecture is parameterized by three constants r, m, $w \in N^+$ which determine the number of general-purpose registers, the memory address space and the word size respectively. There are 2^r general-purpose registers, 2^m memory addresses, and memory words and registers are w-bit words. Informally, the SPM instructions are as follows.

add ra rb rc. $reg[c] := reg[a] + reg[b]$; $pc := pc + 1$.
branch addr. If $reg[0] = 0$ then $pc := pc + addr$ else $pc := pc + 1$.
load ra addr. $reg[a] := md[addr]$; $pc := pc + 1$.
store ra addr. $md[addr] := reg[a]$; $pc := pc + 1$.
set ra val. $reg[a] := val$; $pc := pc + 1$.

The basic format of SPM instructions is shown in Figure 2, from which it can be seen that $w \geq \max(3 + 3r, 3 + r + m)$. The state-spaces for op codes, register and memory addresses, machine words, registers and memory are as follows:

$$OP = W_3 - \{101, 110, 111\}, \quad RI = W_r,$$
$$MAR = W_m, \quad Word = W_w,$$
$$Reg = [RI \rightarrow Word], \quad Mem = [MAR \rightarrow Word].$$

We define SPM in Maude as follows. The [omitted] module BITS defines the sorts Bit (a single bit) and Bits (strings of bits) together with a range of operators on bits. The [omitted] module MACHINE-WORD defines sorts for $OP, \ldots, Word$, together with the operations used to extract bit-fields in instruction decoding. Memory is defined by module MEM (Sect. 3), and registers by module REG (omitted, but similar to MEM). Note in the case of the *set* instruction we assume that the bit field *val* is isomorphic with that of *addr* used in *branch*, *load* and *store* instructions. Also, in practice, it is likely that the bit fied *addr* will overlap with one or more of *ra*, *rb* and *rc*.

---

[4] Using the modified correctness definition of [10, 12, 8] for superscalar processors

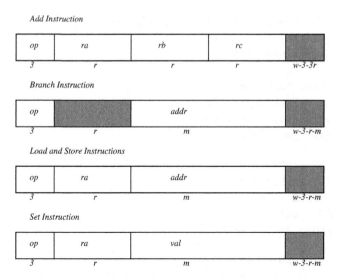

Fig. 2. Format of SPM Instructions.

```
*** State of SPM, together with
*** tupling and projection functions
fmod SPM-STATE is
     protecting BITS .
     protecting MACHINE-WORD .
     protecting MEM .
     protecting REG .

     sort SPMstate .

     *** Create state tuple
     op (_,_,_,_) : Mem Mem Bits Reg -> SPMstate .
     *** Projection functions
     ops mp_ md_ : SPMstate -> Mem .
     op pc_ : SPMstate -> Bits .
     op reg_ : SPMstate -> Reg .

     var S : SPMstate .
     vars MP MD : Mem .
     var PC : Bits .
     var REG : Reg .

     eq mp(MP,MD,PC,REG) = MP .
     eq md(MP,MD,PC,REG) = MD .
     eq pc(MP,MD,PC,REG) = PC .
     eq reg(MP,MD,PC,REG) = REG .
endfm
```

```
*** SPM - the Programmer's Level representation
fmod SPM is
     protecting SPM-STATE .

     *** State function
     op spm : MachineInt SPMstate -> SPMstate .
     *** Next-state function
     op next-spm : SPMstate -> SPMstate .

     var SPM : SPMstate .
     var T : MachineInt .
     vars MP MD : Mem .
     var PC : Bits .
     var REG : Reg .

     eq spm(0,SPM) = SPM .
     ceq spm(T,SPM) = next-spm(spm(T - 1,SPM)) if T > 0 .

     *** Addition
     ceq next-spm(MP,MD,PC,REG) = (MP, MD, PC + 1,
         REG[REG[ra(MP[PC])] ++ REG[rb(MP[PC])] / rc(MP[PC])])
         if op(MP[PC]) == 0 .
     *** [Taken] branch
     ceq next-spm(MP,MD,PC,REG) = (MP, MD, PC + addr(MP[PC]), REG)
         if op(MP[PC]) == 1 and REG[0] == 0 .
     *** [Not-taken] branch
     ceq next-spm(MP,MD,PC,REG) = (MP, MD, PC + 1, REG)
         if op(MP[PC]) == 1 and REG[0] =/= 0 .
     *** Load
     ceq next-spm(MP,MD,PC,REG) = (MP, MD, PC + 1,
         REG[MD[addr(MP[PC])]/ra(MP[PC])]) if op(MP[PC]) == (1 0) .
     *** Store
     ceq next-spm(MP,MD,PC,REG) =
     (MP, MD[REG[ra(MP[PC])]/addr(MP[PC])], PC + 1, REG)
         if op(MP[PC]) == (1 1) .
     *** Set
     ceq next-spm(MP,MD,PC,REG) = (MP, MD, PC + 1,
         REG[addr(MP[PC])/ra(MP[PC])]) if op(MP[PC]) == (1 0 0) .
endfm
```

5   The Pipelined Implementation ACP

The implementation ACP of SPM has a four-stage pipeline (see Figure 3) with the following stages.

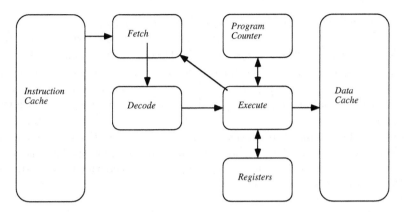

Fig. 3. Structure of ACP.

Fetch. A single instruction is fetched from the instruction cache[5] either using a *fetch program counter fpc*, or a branch target address generated by the execute unit. In normal operation, when the pipeline is full (i.e. a branch has not been taken in the past three cycles), $fpc = pc + 3$, where pc is the architectural program counter. In the event of a read-write conflict (Sect. 5.2), no instruction is fetched. The instruction is stored in an instruction register.

Decode. The contents of the instruction register in the Fetch Unit is decoded into five [overlapping] fields: op, representing the op-code; ra, rb and rc representing three register indices; and $val/addr$ representing a memory address or immediate value. See Figure 2.

Execute. The instruction stored in the Decode Unit is executed, and the results encoded as a triple representing the *result*; the *destination* register/memory address; and the *unit* (if the result is to be stored in memory, register, or program counter; or if outcome is a failed conditional branch, or the pipeline stalls). Some elements of the execution triple are redundant in some circumstances.

Committal. Results from the Execution Unit are written to program counter, registers and/or data cache.

5.1 Pipeline States

The pipeline of ACP has four distinct states, identified by a *reset* counter and the *unit* field of the execution triple (and the relative values of pc and fpc).

Boot. In this state, which can only occur at start up, $reset = 2$, $fpc = pc$, *unit* is set to wait (i.e. no results to be committed), and the instruction register in the Fetch Unit and Decode Unit contain junk.

---

[5] We omit the possibility of cache misses to simplify the example.

After Branch. This state is indistinguishable from one cycle after boot. In this state $reset = 1$, $fpc = pc + 1$, $unit$ is set to `wait`, the instruction register in the Fetch Unit contains $mp[pc]$, and the Decode Unit contains junk.

Stall. This state is indistinguishable from two cycles after boot. In this state, $reset = 0$, $fpc = pc + 2$, $unit$ is set to `wait`, the instruction register in the Fetch Unit contains $mp[pc + 1]$, and the Decode Unit the decoded form of $mp[pc]$.

Pipeline full. In this state, $reset = 0$, $fpc = pc + 3$, $unit$ is set to something other than `wait` (indicating some result to be committed), the instruction register in the Fetch Unit contains $mp[pc + 2]$, and the Decode Unit the decoded form of $mp[pc + 1]$.

5.2   Pipeline Conflicts

There are a number of circumstances in which pairs of consecutive instructions can conflict. If the first instruction is a branch, then there is a *procedural dependency* between the instructions, and the second instruction will need to be discarded if the branch is taken by flushing the pipeline, and switching to the *after branch* state. There may also be *data dependencies* (or *RAW hazards*) between instruction pairs, if the second instruction requires the result of the first. In this case, the pipeline is suspended for one cycle by switching to the *stall* state, to allow the result of the first instruction to be committed before the second executes. The various data dependencies are illustrated in table 1 ('–' indicates no dependency). Note that it is possible to reduce the number of pipeline stalls by providing mechanisms to permit internal *forwarding* of results before they are committed to registers/memory. Such mechanisms also enable execution to provisionally continue in the event of conditional branches where the outcome is unknown. However, we do not include such mechanisms here.

Table 1. Possible data dependencies between instructions: note that data dependencies cannot occur when the first instruction is a branch.

1st Inst.	2nd Instruction				
	add a2 b2 c2	branch a2	load a2 b2	store a2 b2	set a2 b2
add a1 b1 c1	$a2 \neq c1$ $b2 \neq c1$	$c1 \neq 0$	–	$a2 \neq c1$	–
load a1 b1	$a2 \neq a1$ $b2 \neq a1$	$a1 \neq 0$	–	$a2 \neq a1$	–
store a1 b1	–	–	$b2 \neq b1$	–	–
set a1 b1	$a2 \neq a1$ $b2 \neq a1$	$a1 \neq 0$	–	$a2 \neq a1$	–

5.3   Formal Description of ACP

Space precludes including the full formal description of ACP in Maude (about 550 lines of code with comments and whitespace). Instead, we give a partial definition in the notation used in the underlying mathematical model. This description of ACP first appeared in [8].

The state of ACP is the cartesian product of the states of each of its components:

$$State_{ACP} = Icache \times Ftch \times Dec \times Exec \times Reg \times MAR \times Dcache,$$

where

$$Icache = Dcache = Mem,$$
$$Ftch = Word \times MAR,$$
$$Dec = OP \times RI \times RI \times RI \times MAR,$$
$$Exec = Word \times MAR \times Unit \times Ctr,$$
$$Unit = \{\texttt{reg, pc, incpc, dcache, wait}\},$$
$$Ctr = \{0, 1, 2\}.$$

The iterated map $ACP : S \times State_{ACP} \to State_{ACP}$ is defined by

$$ACP(0, \vec{\sigma}) = init_{ACP}(\vec{\sigma}),$$
$$ACP(s + 1, \vec{\sigma}) = next_{ACP}(ACP(s, \vec{\sigma})),$$

where $\vec{\sigma} = (icache, \vec{f}, \vec{d}, \vec{e}, reg, pc, dcache)$, and

$\vec{f} = (ir, fpc)$, representing the instruction register and fetch program counter;
$\vec{d} = (op, ra, rb, rc, addr)$, representing the decoded instruction fields; and
$\vec{e} = (result, dest, unit, reset)$, representing the execution triple and
the pipeline state counter.

It remains to define $init_{ACP}$ and $next_{ACP}$. The definition of $init_{ACP}$ is central to the verification process, and is discussed in Sect. 5.4. Next-state function $next_{ACP} : State_{ACP} \to State_{ACP}$ is defined as follows:

$$next_{ACP}\ (\vec{\sigma}) = (icache, Fetch(\vec{\sigma}), Decode(\vec{\sigma}), Execute(\vec{\sigma}),$$
$$Register(\vec{\sigma}), Counter(\vec{\sigma}), DataCache(\vec{\sigma})).$$

With the exception of $icache$, which does not change state, each unit of ACP has its own next-state function.

Subfunction $Conflict : Dec \times Exec \to \mathbf{B}$ identifies conflicts between instructions, and is defined by

$Conflict(\vec{d}, \vec{e}) =$

$\begin{cases} tt, \text{ if } unit = \texttt{reg} \text{ and } \langle(op = \texttt{branch} \text{ and } dest = 0) \text{ or} \\ \quad (op = \texttt{add} \text{ and } (trim_{m \mapsto r}(dest) = ra \text{ or } (trim_{m \mapsto r}(dest) = rb)) \\ \quad \text{or } (op = \texttt{store} \text{ and } (trim_{m \mapsto r}(dest) = ra))\rangle \\ \quad \text{or } unit = \texttt{dcache} \text{ and } (op = \texttt{load} \text{ and } dest = addr); \\ ff, \text{ otherwise.} \end{cases}$

and $trim_{m \mapsto r}$ truncates a binary word of length m to a word of length r.

The next-state function $Fetch : State_{ACP} \to Ftch$ fetches a new instruction from memory and updates the fetch program counter fpc.

$$Fetch(\overrightarrow{\sigma}) =$$
$$\begin{cases} (ir, fpc), & \text{if } Conflict(\overrightarrow{d}, \overrightarrow{e}); \\ (icache(fpc), fpc + 1), & \text{if } unit \in \{reg, incpc, dcache, wait\} \\ & \text{and not } Conflict(\overrightarrow{d}, \overrightarrow{e}); \\ (icache(result), result + 1), & \text{otherwise.} \end{cases}$$

In the event of a Conflict the state of the Fetch unit remains unchanged. Otherwise, the next instruction is fetched from memory (using either the fetch program counter fpc or the branch target address $result$) and fpc is either incremented or replaced with the [incremented] branch target address.

The next-state function $Decode : State_{ACP} \to Dec$ decodes an instruction into its constituent parts.

$$Decode(\overrightarrow{\sigma}) =$$
$$\begin{cases} \overrightarrow{d}, & \text{if } Conflict(\overrightarrow{d}, \overrightarrow{e}); \\ (op(ir), ra(ir), rb(ir), rc(ir), addr(ir)), & \text{otherwise,} \end{cases}$$

where the [undefined] functions op, ra, rb, rc and $addr$ extract the various bit fields of an operation. Note that as with SPM it is likely that the $addr$ field will overlap with one or more of ra, rb and rc. The state of the decode unit remains unchanged in the event of a conflict: otherwise, the contents of the instruction register are decoded.

The next-state function $Execute : State_{ACP} \to Exec$ generates an execution triple for each instruction, specifying result, destination and unit, together with the value of the $reset$ counter, controlling the state of the pipeline.

$$Execute(\overrightarrow{\sigma}) =$$
$$\begin{cases} exec(\overrightarrow{d}, reg, pc, dcache), & \text{if not } Conflict(\overrightarrow{d}, \overrightarrow{e}) \text{ and } reset = 0; \\ (result, dest, \mathtt{wait}, 0), & \text{if } Conflict(\overrightarrow{d}, \overrightarrow{e}) \text{ and } reset = 0; \\ (result, dest, \mathtt{wait}, reset - 1), & \text{otherwise.} \end{cases}$$

In the event that the pipeline is not full ($reset > 0$), $Execute$ decrements $reset$ and prevents any results from being committed by setting $unit = \mathtt{wait}$. Otherwise, if the pipeline is full ($reset = 0$), then $Execute$ checks for conflicts and either executes an instruction (using $exec$), or stalls the pipeline.

Subfunction $exec : Dec \times Reg \times MAR \times Dcache \to Exec$ executes the current instruction, and is defined as follows.

$$exec(\overrightarrow{d}, reg, pc, dcache) =$$
$$\begin{cases} (reg[ra] + reg[rb], pad_{r \mapsto m}(rc), \mathtt{reg}, 0), \text{ if } op = \mathtt{add}; \\ (result, pc + addr, \mathtt{pc}, 2), & \text{if } op = \mathtt{branch} \text{ and } reg[0] = 0; \\ (result, dest, \mathtt{incpc}, 0), & \text{if } op = \mathtt{branch} \text{ and } reg[0] \neq 0; \\ (dcache[addr], pad_{r \mapsto m}(ra), \mathtt{reg}, 0), & \text{if } op = \mathtt{load}; \\ (reg[ra], addr, \mathtt{dcache}, 0), & \text{if } op = \mathtt{store}; \\ (pad_{m \mapsto w}(addr), pad_{r \mapsto m}(ra), \mathtt{reg}, 0), & \text{if } op = \mathtt{set}; \end{cases}$$

The functions $pad_{a \mapsto b}$ extend binary words of length a to words of length b by adding leading zeros.

The next-state function $Register : State_{ACP} \rightarrow Reg$ updates the data registers.

$$Register(\overrightarrow{\sigma}) =$$
$$\begin{cases} reg[result/dest], \text{ if } unit = \mathtt{reg}; \\ reg, & \text{otherwise.} \end{cases}$$

The registers are unchanged unless the destination of an operation is set to $\mathtt{reg}$. In this case, the result of the operation is written to the register whose index is held in the $dest$ field.

The next-state function $Counter : State_{ACP} \rightarrow MAR$ updates the program counter.

$$Counter(\overrightarrow{\sigma}) =$$
$$\begin{cases} pc, & \text{if } unit = \mathtt{wait}; \\ result, \text{ if } unit = \mathtt{pc}; \\ pc + 1, \text{ otherwise.} \end{cases}$$

The program counter may either remain unchanged (in the event of a pipeline stall); be replaced with the $result$ field of the execute unit (in the event of a taken branch); or be incremented (in all other cases).

The next-state function $DataCache : State_{ACP} \rightarrow DCache$ updates the data cache.

$$DataCache(\overrightarrow{\sigma}) =$$
$$\begin{cases} dcache[result/dest], \text{ if } unit = \mathtt{dmem}; \\ dcache, & \text{otherwise.} \end{cases}$$

The data cache next-state function is very similar in structure and operation to the register next-state function.

5.4 Constructing the Initialization Function

While it is straightforward to define an initialization function that is 'correct', care must be taken if the one-step theorems are to be applied, as the initialization function $init_{ACP}$ must have the property $init_{ACP}(\overrightarrow{\sigma}) = \overrightarrow{\sigma}$, for any

$\vec{\sigma} \in ACP(S \times State_{ACP})$. Given the complexity of the state of pipelined (and superscalar) systems, it is not easy to construct such initialization functions: they must be able to identify and leave unchanged *all* legitimate states of a system.

In the case of ACP, it is possible to systematically construct the initialization function $init_{ACP}$ using the next-state function $next_{ACP}$, employing a technique first described in [8]. This is because the state of the pipeline of ACP is uniquely determined by the instructions currently in the pipeline. This is not always the case. Consider a superscalar processor, with multiple integer function units, where instructions are queued for execution. If instructions are dispatched to function units on the basis of queue lengths, to balance execution loads across units, then the state of the pipeline will depend not only on the current state of the pipeline, but also on the number and distribution of instructions that have now left the pipeline. However, in ACP this is not the case and we can systematically define $init_{ACP}$.

The method used is as follows. Given the current state

$$\vec{\sigma} = (icache, \vec{f}, \vec{d}, result, dest, unit, reset, reg, pc, dcache),$$

we apply a reset function $BootState : State_{ACP} \rightarrow State_{ACP}$

$$BootState(icache, \vec{f}, \vec{d}, result, dest, unit, reset, reg, pc, dcache) =$$
$$(icache, \vec{f}, \vec{d}, result, dest, \texttt{wait}, 2, reg, pc, dcache)$$

that returns ACP to a boot state $\vec{\sigma}_0$ by setting $unit_0 = \texttt{wait}$ and $reset_0 = 2$. Returning the pipeline to a boot state effectively empties it: as the pipeline is refilled we should, at some point, reach the original state $\vec{\sigma}$. If we do not, then $\vec{\sigma}$ was not a legal ACP state.

We can check if $\vec{\sigma}$ is a legal *boot* state, using $Boot : State_{ACP} \rightarrow \mathbf{B}$

$$Boot(\vec{\sigma}) = (unit = BootState(\vec{\sigma})_{unit}$$
$$\text{and } reset = BootState(\vec{\sigma})_{reset}),$$

where $BootState(\vec{\sigma})_x$ projects out element x of $BootState(\vec{\sigma})$.

If ACP is not in a boot state, we can apply $next_{ACP}$ to $BootState(\vec{\sigma})$ and check if $\vec{\sigma}$ is an *after branch* state of ACP with $AfterBranch : State_{ACP} \rightarrow \mathbf{B}$

$$AfterBranch(\vec{\sigma}) = (\vec{f} = next_{ACP}(BootState(\vec{\sigma}))_{\vec{f}}$$
$$\text{and } unit = next_{ACP}(BootState(\vec{\sigma}))_{unit}$$
$$\text{and } reset = next_{ACP}(BootState(\vec{\sigma}))_{reset}).$$

We can determine if ACP is in a *stall* state with $AfterConflict : State_{ACP} \rightarrow \mathbf{B}$ by applying $next_{ACP}$ to $AfterBranch(\vec{\sigma})$:

$$AfterConflict(\vec{\sigma}) = (\vec{f} = next_{ACP}^2(BootState(\vec{\sigma}))_{\vec{f}}$$

$$\text{and } \vec{d} = next^2_{ACP}(BootState(\vec{\sigma}))\vec{d}$$
$$\text{and } unit = next^2_{ACP}(BootState(\vec{\sigma}))_{unit}$$
$$\text{and } reset = next^2_{ACP}(BootState(\vec{\sigma}))_{reset}).$$

Finally, we can determine if the pipeline is *full* with $PipeFull : State_{ACP} \rightarrow \mathbf{B}$:

$$PipeFull(\vec{\sigma}) = (\vec{\sigma} = next^3_{ACP}(BootState(\vec{\sigma}))).$$

We can now define an initialization function that leaves legal initial states unchanged, and returns others to a boot state:

$$init_{ACP} : State_{ACP} \rightarrow State_{ACP},$$
$$init_{ACP}(\vec{\sigma}) = \begin{cases} \vec{\sigma}, & \text{if } AfterBranch(\vec{\sigma}) \text{ or } Afterconflict(\vec{\sigma}) \\ & \text{or } PipeFull(\vec{\sigma}); \\ BootState(\vec{\sigma}), & \text{otherwise.} \end{cases}$$

6   The Correctness Statement and Verification Process

To express the correctness of ACP with respect to SPM, we must construct data abstraction map ψ and retiming λ. The data abstraction map $\psi : State_{ACP} \rightarrow State_{SPM}$ is defined by

$$\psi(\vec{\sigma}) = (icache, dcache, pc, reg).$$

Note that this is an atypically simple data abstraction function: normally, we would expect some transformation of implementation state elements to be necessary. The retiming $\lambda \in URet^{dur}_{ACP}(State_{ACP}, S, T)$ is defined in terms of the following duration function $dur : State_{ACP} \rightarrow S^+$ as described in definition 5.

$$dur(\vec{\sigma}) = \begin{cases} 3, \text{ if } AfterBranch(\vec{\sigma}); \\ 2, \text{ if } AfterConflict(\vec{\sigma}); \\ 1, \text{ if } PipeFull(\vec{\sigma}); \\ 4, \text{ otherwise.} \end{cases}$$

To verify the correctness of ACP with respect to SPM using the one-step theorems 1 and 2, we need to prove the following.

$$next_{ACP}(0, \vec{\sigma}) = init_{ACP}(next_{ACP}(0, \vec{\sigma})) \qquad (1)$$
$$next_{ACP}(dur(\vec{\sigma}), \vec{\sigma}) = init_{ACP}(next_{ACP}(dur(\vec{\sigma}), \vec{\sigma})), \qquad (2)$$
$$next_{SPM}(0, \psi(\vec{\sigma})) = \psi(next_{ACP}(0, \vec{\sigma})) \qquad (3)$$
$$next_{SPM}(1, \psi(\vec{\sigma})) = \psi(next_{ACP}(dur(\vec{\sigma}), \vec{\sigma})), \qquad (4)$$

where $next_{SPM}$ was defined in Maude as next-spm in Sect. 4. Equations 1 and 2 discharge the obligations of theorem 1, and equations 3 and 4 discharge the obligations of theorem 2

6.1   Verification in Maude

Maude is not by itself a theorem proving system, and cannot automatically verify systems without direction. The simplest approach (and the most efficient in terms of number of rewrites) is to manually explore the cases, by defining modules containing sets of constants representing the values of, and relationships between, state elements for each case that needs to be considered. This was the first approach taken with this example. There are 53 cases to consider: three concern correctness and time-consistency when the pipeline is not full; six concern correctness and time-consistency when the pipeline is full at time $t = 0$ (two for branch, one each for the other instructions); and the remaining 44 concern correctness and time-consistency when the pipeline is full at time $t = 1$, and explore all possible cases of conflicting and non-conflicting instruction pairs. The verification process requires approximately 300K rewrites. A previous manual verification had been undertaken in [8]. Verification in Maude uncovered an off-by-one bug in the branch target address calculation, and an error in the initialization function[6]. However, the process of manually defining cases is time-consuming, and potentially error-prone. The example here, with 53 cases, is about the largest that can be sensibly attempted in this way: the superscalar example from [12, 8] will require about 2000 cases.

To address this, we can build proof strategies within the Maude meta-level, that enables Maude modules to be treated as data types, and the rewriting process to be controlled. We can use these properties to construct a range of verification strategies, tailored to classes of example.

In this paper, we employ a simple strategy, and use Maude to automatically construct and check a tree representing all subcases. We do this by defining operations to dynamically extend Maude modules by adding equations, and to direct rewriting.

The following Maude operation, defined within the Maude meta-level, extends a module (of [pre-defined] sort FModule) with a new equation T1 = T2.

```
op AddEq : FModule Term Term -> FModule .
var QI : Qid .
⋮

eq AddEq((fmod QI is IL SD SSDS ODS VDS MAS EqS endfm),T1,T2) =
          fmod QI is IL SD SSDS ODS VDS MAS
          ((eq T1 = T2 .) EqS) endfm .
```

We omit the definitions of most of the variables for brevity. (Variable QI is of sort Qid — *quoted identifier*, representing the name of the module; ODS is of sort OpDeclSet — the set of *operator declarations*, and so on.)

We direct the rewriting process by [recursively] searching for conditional operators within ACP and using the above AddEq operator to add equations, asserting each branch of a conditional in turn. When no more conditionals remain,

---

[6] This was a technical problem in the verification, and would not have affected any implementation.

the correctness of ACP is checked with respect to SPM. We can conclude that ACP is a correct implementation of SPM if it is correct for all possible branches of all conditionals.

7 Modularization and Further Work

The Maude strategy used for this example is experimental, and not particularly efficient. The ACP example can be run in a few tens of seconds on a 700MHz Pentium III and requires less than 128Mbyte of memory, suggesting that there is scope to undertake significantly larger examples on existing hardware without addressing the efficiency of the proof strategy. However, the required resources will grow exponentially with larger examples. Consequently, a number of approaches are being followed to increase the size of examples that can be addressed.

– The existing proof strategy can be made more efficient. For example, there is currently considerable repetitive computations, particularly in computing the *dur* function (Sect. 2.2).
– Different proof strategies can be considered. For example, to automatically generate the sets of case-defining constants used in the manually-directed proof.

However, the main approach being taken is to allow proofs to be modularized. This can be accomplished in a number of ways, all of which exploit the modular nature of iterated map representations:

– modularization by pipeline state;
– modularization by pipeline functionality; and
– modularization by pipeline conflicts.

The first approach, exploiting the pipeline state, has been successfully tried with ACP. Essentially, the verification is partioned according to the potential state of the pipeline. In the case of ACP, the pipeline can be in one of four states (boot, after branch, stall and full: Sect. 5.1), so each of the four proof obligation equations (Sect. 6) is further partitioned into four[7]. Clearly, an example with more pipeline stages would require more partitioning. This approach is simple to apply, and useful, but somewhat inflexible as it is limited by the number of pipeline stages present.

The second approach is analogous to a 'traditional' method of simplifying proofs: by choosing one instruction from each 'class', rather than verifying every instruction. For example, rather than verifying all arithmetic/logical operations, one is chosen. The precise definition of a 'class' of instructions must be made with reference to the structure of the implementation and the architecture. For example, it may be appropriate to regard all instructions executed by a particular functional unit as forming a class. The verification of the functional unit can

---

[7] In practice, there is little point in partioning the two equations for times $t = s = 0$ as their proofs are already short.

be deferred to a later date, when it can be undertaken [semi] independently of its surrounding hardware context, thus simplifying the process. In practice, this lower-level verification would probably be necessary in any case, as generally, arithmetic and logical operators, for example, will be interpreted by exactly the same operations at both SPM and ACP levels of abstraction: or otherwise left uninterpreted. This means the internal structure of the implementation of, say, a functional unit will not be present at the level of abstraction of ACP, meaning the verification of multiple, similar instructions can be of dubious usefulness. This modularization technique can often be a useful approach, though its applicability depends on the structure of the architecture and implementation. It is not applicable to ACP, because of the small number of instructions.

The third approach if taken to its logical conclusion (considering every instruction/conflict case separately) is isomorphic with the original approach taken to verifying ACP, which required 53 cases. However, it is possible to partially specify instruction/conflict cases. For example, consider Table 1. We can partition a verification into five subcases by simply fixing the first instruction, for example. This approach seems promising: ACP is a simple example, with relatively little scope for partitioning the verification, though the technique has been applied successfully. However, realistic processors (and their implementations) would seem on preliminary investigation to admit more complex and fine-grained partitioning schemes. In addition, this approach is flexible, and easy to apply: if verification using one partitioning scheme proves intractable it is relatively straightforward to apply another.

Further work is also needed to simplify debugging of verifications: while it is gratifying to see the output "true", "false" is unhelpful. The current strategy is to output a list of all terms that are not equal together with history information. This is adequate for the ACP example, but does tend to generate output that is voluminous and difficult to interpret. More sophisticated tools will be required for larger examples (and this is of course closely related to modularization strategies).

One of the key strengths of Maude is the ease with which proof strategies can be constructed: together with the efficiency of the underlying term-rewriting engine, this means that effective, specialized verification tools can be built relatively quickly. Further, they can be tailored specifically to the examples at hand, avoiding unnecessary computational overhead. To that end, a well-founded underlying model, such as that outlined here and described in more detail in [18, 11, 12, 8] is extremely useful in identifying and formally justifying simplifications to the proof obligations.

Other future work includes completing verification of ACS, the superscalar implementation of SPM, and addressing the high-level language and compiler examples of [31, 32]. In addition, theoretical work is underway on a model of operating system kernels: starting with a simple system involving multiple communicating processes. The ultimate intention is to build a unified theoretical model of systems from high-level languages to hardware, backed by an implementation in Maude.

References

1. G Birtwistle and B Graham. Verifying SECD in HOL. In J Staunstrup, editor, *Formal Methods for VLSI Design*, pages 129 – 177. North-Holland, 1990.
2. J Burch. Techniques for verifying superscalar microprocessors. In *Design Automation Conference*, 1996.
3. J Burch and D Dill. Automatic verification of pipelined microprocessor control. In D Dill, editor, *Proceedings of the 6th International Conference, CAV'94: Computer-Aided Verification*, pages 68 – 80. Lecture Notes in Computer Science 818, Springer-Verlag, 1994.
4. M Clavel, F Durán, S Eker, P Lincoln, N MartíOliet, J Meseguer, and J Quesada. Maude: Specification and programming in rewriting logic. Technical report, Computer Science Laboratory, SRI International, 1999.
5. A Cohn. A proof of correctness of the Viper microprocessor: the first levels. In G Birtwistle and P A Subrahmanyam, editors, *VLSI Specification, Verification and Synthesis*, pages 27 – 72. Kluwer Academic Publishers, 1987.
6. B Cook, J Launchbury, and J Matthews. Specifying superscalar microprocessors in Hawk. In M Sheeran, editor, *Workshop on Formal Methods for Hardware (Marstrand, Sweden)*, 1998.
7. D Cyrluk, J Rushby, and M Srivas. Systematic formal verification of interpreters. In *IEEE International Conference on Formal Engineering Methods (ICFEM'97*, pages 140 – 149, 1997.
8. A C J Fox. *Algebraic Representation of Advanced Microprocessors*. PhD thesis, Department of Computer Science, University of Wales Swansea, 1998.
9. A C J Fox. *Algebraic Techniques for Verifying Microprocessors in HOL*. Computer Laboratory Report 512, University of Cambridge, 2001.
10. A C J Fox and N A Harman. An algebraic model of correctness for superscalar microprocessors. In *Formal Methods in Computer-Aided Design*, pages 346 – 361. Lecture Notes in Computer Science 1166, Springer-Verlag, 1996.
11. A C J Fox and N A Harman. Algebraic models of correctness for microprocessors. *Formal Aspects of Computer Science*, 12:298 – 312, 2000.
12. A C J Fox and N A Harman. Algebraic models of superscalar microprocessor implementations: A case study. In B Möller and J V Tucker, editors, *Prospects for Hardware Foundations*, pages 138 – 183. Lecture Notes in Computer Science 1546, Springer-Verlag, 1998.
13. A C J Fox and N A Harman. Algebraic models of temporal abstraction for initialised iterated state systems: An abstract pipelined case study. Technical Report CSR 21-98, University of Wales Swansea, 1998, submitted to the *Journal of Algebraic and Logic Programming*.
14. B Graham. *The SECD Microprocessor: a Verification Case Study*. Kluwer, 1992.
15. M J C Gordon and T F Melham. *Introduction to HOL*. Cambridge University Press, 1993.
16. N A Harman. Correctness and verification of hardware systems using Maude. Technical Report Computer Science Report, University of Wales Swansea, 2000.
17. N A Harman and J V Tucker. The formal specification of a digital correlator I: Abstract user specification. In K McEvoy and J V Tucker, editors, *Theoretical Foundations for VLSI Design*, pages 161 – 262. Cambridge University Press Tracts in Theoretical Computer Science 10, 1990.

18. N A Harman and J V Tucker. Algebraic models of microprocessors: Architecture and organisation. *Acta Informatica*, 33:421 – 456, 1996.

19. N A Harman and J V Tucker. Algebraic models of microprocessors: the verification of a simple computer. In V Stavridou, editor, *Proceedings of the 1995 IMA Conference on Mathematics for Dependable Systems*. Oxford University Press, 1997.

20. J L Hennessy and D A Patterson. *Computer Architecture: A Quantative Approach*. Morgan Kaufman, 1996.

21. R Hosabettu, M Srivas, and G Gopalakrishnan. Decomposing the proof of correctness of pipelined microprocessors. In A J Hu and M Y Vardi, editors, *Computer Aided Verification: 10th International Conference*, pages 122 – 134. Springer-Verlag, Lecture Notes in Computer Science 1427, 1998.

22. J Sawada W A Hunt. Processor verification with precise exceptions and speculative execution. In A J Hu and M Y Vardi, editors, *Computer Aided Verification: 10th International Conference*, pages 135 – 147. Springer-Verlag, Lecture Notes in Computer Science 1427, 1998.

23. R B Jones, J U Skakkebæk, and D Dill. Reducing manual abstraction in formal verification of out-of-order execution. In G Gopalakrishnan and P Windley, editors, *Formal Methods in Computer-Aided Design, FMCAD 98*, pages 2 – 17. Springer-Verlag, Lecture Notes in Computer Science 1522, 1998.

24. J U Skakkebæk, R B Jones, and D Dill. Formal verification of out-of-order execution using incremental flushing. In A J Hu and M Y Vardi, editors, *Computer Aided Verification: 10th International Conference*, pages 98 – 109. Springer-Verlag, Lecture Notes in Computer Science 1427, 1998.

25. S Krstic, B Cook, J Launchbury, and J Matthews. A correctness proof of a speculative, superscalar, out-of-order, renaming microarchitecture. Technical report, Oregon Graduate Institute, 1998.

26. F Durán M Clavel, S Eker, and J Meseguer. Building equational proving tools by reflection in rewriting logic. In K Futatsugi, A T Nakagawa and T Tamai, editors, *CAFE: An Industrial-Strength Algebraic Formal Method*, pages 1 – 31, Elsevier, 2000.

27. K Meinke and J V Tucker. Universal algebra. In T S E Maibaum S Abramsky, D Gabbay, editor, *Handbook of Logic in Computer Science*, pages 189 – 411. Oxford University Press, 1992.

28. S Miller and M Srivas. Formal verification of the AAMP5 microprocessor: a case study in the industrial use of formal methods. In *Proceedings of WIFT 95, Boca Raton*, 1995.

29. S Owre, J Rushby, N Shankar, and M Srivas. A tutorial on using PVS. In *Proceedings of TPCD 94*, pages 258–279. Lecture Notes in Computer Science 901, Springer-Verlag, 1994.

30. M Srivas, H Rueß, and D Cyrluk. Hardware verification using PVS. In T Kropf, editor, *Formal Hardware Verification*, pages 156 – 205. Springer-Verlag, Lecture Notes in Computer Science 1287, 1997.

31. K Stephenson. *An Algebraic Approach to Syntax, Semantics and Compilation*. PhD thesis, University of Wales Swansea Computer Science Department, 1996.

32. K Stephenson. Algebraic specification of the Java virtual machine. In B Möller and J V Tucker, editors, *Prospects for Hardware Foundations*. Lecture Notes in Computer Science 1546, Springer-Verlag, 1998.

33. P Windley and J Burch. Mechanically checking a lemma used in an automatic verification tool. In A Camilleri M Srivas, editor, *Formal Methods in Computer-Aided Design*, pages 362 – 376. Lecture Notes in Computer Science 1166, Springer-Verlag, 1996.

34. P Windley and M Coe. A correctness model for pipelined microprocessors. In *Proceedings of the 2nd Conference on Theorem Provers in Circuit Design*, 1994.

35. M Wirsing. Algebraic specification. In J van Leeuwen, editor, *Handbook of Theoretical Computer Science, Volume B: Formal Models and Semantics*, pages 675 – 788. Elsevier, 1990.

Verifying Architectural Specifications

Piotr Hoffman

Warsaw University, Institute of Informatics,
Banacha 2, 02-097 Warszawa, Poland
`piotrek@mimuw.edu.pl`

Abstract. In this paper we develop methods for verifying the correctness of architectural specifications, a mechanism introduced in the CASL specification language. This mechanism offers a formal way to express implementation steps in program development. Each such step states that to implement the unit of interest, one may implement some other units and then assemble them in the prescribed manner. In this paper we define a formal institution-independent semantics of architectural specifications, as well as sound and complete methods for proving them correct, applicable in the case of many institutions, in particular first-order logic.

Introduction

The formal development of a program consists of three phases: writing a requirements specification, implementing it, and proving the correctness of the implementation with respect to the specification (see [ST97]). Architectural specifications are meant to be a tool aiding the developer in breaking down the implementation task into independent subtasks. These subtasks can be either further broken down, or directly coded in a chosen programming language. The benefits of dividing a programming task into independent subtasks are obvious: work on parts of the project can proceed in parallel, the project as a whole is easier to maintain and comprehend. Also, the correctness proof can now be constructed from the correctness proofs for the individual subtasks and a correctness proof of the subdivision (i.e., architectural specification) itself.

In this paper, we first recall the notion of *institution* in Section 1 and then define the syntax and formal semantics of architectural specifications in an institution-independent fashion in Section 2. Both are somewhat simplified versions of what has been first introduced in CASL, the specification language designed by the Common Framework Initative (CoFI). For more information on CASL, see [CASL00,CASL99,CASL]; for an in-depth discussion of architectural specifications, see [BST99].

In Section 3 we precisely define what we understand by verifying architectural specifications. Essentially, we will consider an architectural specification correct, if having completed the subtasks one can always assemble their implementations in the prescribed manner and indeed obtain an implementation of the original task. Our aim is to find an algorithm that, when fed an architectural specification, generates proof obligations equivalent to the correctness of

M. Cerioli and G. Reggio (Eds.): WADT/CoFI 2001, LNCS 2267, pp. 152–175, 2002.

the specification. Of course, we would like these obligations to be of a possibly simple form; we choose what we call *simple obligations*, i.e., statements of the form $\Phi \models_\Sigma \Psi$, where Φ and Ψ are finite sets of sentences of the underlying institution (these "sentences" could actually be structured specifications over some other institution — we would then end up with refinements of specifications). Note that this will *not* provide a decision procedure for the verification of architectural specifications in general — in fact, in typical institutions verification can easily be proven undecidable (since checking simple obligtions, as defined above, is typically undecidable).

In Section 4 we divide the procedure of generating obligations into two stages: first, we want to extract some *intermediate proof obligations* from an architectural specification; next, we want to transform these intermediate proof obligations into simple obligations. We then state our main theorem: under the assumption that the underlying institution has *logical amalgamation*, a notion closely connected to the amalgamation property (see [Tar99] and [SMTKH01]), one can perform stage one of our procedure; that is, one can reduce the problem of checking the correctness of an architectural specification to that of discharging some intermediate proof obligations. The proof of this result can be found in Section 5.

Sections 6 and 7 are devoted to stage two of the procedure of generating proof obligations, i.e., transforming intermediate proof obligations into simple obligations. In Section 6 we simplify the intermediate proof obligations, again under the assumption that the underlying institution has logical amalgamation. In Section 7 we show how to transform the simplified intermediate proof obligations into simple obligations in the case of first-order logic with equality. This section is the only part of the paper that is not institution-independent. Still, it seems that the techniques in use should be applicable to other institutions as well.

Finally, in Section 8 we propose to impose additional restrictions on architectural specifications, thus defining *strict* correctness. This allows us to divide the verification process into two parts: the first one is static and performed automatically, the second one involves non-trivial proof obligations discharged by means of a theorem-prover. The notions of static and strict correctness also form a link between this paper and the *extended static semantics* of [SMTKH01] and [KHTSM01].

1   Institutions

The definition of architectural specifications will be parametrized by an underlying logical system. To formalize this notion, we will use the concept of *institution* (see [GB92]), presented in this section. Before we dive into the details, we introduce some notation. For any function, morphism or functor f, by $dom(f)$ we denote its domain; if f is a morphism or functor, then by $cod(f)$ we denote its codomain. For any category C, by $|C|$ we denote the class of objects of C. Finally, composition is always written in diagrammatical order.

An institution **I** is a quadruple $(Sig, \mathbf{Mod}, Sen, \models)$, where:

- Sig is a category of *signatures*;
- $\mathbf{Mod} : Sig^{op} \to \mathbf{Class}$ is a functor into the quasi-category of all large sets;
- $Sen : Sig \to \mathbf{Set}$ is a functor into the category of all small sets;
- $\models$ is a family $\{\models_\Sigma\}_{\Sigma \in |Sig|}$, where $\models_\Sigma$ is a relation on $\mathbf{Mod}(\Sigma) \times Sen(\Sigma)$.

These data are subject to the so-called *satisfaction condition*, i.e., for any $\sigma : \Sigma \to \Delta$ in Sig, $\phi \in Sen(\Sigma)$ and $M \in \mathbf{Mod}(\Delta)$ the following equivalence holds:

$$M \models_\Delta Sen(\sigma)(\phi) \iff \mathbf{Mod}(\sigma)(M) \models_\Sigma \phi$$

Elements of $\mathbf{Mod}(\Sigma)$ are called Σ-*models* and elements of $Sen(\Sigma)$ are called Σ-*sentences*. We usually denote the (large) function $\mathbf{Mod}(\sigma)$ by $\cdot|_\sigma$ and call it the *(σ-)reduct*. For any $\sigma : \Sigma \to \Delta$ in Sig and $\phi \in Sen(\Sigma)$, by $\sigma(\phi)$ we denote $Sen(\sigma)(\phi)$ and we call it the *translation* of ϕ along σ.

The most prominent example of an institution is many-sorted first-order logic. A possible definition follows:

- a signature is a pair (S, p), where S is a finite set and p is a function with finite domain into $S^* \times S$; the elements of S are called *sorts* and the elements of $dom(p)$ are called *function symbols*;
- a signature morphism $\sigma : (S, p) \to (T, q)$ consists of two functions $\sigma^S : S \to T$ and $\sigma^F : dom(p) \to dom(q)$ such that for any $f \in dom(p)$, if $p(f) = (s_1 \ldots s_n, s)$, then $q(\sigma^F(f)) = (\sigma^S(s_1) \ldots \sigma^S(s_n), \sigma^S(s))$; composition is defined component-wise;
- a sentence over (S, p) is a first-order sentence involving variables of the sorts S, the function symbols $dom(p)$, and equality;
- the translation of a sentence along a signature morphism is defined naturally;
- a model over (S, p) assigns to each $s \in S$ a set called the *carrier* of s and to each $f \in dom(p)$ with $p(f) = (s_1 \ldots s_n, s)$ a function called the *interpretation* of f, whose domain is the product of the carriers of the sorts $s_1, \ldots, s_n$ and codomain is the carrier of the sort s;
- the reduct of a model M over (T, q) along a morphism $\sigma : (S, p) \to (T, q)$ assigns to each sort $s \in S$ the carrier of $\sigma^S(s)$ in M and to each function symbol $f \in dom(p)$ the interpretation of $\sigma^F(f)$ in M;
- the relations $\models_\Sigma$ are defined as usual.

With these definitions the satisfaction condition holds (cf. the equational case in [GB92]) and thus we have an institution; we denote it by FOL.

For any institution **I**, by $|\mathbf{I}|$ we will denote the institution obtained from **I** by removing all the sentences, i.e., by setting $Sen(\Sigma) = \emptyset$ for all signatures Σ.

We will overload the symbols $\mathbf{Mod}$ and $\models$ as follows. For any set Φ of Σ-sentences, $\mathbf{Mod}_\Sigma(\Phi)$ denotes the class of all Σ-models satisfying all sentences from Φ. If additionally Ψ is a set of Σ-sentences, then we write $\Phi \models_\Sigma \Psi$ if $\mathbf{Mod}_\Sigma(\Phi) \subseteq \mathbf{Mod}_\Sigma(\Psi)$.

We parametrize the definition of architectural specifications by an underlying institution $\mathbf{I} = (Sig, \mathbf{Mod}, Sen, \models)$ and a subcategory $\mathcal{J}$ of Sig called the *inclusion subcategory*. We require that $\mathcal{J}$ be a partial order and that $|\mathcal{J}| = |Sig|$. It makes sense to impose further restrictions on the inclusions, but this will not be necessary in the following considerations. An inclusion of Σ into Δ will be denoted by $\iota_{\Sigma,\Delta}$. If this inclusion exists, we will write $\Sigma \leq \Delta$. In the case of FOL, we define a morphism σ to be an inclusion if both σ^S and σ^F are set-theoretic inclusions.

In this paper, we will always assume that signatures, signature morphisms and sentences are representable in a finite way, the translation of sentences is computable, the order $\leq$ is decidable and the binary least upper bound operation in $\leq$ is computable. These assumptions obviously hold in FOL.

2   Architectural Specifications

The language of architectural specifications is parametrized by an institution **I** and inclusion subcategory $\mathcal{J}$, and described by the grammar in Figure 1.

$$
\begin{aligned}
ASP &::= \textbf{units } UDS \textbf{ result } UE \textbf{ implementing } USP \\
UDS &::= \epsilon \mid \\
&\qquad UDS \; ID : USP \\
USP &::= SP \mid \\
&\qquad SP \rightarrow SP \\
UE &::= \bullet\, T \mid \\
&\qquad \lambda \; ID : SP \bullet T \\
T &::= ID \mid \\
&\qquad ID(T) \mid \\
&\qquad T \textbf{ and } T \mid \\
&\qquad \textbf{reduce } T \textbf{ by } \sigma \mid \\
&\qquad \textbf{let } ID = T \textbf{ in } T
\end{aligned}
$$

Fig. 1. Language of architectural specifications.

Here, the *ID*s come from an infinite set of identifiers, the σs are signature morphisms in **I**, and the *SP*s are finite presentations in **I**, i.e. pairs (Σ, Φ), where Φ is a finite set of sentences over Σ.

An architectural specification states that, in order to implement a unit satisfying the specification in the **implementing** section, one can implement the units declared in the **units** section, and then assemble them as described in the **result** section. The denotation of the whole specification is a function taking any implementation of the declared units to a result unit as described by the unit expression *UE*.

Remark 1. The above language of architectural specifications differs somewhat from that used in CASL. One addition we have made is the **implementing** section, which we think makes the idea of architectural specifications more comprehensible; in CASL comments are used for the same purpose. Also, the reduct is more restrictive in CASL. On the other hand, we have left out a lot of constructs: multiparameter units, unit definitions, fitting morphisms in applications, declarations with imports, and translation. Also, we have chosen the specifications to be finite presentations. This last simplification does not mean that structured specifications cannot be dealt with in our framework; all one has to do is formalize them as sentences of an institution, which is indeed possible.

In order to introduce a formal semantics of architectural specifications, we need to provide some definitions. A *parametric signature* is a pair $P\Sigma = (\Sigma, \Delta)$ such that $\Sigma \leq \Delta$. A *unit signature* $U\Sigma$ is either a (regular) signature or a parametric signature. A *parametric unit* over a parametric signature (Σ, Δ) is a partial function U from $\mathbf{Mod}(\Sigma)$ to $\mathbf{Mod}(\Delta)$ which is *persistent*, i.e., for any $M \in dom(U)$ we have $U(M)|_{\iota_{\Sigma,\Delta}} = M$. A *unit* over a unit signature is a model over that signature, if it is a (regular) signature, or a parametric unit over that signature, if it is a parametric signature. By $\mathbf{PUnit}(P\Sigma)$ we denote the set of all parametric units over the parametric signature $P\Sigma$. A *parametric unit specification* over a parametric signature (Σ, Δ) is a pair of presentations $u = ((\Sigma, \Phi), (\Delta, \Psi))$. Then $\mathbf{PUnit}(u)$ is the set of units over (Σ, Δ) with domain $\mathbf{Mod}_\Sigma(\Phi)$ and into $\mathbf{Mod}_\Delta(\Psi)$. A parametric unit specification u is *inconsistent* if $\mathbf{PUnit}(u)$ is empty. A *unit specification* is either a presentation or a parametric unit specification. A *static environment* δ is a pair (B, P), where B maps identifiers to signatures and P maps identifiers to parametric signatures. The domains of B and P are required to be finite and disjoint. A *unit environment* fitting a static environment (B, P) is a map e, sending any identifier $ID \in dom(B)$ to a model in $\mathbf{Mod}(B(ID))$ and any identifier $ID \in dom(P)$ to a parametric unit in $\mathbf{PUnit}(P(ID))$. A *unit context* fitting a static environment δ is any set E of environments fitting δ. A *model function of type* (E, Σ) is a total function F from the context E into $\mathbf{Mod}(\Sigma)$. *Unit functions of type* $(E, U\Sigma)$, denoted UF, are defined analogically.

To deal with the **and**-construct, we will need to define two notions. First, we define the *sum* of signatures Σ, Δ to be their least upper bound $\Gamma = \Sigma \cup \Delta$ with respect to the order $\leq$, if it exists; if so, we call them *compatible*. Then for any models $M \in \mathbf{Mod}(\Sigma)$ and $N \in \mathbf{Mod}(\Delta)$ we define their *amalgamation*, denoted $M \oplus N$, to be a model $P \in \mathbf{Mod}(\Gamma)$ satisfying $P|_{\iota_{\Sigma,\Gamma}} = M$ and $P|_{\iota_{\Delta,\Gamma}} = N$, if it exists and is unique. Models are called *amalgamable* if their amalgamation exists.

The semantics will assign:

- to any unit specification USP, a unit signature $U\Sigma$ and a (large) set $\mathcal{U}$ of units over $U\Sigma$; i.e., either a signature Σ and a (large) set $\mathcal{M}$ of models over Σ or a parametric signature $P\Sigma$ and a (large) set $\mathcal{P}$ of parametric units over $P\Sigma$;
- to any list of unit declarations UDS, a static environment δ and a unit context E fitting δ;
- to any term T, given a static environment δ and a unit context E fitting δ, a signature Σ and a model function F of type (E, Σ);

- to any unit expression UE, given a static environment δ and a unit context E fitting δ, a unit signature $U\Sigma$ and a unit function UF of type $(E, U\Sigma)$;
- to any architectural specification ASP, a static environment δ, a unit signature $U\Sigma$ and a unit function UF of type $(dom(UF), U\Sigma)$, where $dom(UF)$ fits δ.

In the semantics rules below, we use the natural semantics style, with judgments written $\_\_ \vdash \_\_ \rhd \_\_$. We always implicitly add premises stating that objects denoted by δ are static environments, in any pair δ, E, E fits δ, etc. By $f[a/x]$ we denote the function which extends f with the mapping $x \mapsto a$.

$$\boxed{\vdash ASP \rhd \delta, U\Sigma, UF}$$

$$\frac{\begin{array}{c} \vdash UDS \rhd \delta, E \\ \delta, E \vdash UE \rhd U\Sigma, UF \\ \vdash USP \rhd U\Sigma, \mathcal{U} \\ \text{for all } e \in E, UF(e) \in \mathcal{U} \end{array}}{\vdash \textbf{units } UDS \textbf{ result } UE \textbf{ implementing } USP \rhd \delta, U\Sigma, UF}$$

$$\boxed{\vdash UDS \rhd \delta, E}$$

$$\frac{}{\vdash \epsilon \rhd \emptyset, \{\emptyset\}}$$

$$\frac{\begin{array}{c} \vdash UDS \rhd (B, P), E \\ \vdash USP \rhd \Sigma, \mathcal{M} \\ ID \notin dom(B) \cup dom(P) \end{array}}{\begin{array}{c} \vdash UDS \ ID : USP \rhd \\ (B[\Sigma/ID], P), \{ e[M/ID] \mid e \in E, M \in \mathcal{M} \} \end{array}}$$

$$\frac{\begin{array}{c} \vdash UDS \rhd (B, P), E \\ \vdash USP \rhd P\Sigma, \mathcal{P} \\ ID \notin dom(B) \cup dom(P) \end{array}}{\begin{array}{c} \vdash UDS \ ID : USP \rhd \\ (B, P[P\Sigma/ID]), \{ e[U/ID] \mid e \in E, U \in \mathcal{P} \} \end{array}}$$

$$\boxed{\vdash USP \rhd U\Sigma, \mathcal{U}}$$

$$\frac{}{\vdash (\Sigma, \Phi) \rhd \Sigma, \mathbf{Mod}_\Sigma(\Phi)}$$

$$\frac{\Sigma \leq \Delta}{\vdash (\Sigma, \Phi) \to (\Delta, \Psi) \rhd (\Sigma, \Delta), \mathbf{PUnit}((\Sigma, \Phi), (\Delta, \Psi))}$$

$$\boxed{\delta, E \vdash UE \rhd U\Sigma, UF}$$

$$\frac{(B, P), E \vdash T \rhd \Sigma, F}{(B, P), E \vdash \bullet T \rhd \Sigma, F}$$

$$\frac{\begin{array}{c} (B\,[\Sigma/ID]\,, P), \{\, e\,[M/ID] \mid e \in E,\; M \in \mathbf{Mod}_\Sigma(\Phi)\,\} \vdash T \rhd \Delta, F \\ \Sigma \leq \Delta \\ \text{for all } e \in E \text{ and } M \in \mathbf{Mod}_\Sigma(\Phi),\; F(e\,[M/ID])|_{\iota_{\Sigma,\Delta}} = M \end{array}}{\begin{array}{c} (B, P), E \vdash \lambda ID : (\Sigma, \Phi) \bullet T \rhd \\ (\Sigma, \Delta),\, \lambda e \in E \cdot \lambda M \in \mathbf{Mod}_\Sigma(\Phi) \cdot F(e\,[M/ID]) \end{array}}$$

$$\boxed{\delta, E \vdash T \rhd \Sigma, F}$$

$$\frac{ID \in dom(B)}{(B, P), E \vdash ID \rhd B(ID),\, \lambda e \in E \cdot e(ID)}$$

$$\frac{\begin{array}{c} ID \in dom(P) \text{ and } P(ID) = (\Sigma, \Delta) \\ (B, P), E \vdash T \rhd \Sigma, F \\ \text{for all } e \in E,\; F(e) \in dom(e(ID)) \end{array}}{(B, P), E \vdash ID(T) \rhd \Delta,\, \lambda e \in E \cdot e(ID)(F(e))}$$

$$\frac{\begin{array}{c} \delta, E \vdash T_1 \rhd \Sigma_1, F_1 \\ \delta, E \vdash T_2 \rhd \Sigma_2, F_2 \\ \Delta = \Sigma_1 \cup \Sigma_2 \\ \text{for all } e \in E, \text{ the models } F_1(e) \text{ and } F_2(e) \text{ are amalgamable} \end{array}}{\delta, E \vdash T_1 \textbf{ and } T_2 \rhd \Delta,\, \lambda e \in E \cdot F_1(e) \oplus F_2(e)}$$

$$\frac{\delta, E \vdash T \rhd cod(\sigma), F}{\delta, E \vdash \textbf{reduce } T \textbf{ by } \sigma \rhd dom(\sigma),\, \lambda e \in E \cdot F(e)|_\sigma}$$

$$\frac{\begin{array}{c} (B, P), E \vdash T \rhd \Sigma, F \\ ID \notin dom(B) \cup dom(P) \\ (B\,[\Sigma/ID]\,, P), \{\, e\,[F(e)/ID] \mid e \in E\,\} \vdash T' \rhd \Sigma', F' \end{array}}{(B, P), E \vdash \textbf{let } ID = T \textbf{ in } T' \rhd \Sigma',\, \lambda e \in E \cdot F'(e\,[F(e)/ID])}$$

It is easy to see that any given syntactic object has, in a given context, at most one denotation. If this denotation exists, we call the object *correct* in that context. In particular, an architectural specification *ASP* is *correct* if it has a denotation.

3   Correctness

In this paper we investigate ways of checking whether an architectural specification is correct. Obviously, the difficulty of this task depends on the underlying institution **I** and the inclusion category. If we choose **I** to be the institution of first-order logic, checking correctness will of course turn out undecidable, which is a straightforward consequence of first-order logic being undecidable. Generally, there are four conditions in the semantic rules which cannot always be checked statically. These are:

1. "for all $e \in E, UF(e) \in \mathcal{U}$" in the top-most rule;
2. "for all $e \in E, F(e) \in dom(e(ID))$" in the rule for application;
3. "for all $e \in E$, the models $F_1(e)$ and $F_2(e)$ are amalgamable" in the rule for amalgamation;
4. "for all $e \in E$ and $M \in \mathbf{Mod}_\Sigma(\Phi)$, $F(e\,[M/ID])|_{\iota_{\Sigma,\Delta}} = M$" in the rule for λ-definition.

In an institution as powerful as FOL, each of these conditions is sufficient to cause undecidability. We call a statement of the form $\Phi \models_\Sigma \Psi$, where Φ, Ψ are finite sets of Σ-sentences, a *simple obligation* over Σ. Our aim in this paper is to produce an algorithm, for as general a case as possible, which will, given an architectural specification, generate a finite set of *proof obligations* which capture its correctness. More specifically, the algorithm will compute the following two finite sets:

- P, consisting of parametric unit specifications
- S, consisting of simple obligations

We will require that if the algorithm on ASP does not signal failure and produces sets P and S, then ASP is correct if and only if *either* one of the parametric unit specifications in P is inconsistent, *or* all obligations $\Phi \models_\Sigma \Psi$ from S are true. We will additionally require that if any parametric unit specification in P is inconsistent, then the denotation of ASP is the empty function. Of course, the algorithm may signal failure only if ASP is incorrect. The above requirements effectively mean that if we want to prove the correctness of an architectural specification and are not interested in the case that it denotes the empty function, then we are left with checking simple obligations. This is a desirable situation, since proof methods and tools often exist for such obligations (see, e.g., [Pad99], [BCH99], [Borz98]).

4   Intermediate Proof Obligations and Logical Amalgamation

In this section we introduce an intermediate kind of proof obligation and we show that, under certain assumptions, one can reduce the problem of checking the correctness of an architectural specification to discharging such obligations.

Let Σ be a signature. We define Σ-*statements* to be:

- $\sigma \oplus \tau$, where $\sigma : \Delta \to \Sigma$, $\tau : \Gamma \to \Sigma$, Δ and Γ compatible
- $\sigma = \tau$, where $\sigma, \tau : \Delta \to \Sigma$
- $\sigma \neq \tau$, where $\sigma, \tau : \Delta \to \Sigma$

The statement $\sigma \oplus \tau$ is called the *amalgamation statement*, the statements $\sigma = \tau$ and $\sigma \neq \tau$ are called *positive* and *negative equality statements*, respectively. Let M be a model over Σ. We say that:

- $M \models_\Sigma \sigma \oplus \tau$ if $M|_\sigma$ and $M|_\tau$ are amalgamable
- $M \models_\Sigma \sigma = \tau$ if $M|_\sigma = M|_\tau$
- $M \models_\Sigma \sigma \neq \tau$ if $M|_\sigma \neq M|_\tau$

These statements are in general non-trivial, since the morphisms involved need not be inclusions. The *intermediate proof obligations* are *satisfaction* and *amalgamability obligations*. A *satisfaction obligation* is a statement of the form $\Phi \cup \mathcal{E} \models_\Sigma \Psi$, an *amalgamability obligation* takes the form $\Phi \cup \mathcal{E} \models_\Sigma \sigma \oplus \tau$. Here, Φ and Ψ are finite sets of Σ-sentences, $\mathcal{E}$ is a finite set of equality Σ-statements, and $\sigma \oplus \tau$ is a Σ-statement, as introduced above. Obviously, checking satisfaction obligations is computationally at least as hard as checking simple obligations. In particular, in FOL it is undecidable. Though amalgamability obligations seem less complex, this need not be the case — e.g., in FOL for any finite sets Φ and Ψ of sentences over Σ one can compute an amalgamability obligation which is true iff $\Phi \models_\Sigma \Psi$. To see this, let Δ be the signature Σ plus a new sort s and constants a and b of this sort. Now, consider the signature Δ_0 consisting of one sort s and one constant x of this sort, and two morphisms $\sigma, \tau : \Delta_0 \to \Delta$, one sending s to s and x to a, the other sending s to s and x to b. Then the obligation $(\Delta, \{((\bigwedge \Phi \Rightarrow \bigwedge \Psi) \Rightarrow a = b\}) \models_\Delta \sigma \oplus \tau$ is equivalent to $\Phi \models_\Sigma \Psi$. Another interesting example is the logic of CASL. In this logic problems very similar to the problem of checking whether $\models_\Sigma \sigma \oplus \tau$ holds have been shown undecidable, see Section 8 and [KHTSM01]. For any intermediate proof obligation without equality statements one can produce an architectural specification which is correct if and only if that obligation is true. This means that the intermediate obligations are certainly no harder than the original problem. As for the equality statements, they are needed to express the fact that a parametric unit applied to a given argument always yields the same result (actually, they are decompositions of implications between positive equality statements: if the arguments of two applications are equal, then so must be the results). The category $Pres(\mathbf{I})$ of finite presentations in the institution $\mathbf{I}$ is defined as follows. Its objects are finite presentations, and a morphism between (Σ, Φ) and (Δ, Ψ) is simply a morphism $\sigma : \Sigma \to \Delta$ in Sig such that for any $M \in \mathbf{Mod}_\Delta(\Psi)$ we have $M|_\sigma \in \mathbf{Mod}_\Sigma(\Phi)$. This makes $\mathbf{Mod}$ extend to a (contravariant) functor on $Pres(\mathbf{I})$. A *diagram* is a functor $D : I \to Pres(\mathbf{I})$ from a finite category I; the category I is called the *scheme* of the diagram. A family of models $\{M_i\}_{i \in |I|}$ is said to be *consistent* with D if:

- for any $i \in |I|$, $M_i \in \mathbf{Mod}(D(i))$
- for any $m : i \to j$ in I, $M_i = M_j|_{D(m)}$

We say that the institution **I** has *logical amalgamation* if for any diagram D with scheme I one can compute a sink $\{\alpha_i : D(i) \to (\Sigma, \Phi)\}_{i \in |I|}$ such that the set $\{\{M|_{\alpha_i}\}_{i \in |I|} \mid M \in \mathbf{Mod}_\Sigma(\Phi)\}$ is equal to the set of all families consistent with D. This sink is called an *amalgamating sink*. The motivation behind the above notion is as follows. While verifying an architectural specification, we will build diagrams of presentations. Some nodes in these diagrams will correspond to declared non-parametric units, others will correspond to terms. Take a family consistent with the diagram. Then the models at the nodes corresponding to declarations define an environment, while the nodes corresponding to terms define the values of these terms in that environment (actually, to make this intuition precise we would have to add some information on the parametric units as well). In the verification process, two conditions need to be checked. First, whether the model at a given node always satisfies a given set of sentences; second, whether models at two given nodes are always amalagamable. These conditions will give rise to the satisfaction and amalgamability obligations, respectively. Now, suppose we have a diagram D with scheme I and an amalgamating sink $\{\alpha_i : D(i) \to (\Sigma, \Phi)\}_{i \in |I|}$. Then instead of saying "for any family $\{M_i\}_{i \in |I|}$ consistent with D, $M_i \models \phi$", we may say "for any model $M \in \mathbf{Mod}_\Sigma(\Phi)$, $M|_{\alpha_i} \models \phi$"; this is easily equivalent to a satisfaction obligation. Of course, the details are much more complicated, e.g. when applications of parametric units come into play. We now show that the logical amalgamation property is weaker than having finitely composable signatures (see [Tar99]) and computability of colimits:

Proposition 1. *If the signature category of* **I** *is finitely cocomplete, the colimits are computable, and the model functor preserves finite limits, then* $|\mathbf{I}|$ *has logical amalgamation.*

Proof. Let D be a diagram in $|\mathbf{I}|$ with scheme I and let $\{\alpha_i : D(i) \to \Sigma\}_{i \in |I|}$ be its colimit. We claim that α treated as a sink in $Pres(|\mathbf{I}|)$ is an amalgamating sink. Take any M over Σ. Then the family $\{M|_{\alpha_i}\}_{i \in |I|}$ is consistent with D, since α is a cocone. Now take any family $\{M_i\}_{i \in |I|}$ consistent with D. $\{\mathbf{Mod}(\alpha_i)\}_{i \in |I|}$ now forms a limiting cone over $D; \mathbf{Mod}$. Define a cone $\{f_i : A \to \mathbf{Mod}(D(i))\}_{i \in |I|}$ by taking $A = \mathbf{Mod}(\Sigma) \cup \{*\}$, where $*$ is a new element, and f_i to be $\mathbf{Mod}(\alpha_i)$ extended by $f_i(*) = M_i$. Then there exists a (large) function $h : A \to \mathbf{Mod}(\Sigma)$ factoring this cone through $\mathbf{Mod}(\alpha)$. Then we have $M_i = h(*)|_{\alpha_i}$, which proves the proposition. $\qquad\square$

The following nicely generalizes the standard result of [GB92].

Proposition 2. *An institution* **I** *has logical amalgamation if and only if amalgamating sinks in* **I** *can be computed for any diagram* $D : I \to Pres(|\mathbf{I}|)$.

Proof. We will only prove the implication from right to left. Let D be a diagram with scheme I in **I**. Denote by $|D|$ the diagram in $|\mathbf{I}|$ obtained from D by removing all the sentences. We can now compute an amalgamating sink $\{\alpha_i : |D|(i) \to (\Sigma, \Phi)\}_{i \in |I|}$ for $|D|$. Let $\Psi = \bigcup_{i \in |I|} \alpha_i(\pi_2(D(i)))$ and consider the sink obtained by treating the α_i as morphisms in $Pres(\mathbf{I})$ with domain $D(i)$ and codomain $(\Sigma, \Phi \cup \Psi)$. This sink is an amalgamating sink for D.   □

Corollary 1. *For any institution* $\mathbf{I}$, *if* $|\mathbf{I}|$ *has logical amalgamation, then so does* $\mathbf{I}$.   □

Example 1. Take the institution FOL^{der} to be the same as FOL, but with *derived signature morphisms*. That is, a morphism σ from Σ to Δ works as usual on sorts, and may assign to a symbol $f : s_1 \times \cdots \times s_n \to s$ in Σ a term of sort $\sigma^S(s)$ in Δ involving the variables $x_1 : \sigma^S(s_1), \ldots, x_n : \sigma^S(s_n)$. Then the reduct of a model over Δ interprets f as the function which assigns to any tuple $v_1, \ldots, v_n$ of appropriate arguments the value of this term under the valuation $\{x_1 \mapsto v_1, \ldots, x_n \mapsto v_n\}$. As can easily be seen, the signature category of FOL^{der} is not finitely cocomplete. Take Σ to be a signature with a single sort s and a function symbol $f : s \times s \to s$ and take Δ to contain only the sort s. We then may define two morphisms $\sigma, \tau : \Sigma \to \Delta$, which assign to f the terms x_1 and x_2, respectively. The diagram consisting of these two parallel morphisms does not have a colimit (i.e., coequalizer). In fact, no cocone over this diagram exists. However, an amalgamating sink exists: it consists of the signature Δ with the sentence "$\forall x : s \cdot \forall y : s \cdot x = y$" and the morphisms σ, τ, id_Δ.

Proposition 3. *The institutions* $|FOL|$ *and* FOL *have logical amalgamation.*

Proof. We will use the fact that the signature category of FOL is finitely cocomplete and the model functor preserves finite limits (cf. [GB84] for the equational case); it is obvious that the colimits are computable. That $|FOL|$ has logical amalgamation is now a consequence of Prop. 1. Then the fact that FOL has logical amalgamation follows by Corollary 1.   □

Theorem 1. *Assume* $\mathbf{I}$ *has logical amalgamation. Then there exists an algorithm which, given an architectural specification ASP, either signals failure, or computes a finite set of parametric unit specifications* P *and a finite set* S *of satisfaction and amalgamability obligations, and such that:*

ASP is correct $\iff$ the algorithm does not signal failure, and

either some parametric unit specification in P is inconsistent,

or every obligation in S is true

Also, if an element of P is inconsistent, then the denotation of ASP is the empty function.

  Let n be the number of declared parametric units in ASP and k_i, $i = 1, \ldots, n$, the number of times the parametric unit number i is applied in ASP, and let s be the number of amalgamations and applications in ASP. Then the cardinality of S is bounded by $4 + s \cdot 2^{|k_1^2 - 1| + \cdots + |k_n^2 - 1|}$ and the number of equality statements in every obligation is lower than $\frac{1}{2} \sum_{1 \leq i \leq n} k_i(k_i - 1)$.

The above theorem is a major step in verifying architectural specifications. However, the proof is quite technical and we delegate it to the next section. The bounds in the theorem are important, since they show that if all units are applied at most once, then the number of obligations is linear and the equality statements vanish altogether. This is also the reason for explicitly introducing the **let**-construct to the language of architectural specifications, since it allows us to remove common subexpressions and hence also the number of applications of a single parametric unit.

Remark 2. The semantics of architectural specifications considered in this paper is a so-called *non-generative* semantics, meaning that repeated application of a parametric unit to the same arguments always yields the same result. One could introduce a *generative* semantics (as that of functors in SML [Pau96]) in either of two equivalent ways. One method would be to define parametric units as multi-valued functions (i.e., relations) and rewrite the semantics appropriately. Another method would be to treat any declaration of a parametric unit F with some type as a sequence of declarations $F_1, \ldots, F_k$ with the same type and in the terms substitute F_i for the ith application of F. The above theorem implies that with generative semantics (defined either way), the cardinality of the set of obligations S will be bounded by $4 + s$ and the obligations themselves will not involve equality statements.

Example 2. The architectural specification in Fig. 2 is an example of a correct specification which is not correct with respect to a generative semantics. The diagram associated with this specification is depicted below (here, σ is the signature morphism used in the definition of the unit C''):

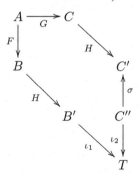

The correctness of this specification can be established as follows. First, all the applications are obviously correct. Second, the amalgamation defining T is correct because the carriers of the sort s in the units B' and C'' are identical, since they both come from the unit A. So, what remains to be checked is that the unit T always satisfies the sentence $a = a' \Rightarrow b = b'$. We will consider two cases. If the units B and C are not equal, then the interpretations of the constant a in B and the constant a in C must be different. Thus, in the unit T the constants a and a' have different interpretations and the sentence $a = a'$ does not hold in T. If, on the other hand, the units B and C are equal, then so are B' and C'

(this is where the non-generativity comes into play). Thus the constants b and b' must have equal interpretations in T, and so $b = b'$ holds in T. In either case, the implication $a = a' \Rightarrow b = b'$ holds in T.

It is easy to see that the specification in Fig. 2 is not correct with respect to a generative semantics. What is more interesting is that there seems to be no obvious way of making it generatively correct, e.g. by using **let**-constructs for common subexpressions.

> **units**
> > $A : \{\ \textbf{sort}\ s;\ \}$
> > $F : \{\ \textbf{sort}\ s;\ \} \to \{\ \textbf{sort}\ s;\ \textbf{op}\ a : s;\ \}$
> > $G : \{\ \textbf{sort}\ s;\ \} \to \{\ \textbf{sort}\ s;\ \textbf{op}\ a : s;\ \}$
> > $H : \{\ \textbf{sort}\ s;\ \textbf{op}\ a : s;\ \} \to \{\ \textbf{sort}\ s;\ \textbf{ops}\ a, b : s;\ \}$
> > **result**
> > > • **let**
> > > > $B = F(A)$
> > > > $C = G(A)$
> > > > $B' = H(B)$
> > > > $C' = H(C)$
> > > > $C'' = \textbf{reduce}\ C'\ \textbf{by sort}\ s \mapsto s,\ \textbf{ops}\ a' \mapsto a,\ b' \mapsto b$
> > > > $T = B'\ \textbf{and}\ C''$
> > > > **in**
> > > > > T
> > > **implementing**
> > > > $\{\ \textbf{sort}\ s;\ \textbf{ops}\ a, b, a', b' : s;\ \textbf{axioms}\ a = a' \Rightarrow b = b';\ \}$

Fig. 2. A correct architectural specification which is incorrect with respect to a generative semantics.

5   Proof of Theorem 1

An *extended environment* is a quadruple $e\delta = (D, b, p, lp)$, where:

– D is a presentation diagram with scheme I
– b is a partial injective function from identifiers to $|I|$
– p is a partial function from identifiers to parametric unit specifications
– the domains of b and p are disjoint
– lp labels some edges in I with elements of $dom(p)$ in such a way that, for any morphism m in its domain with $p(lp(m)) = ((\Sigma, \Phi), (\Delta, \Psi))$, $D(m)$ is an inclusion from Σ into Δ.

Intuitively, an extended environment consists of a diagram D whose meaning is as in the previous section, a map b which connects identifiers of non-parametric units to nodes in the diagram, a map p which gives the specifications of parametric units and a map lp such that $lp(m) = ID$ means that the edge m in the diagram D represents an application of the parametric unit ID.

Any extended environment $e\delta = (D, b, p, lp)$ defines a static environment (B, P), where B sends any identifier ID in $dom(b)$ to the signature of the presentation $D(b(ID))$, and P sends any identifier ID in $dom(p)$ to the parametric signature obtained by removing the sets of sentences from $p(ID)$. This environment is denoted by $st(e\delta)$.

Let $e\delta = (D, b, p, lp)$ be an extended environment. We say that a family of models $\{N_i\}_{i \in |I|}$ consistent with D satisfies the lp-*condition*, if for any morphisms $m : i \to j$ and $n : k \to l$ with $lp(m) = lp(n)$, $N_i = N_k$ implies $N_j = N_l$. Thus, the lp-condition states that applying the same parametric unit to the same argument must yield the same result.

Let $\mathcal{B}$ send every ID in $dom(b)$ to a model in $\mathbf{Mod}(D(b(ID)))$ and $\mathcal{P}$ send every ID in $dom(p)$ to a unit in $\mathbf{PUnit}(p(ID))$ (they *fit* $e\delta$). Now, let $\{N_i\}_{i \in |I|}$ be a model family consistent with D, such that $N_{b(ID)} = \mathcal{B}(ID)$ for $ID \in dom(b)$ and that for any m in $dom(lp)$ we have $\mathcal{P}(lp(m))(N_{dom(m)}) = N_{cod(m)}$. If such a family N exists and is unique, we denote it by $F_{e\delta}(\mathcal{B}, \mathcal{P})$. The interpretation here is that $F_{e\delta}(\mathcal{B}, \mathcal{P})$ extends the valuation of the declarations given by $\mathcal{B}$ and $\mathcal{P}$ to a valuation of all the terms represented in D (we note here that by definition it will always satisfy the lp-condition).

An extended environment $e\delta$ is *good* if $F_{e\delta}(\mathcal{B}, \mathcal{P})$ is defined for any $\mathcal{B}, \mathcal{P}$ fitting $e\delta$, and if, for any family $\{N_i\}_{i \in |I|}$ consistent with D and satisfying the lp-condition and any $m : i \to j$ in the domain of lp with $p(lp(m)) = ((\Sigma, \Phi), (\Delta, \Psi))$, we have $N_i \models_\Sigma \Phi$ and $N_j \models_\Delta \Psi$. Thus, an environment is good if all valuations on the declared units extend properly *and* if the diagram and lp-condition themselves ensure that all applications are correct.

Let $e\delta = (D, b, p, lp)$ be good. Then by $ctx(e\delta)$ we denote the set of environments of the form $\mathcal{B} \cup \mathcal{P}$, where $\mathcal{B}$ and $\mathcal{P}$ fit $e\delta$. This set is a context fitting $st(e\delta)$. Pick some node $i \in |dom(D)|$. Then by $mf_{e\delta,i}$ we denote the model function of type $(ctx(e\delta), \Sigma)$, where $D(i) = (\Sigma, \Phi)$, which sends any environment $e \in ctx(e\delta)$ to $(F_{e\delta}(\mathcal{B}, \mathcal{P}))_i$, where $\mathcal{B} = e|_{dom(b)}$ and $\mathcal{P} = e|_{dom(p)}$.

We say that an extended environment $e\delta' = (D', b, p, lp')$ *extends* $e\delta = (D, b, p, lp)$, if D' extends D and lp' extends lp.

Let $e\delta = (D, b, p, lp)$ and let $\alpha = \{\alpha_i : D(i) \to (\Sigma, \Phi)\}_{i \in |dom(D)|}$ be a sink. For any $ID \in dom(p)$, let V_{ID} be the set of vertices of the form $dom(m)$ for m with $lp(m) = ID$. Let $P = \{P_{ID}\}_{ID \in dom(p)}$ be a family where P_{ID} is a partition of V_{ID} (we denote the set of all such families by $part(lp)$). By $\mathcal{E}_{\alpha,P}$ we denote the following set of equality statements: for any pair (m, n) with $lp(m) = lp(n)$, if $dom(m)$ and $dom(n)$ are in one partitioning set of $P_{lp(m)}$ and $cod(m) \neq cod(n)$, then the statement $\alpha_{cod(m)} = \alpha_{cod(n)}$ is in $\mathcal{E}_{\alpha,P}$; if $dom(m)$ and $dom(n)$ are in different partitioning sets, then the statement $\alpha_{dom(m)} \neq \alpha_{dom(n)}$ is in $\mathcal{E}_{\alpha,P}$.

The following lemma is the reason for introducing the above sets of equality statements.

Lemma 1. *Let $e\delta = (D, b, p, lp)$ be good, $\alpha = \{\alpha_i : D(i) \to (\Sigma, \Phi)\}_{i \in |dom(D)|}$ be an amalgamating sink for D, i and j be nodes in the scheme of D such that $D(i)$ and $D(j)$ are presentations over compatible signatures and Ψ be a set of sentences over the signature Δ of $D(i)$. Then:*

for all $e \in ctx(e\delta)$, $mf_{e\delta,i}(e) \models_\Delta \Psi \iff$

   either some parametric unit specification in p is inconsistent,

   or for all $P \in part(lp)$ we have $\Phi \cup \mathcal{E}_{\alpha,P} \models_\Sigma \alpha_i(\Psi)$

and

for all $e \in ctx(e\delta)$, $mf_{e\delta,i}(e)$ and $mf_{e\delta,j}(e)$ are amalgamable $\iff$

   either some parametric unit specification in p is inconsistent,

   or for all $P \in part(lp)$ we have $\Phi \cup \mathcal{E}_{\alpha,P} \models_\Sigma \alpha_i \oplus \alpha_j$

$\square$

Lemma 2. *Let $e\delta = (D, b, p, lp)$ be an extended environment. Let k_{ID} be the cardinality of the set $lp^{-1}(ID)$ for each identifier $ID \in dom(p)$. Then the cardinality of the set $part(lp)$ is bounded by $2^{\sum_{ID \in dom(p)} |k_{ID}^2 - 1|}$. Also, for any sink α over D and any $P \in part(lp)$, the cardinality of the set $\mathcal{E}_{\alpha,P}$ is bounded by $\frac{1}{2} \sum_{ID \in dom(p)} k_{ID}(k_{ID} - 1)$.* $\square$

We say that some parametric unit specification *appears* in the extended environment (D, b, p, lp) if it appears in p. We are now ready to state the main lemma needed for proving our theorem.

Lemma 3. *Assume that $\mathbf{I}$ has logical amalgamation. Let $e\delta$ be an extended environment such that if no unit specification in it is inconsistent, then it is good, and T a unit term. There exists an algorithm which either signals failure, or computes an extended environment $e\delta'$ extending $e\delta$, together with an object i of the scheme of the diagram of $e\delta'$, and a finite set A of intermediate proof obligations, and such that:*

T is correct $\iff$ the algorithm does not signal failure, and

   either some parametric unit specification in $e\delta$ is inconsistent,

   or all obligations in A are true

Moreover, if T is correct and no parametric unit specification in $e\delta$ is inconsistent, then $e\delta'$ is good and the denotation of T in the context $st(e\delta)$, $ctx(e\delta)$ is $\pi_2(D(i))$, $mf_{e\delta',i}$, where D is the diagram of $e\delta'$.

Proof. The proof is by induction on the structure of T. We show the algorithm only, its correctness is easy but lengthy to check. We silently assume that failure propagates as if it were an exception. Let $e\delta = (D, b, p, lp)$.

The first case is $T = $ "ID". If $ID \notin dom(b)$, signal failure. Otherwise return $e\delta' = e\delta$, $i = b(ID)$, and $A = \emptyset$.

The second case is $T = $ "$ID(T_0)$". Let $e\delta_0 = (D_0, b, p, lp_0)$, i_0, A_0 be the result of the computation on $e\delta$ and T_0. If $ID \notin dom(p)$, signal failure. Otherwise, let $p(ID) = ((\Sigma, \Phi), (\Delta, \Psi))$ and $D_0(i_0) = (\Sigma', \Theta)$. If $\Sigma \neq \Sigma'$, signal failure.

Otherwise add an edge labeled $\iota_{\Sigma,\Delta}$ with source i_0 and the target a new node i labelled by $(\Delta, \Psi \cup \iota_{\Sigma,\Delta}(\Theta))$ to D_0, obtaining D'. Let lp' be lp_0 with an additional label ID on this new edge. Let $\{\alpha_j : D_0(j) \to (\Gamma, \Omega)\}_{j \in |dom(D_0)|}$ be an amalgamating sink for D_0. Now return $e\delta' = (D', b, p, lp')$, i and $A = A_0 \cup \{\Omega, \mathcal{E}_{\alpha,P} \models_\Gamma \alpha_{i_0}(\Phi) \mid P \in part(lp_0)\}$.

The third case is $T = "T_1 \text{ and } T_2"$. Let $e\delta_j = (D_j, b, p, lp_j)$, i_j, A_j be the result of the computation on $e\delta$ and T_j, $D_j(i_j) = (\Sigma_j, \Phi_j)$ for $j = 1, 2$. If Σ_1 and Σ_2 are incompatible, signal failure; otherwise, let $\Sigma = \Sigma_1 \cup \Sigma_2$. We may assume that the intersection of the schemes of D_1 and D_2 is the scheme of D. Let D_0 be the sum of D_1 and D_2. Let D' be D_0 with an additional node i labelled by $(\Sigma, \iota_{\Sigma_1,\Sigma}(\Phi_1) \cup \iota_{\Sigma_2,\Sigma}(\Phi_2))$ and edges from i_1 and i_2 to i, labelled by inclusions. Let $\{\alpha_j : D_0(j) \to (\Delta, \Psi)\}_{j \in |dom(D_0)|}$ be an amalgamating sink for D_0. Now, let $lp' = lp_1 \cup lp_2$, and return $e\delta' = (D', b, p, lp')$, i and $A = A_1 \cup A_2 \cup \{\Psi, \mathcal{E}_{\alpha,P} \models \alpha_{i_1} \oplus \alpha_{i_2} \mid P \in part(lp')\}$.

The fourth case is $T = "\mathbf{reduce} \ T_0 \ \mathbf{by} \ \sigma"$. Let $e\delta_0 = (D_0, b, p, lp_0)$, i_0, A_0 be the result of the computation on $e\delta$ and T_0. Let $D_0(i_0) = (\Sigma, \Phi)$. If $cod(\sigma) \neq \Sigma$, signal failure, else construct D' by adding to D_0 a node i labelled by $(dom(\Sigma), \emptyset)$ and an edge from i to i_0 labelled by σ. Return $e\delta' = (D', b, p, lp_0)$, i and A.

The last case is $T = "\mathbf{let} \ ID = T_1 \ \mathbf{in} \ T_2"$. If $ID \in dom(b)$ or $ID \in dom(p)$, signal failure. Let $e\delta_1 = (D_1, b, p, lp_1)$, i_1, A_1 be the result of the computation on $e\delta$ and T_1 and $e\delta_2 = (D_2, b[i_1/ID], p, lp_2)$, i_2, A_2 its result on $(D_1, b[i_1/ID], p, lp_1)$ and T_2. Now, return $e\delta' = (D_2, b, p, lp_2)$, $i = i_2$ and $A = A_1 \cup A_2$. $\qquad \square$

Lemma 4. *For any correct sequence UDS of unit declarations, one can compute a good extended environment $e\delta$ such that the denotation of UDS is equal to $st(e\delta)$, $ctx(e\delta)$.* $\qquad \square$

Proof (Theorem 1). We will show the proof for the non-parametric case only, as the parametric case reduces to it easily. Let ASP be an architectural specification of the form "$\mathbf{units} \ UDS \ \mathbf{result} \ \bullet T \ \mathbf{implementing} \ (\Sigma, \Phi)$". Define UDS' to be UDS extended by the declaration $F : (\Sigma, \Phi) \to (\Sigma, \Phi)$. Now compute an extended environment $e\delta$ for UDS' (Lemma 4). Next compute $e\delta'$, i, A for the term "$F(T)$" and $e\delta$, as in Lemma 3. Finally, as P return the set of parametric unit specifications appearing in $e\delta$ and as S the set A. That the bounds are as stated is a corollary of Lemma 2. $\qquad \square$

6   Simplifying Intermediate Proof Obligations

In this section we show that positive equality statements in intermediate proof obligations are superfluous in an institution which has logical amalgamation. Essentially, the idea here is that any positive equality statement $\sigma = \tau$ can be substituted by a translation along a morphism which equalizes them; this idea can be generalized to work in any institution which has logical amalgamation, even if it has no coequalizers.

Let $\mathcal{E}$ be a finite set of positive Σ-statements. By $\eta_{\mathcal{E}} : (\Sigma, \emptyset) \to (\Gamma_{\mathcal{E}}, \Phi_{\mathcal{E}})$ we will denote a morphism in $Pres(\mathbf{I})$ such that the map $\cdot|_{\eta_{\mathcal{E}}}$ (with domain $\mathbf{Mod}_{\Gamma_{\mathcal{E}}}(\Phi_{\mathcal{E}})$) is in and onto $\mathbf{Mod}_{\Sigma}(\mathcal{E})$.

Proposition 4. *If the institution $\mathbf{I}$ has logical amalgamation, then the morphism $\eta_{\mathcal{E}}$ exists and can be computed for any finite set $\mathcal{E}$ of positive equality statements.*

Proof. Let $\mathcal{E} = \{\sigma_x = \tau_x\}_{x \in X}$. Take D to be a diagram with nodes v_x and a special node V, and morphisms $m_x, n_x : v_x \to V$ for any $x \in X$. The morphisms m_x are labeled with σ_x and n_x with τ_x, and the nodes are labelled accordingly. All sentence sets in the labels are empty. Now take the amalgamating sink of D and let $\eta_{\mathcal{E}} : D(V) \to (\Gamma_{\mathcal{E}}, \Phi_{\mathcal{E}})$ be the morphism in it associated with the node V. It is easy to see that this morphism induces a reduct which is in and onto $\mathbf{Mod}(\mathcal{E})$, since $\mathbf{Mod}(\mathcal{E})$ is in bijective correspondence with the set of all families consistent with D. □

For any Σ-statement γ and morphism $\eta : \Sigma \to \Gamma$ the *translation* of γ by η, denoted $\eta(\gamma)$, is defined by composing it with the elements of γ. Clearly, $M \models_{\Gamma} \eta(\gamma) \iff M|_{\eta} \models_{\Sigma} \gamma$, for any $M \in \mathbf{Mod}(\Gamma)$. This holds both for Σ-sentences and Σ-statements and we will call it the *extended satisfaction condition* (e.s.c.).

Proposition 5. *Let $\mathcal{E}$ and $\mathcal{F}$ be finite sets of positive and negative Σ-statements, respectively. Let γ be a Σ-sentence or Σ-statement and Φ a set of Σ-sentences. Then:*

$$\Phi \cup \mathcal{E} \cup \mathcal{F} \models_{\Sigma} \gamma \iff \Phi_{\mathcal{E}} \cup \eta_{\mathcal{E}}(\Phi \cup \mathcal{F}) \models_{\Gamma_{\mathcal{E}}} \eta_{\mathcal{E}}(\gamma)$$

Proof. For the implication from left to right, take any $M \models_{\Gamma_{\mathcal{E}}} \Phi_{\mathcal{E}} \cup \eta_{\mathcal{E}}(\Phi \cup \mathcal{F})$. By the e.s.c., $M|_{\eta_{\mathcal{E}}} \models_{\Sigma} \Phi \cup \mathcal{F}$ and by the definition of $\eta_{\mathcal{E}}$, $M|_{\eta_{\mathcal{E}}} \models_{\Sigma} \mathcal{E}$. Thus $M|_{\eta_{\mathcal{E}}} \models_{\Sigma} \gamma$ and, by the e.s.c., $M \models_{\Gamma_{\mathcal{E}}} \eta_{\mathcal{E}}(\gamma)$.

For the other implication, take any $M \models_{\Sigma} \Phi \cup \mathcal{E} \cup \mathcal{F}$. Then, by the definition of $\eta_{\mathcal{E}}$, there exists a model $N \models_{\Gamma_{\mathcal{E}}} \Phi_{\mathcal{E}}$ such that $M = N|_{\eta_{\mathcal{E}}}$. By the e.s.c., $N \models_{\Gamma_{\mathcal{E}}} \eta_{\mathcal{E}}(\Phi \cup \mathcal{F})$. Hence, $N \models_{\Gamma_{\mathcal{E}}} \eta_{\mathcal{E}}(\gamma)$, and, again by the e.s.c., $M \models_{\Sigma} \gamma$. □

Thus we have reduced the problem of checking intermediate proof obligations to checking obligations that do not involve any positive equality statements.

7   Transforming Simplified Intermediate Proof Obligations in *FOL*

In this section we aim at transforming the intermediate proof obligations generated in Theorem 1 and simplified in the previous section into simple obligations. We do this in the framework *FOL* of first-order logic, although we do conjecture that the approach is quite universal and would be successful in many similar logics as well. In the case of $|FOL|$ we obtain a decidability result.

Our transformations of intermediate proof obligations into simple obligations will be based on the following lemma. This lemma states that, in first-order logic with negative equalities, the only way to ensure that two different sorts have always equal carriers is to force these carriers to be empty.

Lemma 5. *Let $\mathcal{E}$ be a set of negative Σ-statements in FOL. Take a model $M \models_{\Sigma} \mathcal{E}$. There exists a model $N \models_{\Sigma} \mathcal{E}$ satisfying precisely the same Σ-sentences as M such that, for any pair of different sorts s, s' in Σ, either the carriers of both of them are empty in M or they are different in N.*

Proof. As N one can take a model where elements of the carrier of a sort s are pairs (s, x), with x an element of its carrier in M. The interpretations of function symbols are defined straightforwardly. Clearly, N is isomorphic to M, thus it satisfies the same Σ-sentences. Also, non-empty carriers are obviously different. It is clear that any negative statement true in M is true in N as well.   □

Theorem 2. *Any amalgamability obligation in FOL can be transformed into an equivalent satisfaction obligation.*

Proof. It suffices to consider an amalgamability obligation of the form $\Phi \cup \mathcal{E} \models_{\Sigma} \sigma \oplus \tau$ where $\mathcal{E}$ consists of negative statements only ($\sigma : \Delta \to \Sigma, \tau : \Gamma \to \Sigma, \Delta$ and Γ are compatible). Obviously, amalgamability in FOL means that for any sort s in the intersection of Δ and Γ, the carriers of $\sigma^S(s)$ and $\tau^S(s)$ in any model of $\Phi \cup \mathcal{E}$ are equal. The same applies for any function symbol $f : s_1 \times \cdots \times s_n \to s_{n+1}$. In the proof we formalize the observation that in the case of sorts, this can happen either if $\sigma^S(s) = \tau^S(s)$ or if both of these sorts have empty carriers; in the case of function symbols, we have two subcases. First, if $\sigma^S(s_i) \neq \tau^S(s_i)$ for some $1 \leq i \leq n+1$, then by the above observation, the carriers of both of these sorts are empty and, hence, both functions $\sigma^F(f)$ and $\tau^F(f)$ are empty and equal. If, on the other hand, the respective pairs of sorts are always equal, then we may write an axiom which states that $\sigma^F(f)$ and $\tau^F(f)$ are equal.

Now, let Ψ be the following set of Σ-sentences:

- for any sort s appearing in both Δ and Γ such that $\sigma^S(s) \neq \tau^S(s)$, the sentence ψ_s of the form "$(\forall x : \sigma^S(s) \cdot \neg(x = x)) \wedge (\forall x : \tau^S(s) \cdot \neg(x = x))$";
- for any function symbol f appearing in both Δ and Γ with profile $s_1 \times \ldots \times s_n \to s_{n+1}$ and such that $\sigma^S(s_i) = \tau^S(s_i)$ for $1 \leq i \leq n+1$, but $\sigma^F(f) \neq \tau^F(f)$, the sentence ψ_f of the form "$\forall x_1 : \sigma^S(s_1) \cdot \ldots \forall x_n : \sigma^S(s_n) \cdot \sigma^F(f)(x_1, \ldots, x_n) = \tau^F(f)(x_1, \ldots, x_n)$".

We claim that the following equivalence holds:

$$\Phi \cup \mathcal{E} \models_{\Sigma} \sigma \oplus \tau \iff \Phi \cup \mathcal{E} \models_{\Sigma} \Psi$$

It remains now to prove the above claim.

Assume the right side of the equivalence holds. Take $M \models_{\Sigma} \Phi \cup \mathcal{E}$. Then $M \models_{\Sigma} \Psi$. We will prove that $M|_{\sigma}$ and $M|_{\tau}$ are amalgamable. Toward this end we will check that common sorts and function symbols are interpreted identically in both models.

Let s be a sort appearing in both Δ and Γ. We are to prove that the interpretations of $\sigma^S(s)$ and $\tau^S(s)$ in M are identical. If $\sigma^S(s) = \tau^S(s)$, then this is trivial. Assume $\sigma^S(s) \neq \tau^S(s)$. Then $\psi_s \in \Psi$ and $M \models_{\Sigma} \psi_s$. Then the carriers of $\sigma^S(s)$ and $\tau^S(s)$ are indeed equal in M, since they are both empty.

Let f be a function symbol with profile $s_1 \times \ldots \times s_n \to s_{n+1}$ in both Δ and Γ. If for some $1 \leq i \leq n+1$ we have $\sigma^S(s_i) \neq \tau^S(s_i)$, then, as above, the carriers of sorts $\sigma^S(s_i)$ and $\tau^S(s_i)$ must be empty. Hence, the interpretations of $\sigma^F(f)$ and $\tau^F(f)$ must be empty functions and are thus equal. If, on the other hand, $\sigma^S(s_i) = \tau^S(s_i)$ for all $1 \leq i \leq n+1$, then we have $M \models_\Sigma \psi_f$, which ensures that the interpretations of $\sigma^F(f)$ and $\tau^F(f)$ are equal.

This proves the implication to the left. For the converse, assume $\Phi \cup \mathcal{E} \models_\Sigma \sigma \oplus \tau$ and take $M \models_\Sigma \Phi \cup \mathcal{E}$. We are to prove that $M \models_\Sigma \Psi$.

First take some $\psi_s \in \Psi$. According to Lemma 5, there exists a model $N \models_\Sigma \Phi \cup \mathcal{E}$ such that either the carriers of both $\sigma^S(s)$ and $\tau^S(s)$ are empty in M, or they are different in N. Since $N \models_\Sigma \Phi \cup \mathcal{E} \models_\Sigma \sigma \oplus \tau$, they cannot be different. Thus, they are empty. That is, $M \models_\Sigma \psi_s$.

Now take $\psi_f \in \Psi$. It is apparent that, since the interpretations of $\sigma^F(f)$ and $\tau^F(f)$ coincide in M, we must have $M \models_\Sigma \psi_f$. This finishes the proof. $\qquad \square$

Theorem 3. *Any satisfaction obligation in FOL can be transformed into an equivalent simple obligation.*

Proof. It is sufficient to consider an obligation of the form $\mathcal{E} \models_\Sigma \psi$, where $\mathcal{E}$ consists of negative statements only. This is because any obligation of the form $\Phi \cup \mathcal{E} \models_\Sigma \Psi$ can be transformed into $\mathcal{E} \models_\Sigma \bigwedge \Phi \to \bigwedge \Psi$.

The idea of this proof is that a negative equality obligation consisting of morphisms $\sigma, \tau : \Delta \to \Sigma$ can influence satisfaction of first-order sentences only as follows: if all sorts in Δ which are mapped to different sorts in Σ by σ and τ have empty carriers, then for some function symbol f the sorts in whose profile are mapped in the same way by σ and τ, the interpretations of $\sigma^F(f)$ and $\tau^F(f)$ are different. This intuition is formalized below.

For any statement e of the form $\sigma \neq \tau$, $\sigma, \tau : \Delta \to \Sigma$, define the following Σ-sentences:

- for any sort s in Δ such that $\sigma^S(s) \neq \tau^S(s)$, $\rho_{e,s} = $"$(\forall x : \sigma^S(s) \cdot \neg(x = x)) \wedge (\forall x : \tau^S(s) \cdot \neg(x = x))$"; the set of such sorts is denoted by S_e;
- for any function symbol f in Δ with profile $s_1 \times \ldots \times s_n \to s_{n+1}$, if for all $1 \leq i \leq n+1$ we have $\sigma^S(s_i) = \tau^S(s_i)$, then $\rho_{e,f} = $"$\exists x_1 : \sigma^S(s_1) \cdot \ldots \exists x_n : \sigma^S(s_n) \cdot \neg(\sigma^F(f)(x_1, \ldots, x_n) = \tau^F(f)(x_1, \ldots, x_n))$"; the set of such function symbols is denoted by F_e;
- $\omega_e = $"$\bigwedge_{s \in S_e} \rho_{e,s} \implies \bigvee_{f \in F_e} \rho_{e,f}$".

Let Ω be the set of all sentences ω_e for $e \in \mathcal{E}$. We claim that ψ the following equivalence holds:

$$\mathcal{E} \models_\Sigma \psi \iff \Omega \models_\Sigma \psi$$

The proof of the above claim proves the whole theorem.

To prove the implication from right to left, we will prove that $\mathcal{E} \models_\Sigma \Omega$. Assume $M \models_\Sigma \mathcal{E}$. Take any $e \in \mathcal{E}$ of the form $\sigma \neq \tau$ and $\omega_e \in \Omega$. If M satisifies the premise of ω_e then every pair of different sorts $\sigma^S(s)$, $\tau^S(s)$ has empty carriers in M. Thus, there must be some function symbol f such that $\sigma^F(f)$ and

$\tau^F(f)$ have different interpretations in M. If a sort s is in the profile of f then it must be mapped identically by σ and τ; otherwise, both of the mapped sorts would be empty, hence the interpretations of $\sigma^F(f)$ and $\tau^F(f)$ would be empty and equal. But the existence of such a symbol f means precisely that $\rho_{e,f}$ is satisfied in M. So, $M \models_\Sigma \Omega$ and then $M \models_\Sigma \psi$.

Now we prove the implication from left to right. Assume $\mathcal{E} \models_\Sigma \psi$ and take a model $M \models_\Sigma \Omega$. According to Lemma 5, there exists a model N satisfying the same Σ-sentences as M and such that for any two sorts having non-empty carriers in M, their carriers in N differ. If we had $N \models_\Sigma \mathcal{E}$, then we would also have $M \models_\Sigma \psi$, which would end the proof. So suppose $N \not\models_\Sigma \mathcal{E}$. Thus there exists a statement $e \in \mathcal{E}$ of the from $\sigma \neq \tau$ such that $N \models_\Sigma \sigma = \tau$. To get a contradiction, it suffices to prove that then $N \not\models_\Sigma w_e$. Take any $s \in S_e$. The carriers of $\sigma^S(s)$ and $\tau^S(s)$ in N must be equal, hence they are empty. Thus the premise of w_e is satisfied in N. If w_e were satisfied in N, there would exist a function symbol $f \in F_e$ such that $N \models_\Sigma \rho_{e,f}$. But this would mean that the symbols $\sigma^F(f)$ and $\tau^F(f)$ have different interpretations in N, which contradicts $N \models_\Sigma \sigma = \tau$. Thus we conclude that the initial assumption that $N \not\models_\Sigma \mathcal{E}$ was false, which finishes the proof. $\square$

It is perhaps worth pointing out that intermediate proof obligations in $|FOL|$ are decidable, making correctness in $|FOL|$ decidable. Because in $|FOL|$ we have no sentences at all, the satisfaction obligations are all trivially true. We also have:

Proposition 6. *Amalgamability obligations in $|FOL|$ are decidable.*

Proof. We may assume the amalgamability obligation to be of the form $\mathcal{E} \models_\Sigma \sigma \oplus \tau$, where $\mathcal{E}$ consists of negative statements only. This obligation is true if and only if:

- *either* there exists a statement of the form $\rho \neq \rho$ in $\mathcal{E}$,
- *or* the morphisms σ and τ coincide on the intersection of their domains.

It is obvious that the above conditions are sufficient for the obligation to be true. On the other hand, if σ and τ do not coincide on the intersection of their domains, then $\mathcal{E}$ must be inconsistent. This means that some statement of the form $\rho \neq \rho$ is an element of $\mathcal{E}$. $\square$

8   Static and Strict Correctness

In Section 3 we listed the four conditions that have to be checked during the verification of an architectural specification. It is easy to see that in the case of FOL checking each of these conditions is undecidable. This is not at all unexpected in the case of conditions 1 (does the result unit satisfy the **implementing** specification?), and 2 (does a unit to which a parametric unit is applied satisfy the specification which defines the domain of this parametric unit?). However,

conditions 3 and 4 have to do with amalgamability and do not involve any specifications or axioms. So, one would want to have these conditions checked automatically, during static analysis, as in the intuitively analogous case of checking
sharing conditions in the module system of the SML programming language.

Toward this end, define, for any architectural specifiation ASP over $\mathbf{I}$, the
architectural specification $|ASP|$ over $|\mathbf{I}|$, obtained by removing all sentences
from ASP; thus, $|ASP|$ is the "static" version of ASP — meaning that it involves
signatures and signature morphisms only. Let us say that ASP is *statically correct*
if $|ASP|$ is correct in the usual sense (as a matter of fact, generativity problems
aside, an architectural specification is statically correct if and only if it has a
denotation with respect to the *extended static semantics* of [SMTKH01]).

Of course, checking static correctness is no harder than checking regular
correctness. In particular, checking static correctness in FOL is decidable, thanks
to Prop. 6. Similar results are true for many institutions of interest. Moreover,
the following holds:

Proposition 7. *If ASP is statically correct, then in the elaboration of ASP
conditions 3 and 4 will always be satisfied.*   □

> units
> $A : \{$**sort** $s;$ **op** $a : s; \}$
> $F : \{$**sort** $s;$ **op** $a : s; \} \to \{$**sort** $s;$ **ops** $a, b : s;$ **axiom** $a = b; \}$
> $G : \{$**sort** $s;$ **op** $a : s; \} \to \{$**sort** $s;$ **ops** $a, b : s;$ **axiom** $a = b; \}$
> result
> $\bullet\ F(A)$ **and** $G(A)$
> implementing
> $\{$**sort** $s;$ **ops** $a, b : s; \}$

Fig. 3. A correct architectural specification which is not statically correct.

Call an architectural specification *strictly correct* if it is both correct and statically correct. We propose to require architectural specifications to be strictly
correct. Then verifying a specification would consist of first checking (hopefully
automatically) its static correctness, and then discharging only the *satisfaction*
proof obligations generated by the algorithm presented in this paper. This procedure would be sound, since static correctness ensures that all the amalgamability
obligations generated will be satisfied. However, it is not complete, in the sense
that not all correct architectural specifications are statically correct. In Fig. 3
we have an example of an architectural specification which is correct, but is not
statically correct (and hence not strictly correct). In the following discussion we
use the notation $T.x$ for the component x of the model denoted by the term T
in a given environment. In our example, we know that $A.s = F(A).s = G(A).s$
and $A.a = F(A).a = G(A).a$, and the axioms enforce $F(A).b = F(A).a$ and
$G(A).b = G(A).a$. This causes the amalgamation $F(A)$ **and** $G(A)$ to be correct,
since $F(A).a = G(A).a$ and $F(A).b = G(A).b$. If we remove the axioms, then
$F(A).b$ and $G(A).b$ may be unrelated and hence the models $F(A)$ and $G(A)$

will not be amalgamable. In fact, we consider this incompleteness valuable, since we feel that architectural specifications like the one in Fig. 3 are really flawed anyway.

As has already been said, the problem of checking static correctness is easily decidable for FOL and many other institutions. Unfortunately, the institution of CASL is an exception. The logic of CASL involves a form of subsorting where the inclusion of a subsort element into a supersort is performed by means of an (almost) arbitrary injective map between the respective carriers. These injective maps are required to satisfy only one simple compatability axiom: the injection of s into s' composed with the injection of s' into s'' must be equal to the injection of s into s''. These compatibility conditions are the reason of two related phenomena: lack of logical amalgamation (cf. [SMTKH01][1]), and undecidability of static correctness (see [KHTSM01], where this is proved and algorithms for checking static correctness in practically relevant cases are developed).

9   Conclusion and Future Work

In this paper we have shown — for a very general case — how to reduce the problem of verifying an architectural specification to the problem of checking some intermediate proof obligations. We have also shown — for the case of first-order logic — how to transform these intermediate proof obligations into simple obligations.

However, some work still needs to be done, if we aim at verifying architectural specifications in CASL. First, we need to generalize the above results to partial logic with sort generation constraints. Next, we need to check that the whole construction works for structured specifications as well. These steps should be fairly straightforward. Some problems even turn out simpler in CASL, since non-empty carriers are not admitted in its model theory.

The problems that arise with the introduction of subsorts require more work. The main obstacle here is the lack of logical amalgamation, which causes verification for CASL *without* any sentences (i.e., static correctness) to be undecidable already. What is needed here is an embedding of the CASL logic into a richer logic (*enriched* CASL) which does have the logical amalgamation property. This embedding will probably go along the lines of [SMT01]. A drawback of using such an embedding would be that as a result one would not be able to reason about CASL architectural specifications in pure CASL.

All this would allow us to generate obligations in enriched CASL for the sublanguage of architectural specifications presented in this paper and for the full CASL logic. In order to support full CASL architectural specifications, the additional architectural constructs, such as translation or fitting-morphisms in applications, will have to be taken care of.

---

[1] Actually, only lack of amalgamation has been discussed there, but the same argument can be used to show lack of logical amalgamation.

Acknowledgments

I would like to thank Andrzej Tarlecki for introducing me to the subject and for collaboration on architectural specifications in general. I am also grateful to Bartek Klin, as well as Till Mossakowski and Lutz Schröder, for many discussions and joint work on related topics. This research has been partially supported by Polish State Committee for Research (KBN) grant no. 7 T11C 002 21 and by ESPRIT working group 29432 (CoFI WG).

References

[BCH99] M. Bidoit, M. V. Cengarle, R. Hennicker: Proof Systems for Structured Specifications and Their Refinements. In E. Astesiano, H.-J. Kreowski, B. Krieg-Brückner (eds.): Algebraic Foundations of Systems Specification, pp. 385–434. Springer, 1999.

[Borz98] T. Borzyszkowski: Completeness of a logical system for structured specifications. In F. Parisi Presicce (ed.): Recent Trends in Algebraic Development Techniques, Selected Papers, 12th Intl. Workshop, Proc. WADT'97, LNCS 1376, pp. 107–121. Springer, 1998.

[BST99] M. Bidoit, D. Sannella, A. Tarlecki. Architectural Specifications in Casl. In: Proc. 7th Intl. Conference on Algebraic Methodology and Software Technology (AMAST'98), LNCS 1548, pp. 341–357. Springer, 1999; final version to appear in Formal Aspects of Computing.

[CASL] E. Astesiano, M. Bidoit, H. Kirchner, B. Krieg-Brückner, P. Mosses, D. Sannella, A. Tarlecki. Casl: The Common Algebraic Specification Language. Theoretical Computer Science (to appear).

[CASL99] CoFI: Casl — Summary, version 1.0. In: Documents/CASL/Summary, [CoFI]. 1999.

[CASL00] CoFI: Casl — Semantics, version 1.0. In: Documents/CASL/Semantics, note S-9, [CoFI]. 2000.

[CoFI] CoFI, The Common Framework Initiative for Algebraic Specification and Development. Documents accessible at http://www.brics.dk/Projects/CoFI and ftp://ftp.brics.dk/Projects/CoFI.

[GB84] J. Goguen, R. Burstall: Some fundamental algebraic tools for the semantics of computation. Part 1: Comma categories, colimits, signatures, and theories. Theoretical Computer Science 31(2), pp. 175–209 (1984)

[GB92] J. Goguen, R. Burstall: Institutions: abstract model theory for specification and programming. Journal of the ACM 39, pp. 95–146 (1992).

[KHTSM01] B. Klin, P. Hoffman, A. Tarlecki, L. Schröder, T. Mossakowski: Checking Amalgamability Conditions for Casl Architectural Specifications. In J. Sgall, A. Pultr, P. Kolman (eds.): Proc. 26th Intl. Symp. on Mathematical Foundations of Computer Science (MFCS'01), LNCS 2136, pp. 451–463. Springer, 2001.

[Pad99] P. Padawitz: Proof in Flat Specifications. In E. Astesiano, H.-J. Kreowski, B. Krieg-Brückner (eds.): Algebraic Foundations of Systems Specification, pp. 321–384. Springer, 1999.

[Pau96] L. Paulson: ML for the Working Programmer. Cambridge University Press, 1991.

[SMT01] L. Schröder, T. Mossakowski, A. Tarlecki: Amalgamation in Casl via Enriched Signatures. In F. Orejas, P. Spirakis, J. van Leeuwen (eds.): Proc. 28th Intl. Coll. on Automata, Languages and Programming (ICALP'01), LNCS 2076, pp. 993–1004. Springer, 2001.

[SMTKH01] L. Schröder, T. Mossakowski, A. Tarlecki, B. Klin, P. Hoffman: Semantics of Architectural Specifications in CASL. In H. Hussmann (ed.): Proc. 4th Intl. Conf. on Fundamental Approaches to Software Engineering (FASE'01), LNCS 2029, pp. 253–268. Springer, 2001.

[ST88] D. Sannella, A. Tarlecki: Specifications in an Arbitrary Institution. Information and Computation vol. 76, 2/3, pp. 165–210 (1988).

[ST97] D. Sannella, A. Tarlecki: Essential Concepts of Algebraic Specification and Program Development. Formal Aspects of Computing 9, pp. 229–269 (1997).

[Tar99] A. Tarlecki: Institutions: An Abstract Framework for Formal Specifications. In E. Astesiano, H.-J. Kreowski, B. Krieg-Brückner (eds.): Algebraic Foundations of Systems Specification, pp. 105–130. Springer, 1999.

Geometric Modelling with CASL*

Franck Ledoux[1], Agnès Arnould[2], Pascale Le Gall[1], and Yves Bertrand[2]

[1] LaMI, UMR 8042, Evry 2, 523 Place des Terrasses,
91000 ÉVRY , France
{fledoux,legall}@lami.univ-evry.fr
[2] IRCOM-SIC, UMR 6615, Université de Poitiers, SP2MI,
86962 Futuroscope Cedex, France.
{arnould,bertrand}@sic.sp2mi.univ-poitiers.fr

Abstract. This paper presents an experiment that demonstrates the feasibility of successfully applying CASL to design 3D geometric modelling software. It presents an abstract specification of a 3D geometric model, its basic constructive primitives together with the definition of the rounding high-level operation. A novel methodology for abstractly specifying geometric operations is also highlighted. It allows one to faithfully specify the requirements of this specific area and reveals new mathematical definitions of geometric operations. The key point is to introduce an inclusion notion between geometric objects, in such a way that the result of an operation is defined as the smallest or largest object satisfying some pertinent criteria. This work has been made easier by using different useful CASL features, like first-order logic, free types or structured specifications. Some assets of this specification are to be abstract, readable by researchers in geometric modelling and to simplify the programming process.

Keywords. formal specifications, algebraic specifications, CASL language, geometric modelling, specification methodology.

1 Introduction

3-dimensional geometric modelling software, or modellers, allow one to display and interactively handle complex 3-dimensional geometric objects made up of several thousands of faces. They are at the centre of domains such as Computer-Aided Design (CAD) for manufactured objects, geophysics when the considered objects are geological layers, or botany for modelling the volumic structure of plants. For some years, they have become large-sized and sophisticated software. They should be efficient in calculating time and memory capacity for manipulating huge data structures. They also manipulate subtle algorithms, advanced visualisation features, many interface facilities and several libraries of predefined

---

* This project is partially supported by a French national project (Plan Pluri Formation 1998-2001) between the universities of Poitiers, Strasbourg and Évry, and by the ESPRIT Working Group 29432 (CoFI WG).

M. Cerioli and G. Reggio (Eds.): WADT/CoFI 2001, LNCS 2267, pp. 176–201, 2002.

objects. Moreover, a deep mathematical understanding of specific and/or complex structures and operations is needed for this kind of software. These facts lead to use formal methods in computer graphics which is a non safety but competitive area [BD94,FBBD96,Duf97]. A kernel of a geometric modeller has successfully been developed using algebraic specifications [BD94]. The resulting modeller was more efficient and reliable than most existing ones. It benefited of readable concise code together with clear data structures which made easier its development and maintenance. Because most existing algebraic languages lacked of expressiveness at that time, the authors of [BD94] built an ad-hoc formalism gathering the necessary features (subsorts, partiality, iterators, ...) with the result of having no semantics and no tools available. This was a major drawback of their approach. Actually, they had just written low-level systematic descriptions within a specification style. The [BD94] experiment being conclusive from a software engineering point of view, our aim is to get on this experiment by providing a complete framework of formal design dedicated to geometric modelling. To correct the drawback given above, we want to get an easy formal framework for geometric modelling including a language, a methodology and tools. To carry on the algebraic approach, we have decided to use CASL, the emergent common algebraic specification language recently introduced by CoFI [oLD00b,oLD00a]. CASL is a relevant algebraic language to solve the problems of the first experiment: it offers a clear and readable syntax, most of the necessary features with well-defined semantics and tools as a parser. A prover taken into account structured specifications should be available soon.

Fig. 1. A car body obtained by rounding a rough object.

To validate this choice, we not only specify a kernel provided with classical basic operations, but we also focus on a particular complex operation known in CAD software as the rounding operation. This operation is one of the most useful (but complex) operations in a geometric modeller. It is also representative of algorithmic difficulties of geometric modelling area. From a technical point of view, it simply consists in replacing a vertex or a sharp edge (or a collection of vertices or sharp edges) by a flat or curved surface (or a collection of flat or curved surfaces), while preserving some constraints (for instance, continuity constraints). It is very frequently used to easily construct manufactured objects like TV sets or telephones from a few elementary volumes, where most edges are rounded. This operation quickly provides a realistic shape to geometric objects. For example, let us be keen on building a realistic car body (see Fig. 1). In practice, we never build it straight but we rather build a rough shape just made up of a few plane faces. Then all its vertices and edges are rounded to get a

nice car body with no more plane faces. Obviously to get such a result, you deal
with many well-chosen parameters indicating how to round an object. Due to the
intrinsic complexity of the rounding operation, the result given by most modellers
is often not the expected one in practice. Moreover, it may be surprising but the
rounding operation is actually mismanaged by most "commercial" modellers
[VMV94]. There are two main reasons for this:

- Even though smooth shape is the trademark of CAD systems, *most modellers don't achieve a satisfactory smooth shape*. It may be explained by the
  fact that the underlying geometric model is not adapted. Often modellers
  only handle the geometry of 3-dimensional objects, i.e. the curves and the
  surfaces which make up the shape of the objects, and don't consider the
  structure of the objects. With such representations, many configurations
  must be considered for geometric operations, particularly for the rounding
  one, and you have to deal with important numerical approximations. For
  example, in the rounding case there are many continuity problems between
  adjacent surfaces.
- *Usual algorithms are implemented case by case*. Therefore, as soon as a new
  particular case appears, algorithms become obsolete. The rounding operation, then, is not applicable to all object configurations.

To solve these problems, we have investigated a general and systematic approach.
From the geometric point of view, a well-founded mathematical model based
on topology has been chosen. It emphasises the structure of the objects, i.e.
their *topology*, and not only their shape, i.e. their *geometry*. For example, we
emphasis that a 3-dimensional object is made up of volumes, faces, edges and
vertices. Being aware of the topology of an object allows one to reason about its
general structure and not about several particular details. Such a representation
has already been proved useful in geometric modelling: this is the best way to
get general definitions and efficient associated algorithms. From the software
design side, we want to get an abstract general specification of the rounding
operation to avoid the listing of useless particular cases. For that, we have led a
reverse-engineering activity by using formal methods.

After reviewing the CASL bases in Section 2, we begin by specifying a geometric model, the n-G-map one [Lie94], which allows one to mathematically
represent geometric objects in Section 3. Then in Section 4 we specify a basic
operation, the sewing one, which corresponds to the well-known operation of
sticking between two objects in the n-G-map model. By introducing this specification, we explain our methodology to abstractly specify geometric operations.
This operation is easiest than the rounding one and suffices to fully present our
specification methodology. Afterward in Section 5, we focus on the rounding
operation and more precisely on how we apply our methodology for specifying
rounding operation. The modelling issues and the CASL contributions are finally
discussed in Section 6.

2   CASL

The Common Algebraic Specification Language, or CASL, is a property-oriented language for specifying the requirements and the design of software systems. In other words, CASL is based on algebraic and logic techniques, and a set of axioms specify the intended properties of systems. CASL has been recently introduced by CoFI, the Common Framework Initiative for algebraic specifications and development, to get and promote a common agreement on some basic concepts. That's why CASL gathers many features found in the different previous algebraic approaches. For example, it allows the declaration of both partial and total functions and provides some constructs for subsorting and for expressing sort generation constraints. CASL is structured as follows: basic specifications, structured specifications and architectural specifications. It also offers libraries of specifications. Basic specifications consist of definitions of sorts, types, predicates or operations and axioms, that is, all the classical algebraic concepts. Structured specifications are formed in various familiar ways (union, extension, translation, reduction, etc...) starting from basic specifications. This possibility allows one to greatly improve the readability and the reusability of specifications. Finally, architectural specifications reflect the software architecture and libraries allow one to store and retrieve specifications. In this paper, as we only consider an abstract level, we rely on basic and structured specifications, together with some libraries described in [RMS00]. We will see some CASL constructs in the next Sections, but to know more about CASL, its syntax and its semantics, you can look at [oLD00b,Mos99].

3   The n-Dimensional Generalized Map Specification

Most of new modellers are based on an underlying mathematical model which allows one to represent 3-dimensional objects and geometric operations. In this paper, we are interested in topology-based models [Lie91] which distinguish the *topology*, i.e. the structure of an object from its *embedding*, i.e. its shape. Computations of topologic properties are simpler and algorithms are more efficient in this way. To represent the topology of an object, we use a very representative mathematical model, the one of n-dimensional generalized maps (or n-G-maps)[Lie94,Man83], which allows to represent n-dimensional manifolds and non-manifolds, opened or not, and orientable or not (see 3.2 for more details).

3.1   The n-G-Map Data Type

The main idea of the n-G-map model is that a complex n-dimensional object is made up of a finite set of basic elements, named *darts* and different links between them. In fact, a n-G-map represents an n-dimensional object by considering its structure. Let us consider the example of a 2-dimensional house (see Fig. 2). Fig. 2(a) shows the initial object. To get basic elements, we begin by breaking it down into two parts, a square and a triangle (see Fig. 2(b)). We have then two distinct 2D objects that we can break down into 1D objects to obtain the result of Fig. 2(c). These 1D objects are called *edges*. Once more, each edge can be broken down into 0D objects called *darts* corresponding to the 2 extremity vertices of an edge (see Fig. 2(d)). To simplify the different figures, a dart is represented by an half-edge and not just by a point.

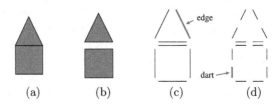

(a)   (b)   (c)   (d)

Fig. 2. A 2-dimensional house is broken down into a finite set of darts.

Now, we have totally broken down the initial object into a finite set of darts. As our aim is to specify n-G-maps, we begin by declaring a sort Dart which provides anonymous darts without any particular characteristic (see Spec. 1). Let us remind that a n-G-map represents the topology of a geometric. So a dart carries no geometric information as coordinates.

Remark. Each piece of specification presented in this paper is extracted from specification libraries [LA01] that we have developed. These libraries consist of specifications about topology and pure geometry. Among the 2000 lines of specifications, around 900 lines are devoted to topology. These libraries have been successfully parsed and statically checked with CATS version 0.73 [MRSS01a]. However this version does not allow to get a pretty syntax in LaTeX currently. So to make easier the reading, the specifications presented in this paper have been manually written and look as they would be with the pretty-printing feature of CASL. Note that as the pretty-printing feature was managed by previous versions of CATS, the LaTeX version given here is close to the LaTeX outputs given by the previous versions of CATS.

---

spec DART =
   sort *Dart*   .
end

Spec. 1. Specification of the darts in CASL.

But the initial object is not totally represented by a finite set of darts. To get it, we introduce some relations between the darts by getting up dimension after dimension. First, we begin by rebuilding the 1D objects, namely the edges, then the faces, ... Fig. 3 shows this process. In Fig. 3(a), the geometric object

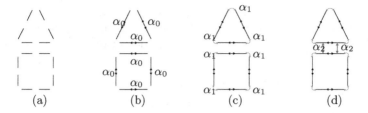

Fig. 3. How to retrieve a 2-dimensional house from a finite set of darts.

is just made up of a finite set of darts. In Fig. 3(b), we retrieve the edges of the initial object by introducing a function on the darts that we call α_0 because it links 0-dimensional objects. To specify what α_0 is, we have introduced the type $SFiniteMap$ which is a particular case of the type $FiniteMap$ provided in the CASL libraries [RMS00]. The type $SFiniteMap$ provides finite maps whose source and target sorts are the same. With this restriction, we easily specify operations as map composition [LA01]. Therefore the type of α_0 is $SFiniteMap[Dart]$. But α_0 represents a physical link between two darts: if a dart d is linked to a dart d', we have also d' linked to d. Mathematically a function verifying this property is called an involution[1]. We specify it by a predicate called *involution* (see Spec. 2).

---

spec SFiniteMap [**sort** S] =

. . .

    pred *involution* : $SFiniteMap[S] \times FinSet[S]$
    vars f : $SFiniteMap[S]$; X : $FinSet[S]$
  • $involution(f, X) \Leftrightarrow (f \circ f)|_X = id(X)$

. . .

end

Spec. 2. Specification of involution in CASL.

Note that the predicate *involution* is defined by using the operations $\_ \circ \_$, $\_|\_$ and id which are all provided in the specification SFiniteMap [LA01]. They are specified according to the classical meaning associated to these symbols. So α_0 is an involution and an axiom specifying this property will appear in our specification (see Spec. 4).

The edges may now be rebuilt. To rebuild the faces of the house, we introduce a new link between edges, named α_1 (see Fig. 3(c)). Like α_0, the function α_1 is an involution. Finally, we finish to rebuild the whole object by introducing the α_2 involution (see Fig. 3(d)) which intuitively represents a link between faces.

---

[1] An application $\alpha_i : D \to D$ is said to be an involution iff $\forall d \in D$ $(\alpha_i \circ \alpha_i)(d) = d$ with $(f \circ g)(d) = f(g(d))$.

Note that, we have just introduced two links α_2 in Fig. 3(d). That does not mean that only four darts are in the domain of α_2. As α_0 and α_1, the range of α_2 is the whole set of darts of the n-G-map. But we do not represent fix points[2] of α_2 in Figure 3(d).

Fig. 4. Why is $\alpha_0 \circ \alpha_2$ an involution ?

At this stage, the initial object is fully rebuilt. However, a condition is missed. We have seen that α_0, α_1 and α_2 are involutions. But, with these conditions only, the object of Fig. 4 can also be made up. As such an object is unauthorised in the n-G-map model, an other property is introduced:

$$(\alpha_0 \circ \alpha_2)^2 = id$$

This property ensures that the geometric object is properly built. In other words, it ensures that two n-dimensional objects are always stuck along isomorphic $(n-1)$-dimensional objects. For instance, the house is made up of two 2-dimensional objects, a square and a triangle stuck along an edge. They are stuck by α_2 and $\alpha_0 \circ \alpha_2$ is really an involution. Fig. 5 focuses on the edge shared by the square and the triangle. The dart d is linked to d_1 by α_0, d_1 is linked to d_2 by α_2, then d is linked to d_2 by $\alpha_2 \circ \alpha_0$ and finally d is linked to itself by $(\alpha_2 \circ \alpha_0)^2$.

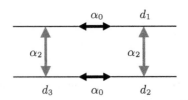

Fig. 5. Why is $\alpha_0\alpha_2$ an involution ?

We have considered a 2-dimensional object made up with a 2-G-map. Generally speaking, a n-dimensional object is made up with a n-G-map. The involved functions, then, are $\alpha_0, \alpha_1, \ldots, \alpha_n$. As we want to specify n-G-maps whichever number n we choose, we must introduce a generic operation α providing all the functions α_i, $0 \leq i \leq n$. This operation α is introduced in Spec. 4. To get the

---

[2] A dart d is a fix point of α_2 if and only if $\alpha_2(d) = d$.

---

spec DIMENSION [**op** n : Nat] **given** NAT =
  sort $Dim[n] = \{i : Nat \bullet i \leq n\}$
end

Spec. 3. Specification of natural numbers less or equal to n in CASL.

---

natural numbers from 0 to n, a type $Dim[n]$ is specified as a subsort of Nat corresponding to the natural numbers less or equal to n (see Spec. 3). Thus we exactly have the right number of links α_i. To sum up, we have the following definition:

Definition 1. A n-G-map $G = (D, \alpha_0, \ldots, \alpha_n)$ is a combinatorial structure such that

- D is a finite set of elements, named *darts*,
- $\alpha_0, \ldots, \alpha_n$ are functions on D such that
  - α_i, $0 \leq i \leq n$, is an involution,
  - $\alpha_i \circ \alpha_j$, $(0 \leq i < i+1 < j \leq n)$, is an involution.

In CASL, we specify n-G-maps as shown in specification 4. This specification asks for some explanations:

- As a n-G-map consists of a finite set of darts D and $n+1$ functions $\alpha_0, \ldots, \alpha_n$, we define a type $Gmap$ with two selectors, D and α. The selector D provides the finite set of darts of the n-G-map and α is a generic operation which provides all the necessary functions α_i. A constructor $gmap$ providing a n-G-map from a finite set of darts and $n + 1$ functions is also declared. Since the finite set of darts and the functions must verify some conditions, $gmap$ is a partial function (keyword "?").
- These conditions are introduced into the definedness axiom of the constructor $gmap$. It specifies that to be well-founded, a n-G-map verifies that the α_i functions have their domain and range included in D and α_i $(0 \leq i \leq n)$, $\alpha_i \circ \alpha_j$ $(0 \leq i < i+1 < j \leq n)$ are involutions.
- The type $Gmap$ is parametrised by a natural number n indicating its dimension. This parameter is also used to define $Dim[n]$.
- Finally, we declare the specification as being free in order to only consider its initial models. In this way, each element of the carrier set corresponding to $Gmap$ is exactly what we are waiting for.
- At first sight, it is not obvious that this specification is consistent (i.e., has at least a non trivial model). Nevertheless, the specification is an accurate translation of the mathematical definition 1 written by researchers in geometric modelling. Using a free constructor excludes the models not directly built from the $gmap$ constructor. In particular, all the models verifying additional properties w.r.t. the involution conditions for the α_i and $\alpha_i \circ \alpha_j$ functions, are excluded. Graphics designers have a strong confidence in the

definition 1. As Spec. 4 is a straight formalization of the definition 1, we can reasonably think our specification is consistent even if we have not yet proven it. Later on, we will lead a formal stepwise refinement towards a concrete specification (or final code) using the forthcoming prover[3]. The success of such an activity will then ensure the consistency of the original specification.

---

spec GMAP [**op** n : *Nat*] **given** NAT =
   ...
then
  free
  {   **type** *Gmap*[n] ::=
          $gmap(\_\_.D : FinSet[Dart]; \_\_.\alpha : FiniteMap[Dim[n], SFiniteMap[Dart]])$?
     var G : *Gmap*[n]
     • $def\,gmap(G.D, G.\alpha) \Leftrightarrow$
        $\forall i, j : Dim[n]$ • $(G.\alpha(i) :: G.D \rightarrow G.D \ \wedge\ involution(G.\alpha(i), G.D) \ \wedge$
               $(j > i + 1 \Rightarrow involution(G.\alpha(i) \circ G.\alpha(j), G.D)))$
  }
end

Spec. 4. Specification of the n-G-map model in CASL.

Remarks. First, the previous experiment [BD94] specifies n-G-map by introducing a data type GMAP through a very complex hierarchy of sorts, each intermediate sort adding a new constraint. So our approach is really more concise and readable. And these two experiments can not be compared. Second, we have chosen a mixfix notation for the two selectors in the style of object oriented languages. A n-G-map is a combinatorial structure whose basic element is the dart. But it is natural to want to retrieve the edges, the faces or volumes which make up the handled object. Therefore, the concept of *orbit* is introduced and we will see that it generalises the expected notions. Let us consider Fig. 6(a). The edge owning the dart d is made up of four darts: the two darts of the edge of the cube and the two darts of the edge of the triangle. These four darts can be reached from d by composition of the involutions α_0 and α_2. They are d, $\alpha_0(d)$, $\alpha_2(d)$ and $(\alpha_0 \circ \alpha_2)(d)$ (see Fig. 6(c)). We define the notion of orbit as follows:

Definition 2. Let $G = (D, \alpha_0, ..., \alpha_n)$ be a n-G-map, $d \in D$ a dart and $P = \{p_1, ..., p_m\}$ a set of permutations on D. The **orbit** of d compared to P is the unit $\{d' \in D \mid \exists\, p'_1 \in P, ..., \exists p'_k \in P, \ with \ d' = (p'_1 \circ \cdots \circ p'_k)(d)\}$. ($p'_i$ may be equal to p'_j with $i \neq j$). And we note it

$$< p_1, ..., p_m > (d).$$

---

[3] There is currently a first prover capable of dealing with small specifications[MRSS01b], but this prover still remains not sufficient to handle with specifications with a strong structuring.

In Fig. 6(b) all the darts of the vertex owning d are in bold features. They are reachable from d by composing α_1 and α_2 (orbit $< \alpha_1, \alpha_2 > (d)$). In Fig. 6(c), the darts of an edge are represented in bold features and they are reachable from d by composition of α_0 and α_2 (orbit $< \alpha_0, \alpha_2 > (d)$). In Fig. 6(d), the darts of a face are represented in bold features and they are reachable from d by composition of α_0 and α_1 (orbit $< \alpha_0, \alpha_1 > (d)$). And Fig. 6(e) shows all the darts reachable from d by composition of α_0, α_1 and α_2 (orbit $< \alpha_0, \alpha_1, \alpha_2 > (d)$) . The specification of the orbit is partially[4] given in Spec. 5.

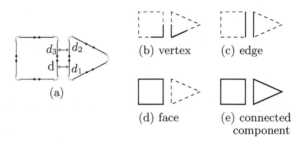

(a)

(b) vertex   (c) edge

(d) face   (e) connected
                   component

Fig. 6. A dart d and its cells: vertex (b), edge (c), face (d) and connected component (e).

3.2   Objects Represented with the n-G-Map Model

A first relevant question about n-G-map is: to what kind of modelling does the n-G-map model belong ? Two classical main approaches can be distinguish: polygonal modelling and parametric surface modelling. A model is generally owned by one of them. By separating topology and shape, the n-G-map model belongs to both polygonal modelling and parametric surface modelling. Everything depends on how the topology is embedded within. In other words, in a topological model, a shape is associated with each topological cell. A 0-cell, namely a vertex, is associated with a point; a 1-cell, namely an edge, is associated with a curve; a 2-cell, namely a face, is associated with a surface... If we work with parametric surfaces, the model belongs to parametric surface modelling. But if only 0-cells and 1-cells are respectively associated with points and straight lines, the model belongs to polygonal modelling.

This richness is offered by any topologic model. We have chosen the n-G-map model because it can represent most of the geometric objects. The n-dimensional manifolds and the non-manifolds can be represented. They can be opened or not, and orientable or not. For example, a 2-dimensional non-manifold is built with a 3-G-map in Fig. 7. Generally, a n-dimensional manifold is built with a n-G-map and a n-dimensional non-manifold is built with a $(n+1)$-G-map. A non-orientable object, as the Moebius ribbon, can also be represented with a 2-G-map.

---

[4] In fact this specification is written according to the methodology described in the sequel. In particular, it involves a notion of inclusion between n-G-maps enabling us to easily specify topological operations. It is partially given because we do not respect the pattern introduced in the sequel.

spec CELL [**op** n : Nat] **given** NAT =

...

   ops __.$darts\_of\_orbit$: $Gmap[n] \times Dart \times FinSet[Dim[n]] \rightarrow FinSet[Dart]$;
      __.$orbit$: $Gmap[n] \times Dart \times FinSet[Dim[n]] \rightarrow Gmap[n]$

   vars G, G' : $Gmap[n]$; d, d' : $Dart$; i : $Dim[n]$; I : $FinSet[Dim[n]]$

    • $d \in G.darts\_of\_orbit(d, I)$
    • $d' \in G.darts\_of\_orbit(d, I) \Leftrightarrow$
       $(\exists\, j : Dim[n] \bullet j \in I \wedge G.\alpha(j)(d') \in G.darts\_of\_orbit(d, I))$
    • $G.orbit(d, I) \subseteq G$
    • $G.orbit(d, I).D = G.darts\_of\_orbit(d, I)$
    • $\forall\, G' : Gmap[n] \bullet$
       $G'.D = G.darts\_of\_orbit(d, I) \wedge G' \subseteq G \Rightarrow G' \subseteq G.orbit(d, I)$

   ...

end

Spec. 5. Specification of the orbit in CASL.

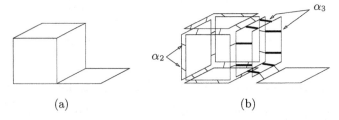

(a)   (b)

Fig. 7. A 2-dimensional non-manifold (a) represented thanks to a 3-G-map (b).

4   Algebraic Specification of Geometric Operations

Writing of a requirement specification is not an easy task. To simplify it, we introduce a methodology allowing one to systematically and abstractly specify geometric operations with the n-G-map model. To present this methodology, let us focus on the basic operation of sewing, which is the basic operation used to stick two objects in the n-G-map model[5]. Once our methodology will be presented, we shall apply it to other basic operations. Before going any further, note that all the basic operations are gathered in the same named specification, BASICOPERATIONS [LA01].

4.1   The Sewing Operation

The aim of this part is to introduce the abstract specification of the sewing operation in CASL. First, the intuitive description of the sewing operation is presented. Then the classical mathematical description of the sewing operation

---

[5] From a geometric modelling point of view, it corresponds to the classical operation of edge identification in boundary representation modelling.

is introduced. Finally, the CASL specification is introduced. And we discuss of our approach.

An Intuitive Description. Let us consider two cubes that we want to stick up (see Fig. 8). Intuitively, we would point at two faces (the two grey faces in Fig. 8(a)) to stick them up. But since we work with the n-G-map model, we only have to choose a couple of darts to link together and give the dimension by which they will be linked (see Fig. 8(b)). In this example, the dimension is 3 as we want to stick 3-dimensional objects. Directly pointing at darts and not faces is not a constraint but more an asset simplifying the user job. In Fig. 8 we have chosen two darts face to face, hence the result is the expected one: all the darts of the two faces we wanted to stick together are now linked by α_3. But if the two darts had not been face to face, the object would have been distorted (and/or moved) to link the two darts together.

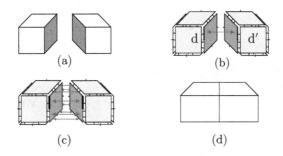

Fig. 8. Two cubes get one object by using sewing operation.

Now, let us explain why two darts and one link are sufficient to characterize the sewing operation. Let us focus on the two faces we want to stick up (see Fig. 9). Fig. 9(a) zooms in the two initial faces with the darts d and d', and the α_3 link between d and d'. By definition of the n-G-map model, $\alpha_0 \circ \alpha_3$ must be an involution. To respect the sewing constraint ($\alpha_3(d) = d'$) and the n-G-map one ($\alpha_0 \circ \alpha_3$ is an involution), the two darts $\alpha_0(d)$ and $\alpha_0(d')$ must be linked by α_3 (see Fig. 9(b)). Likewise, as $\alpha_1 \circ \alpha_3$ is an involution, the two darts $\alpha_1(d)$ and $\alpha_1(d')$ are linked by α_3 too (see Fig. 9(c)). And so on, all the darts of the two faces are linked by α_3 (see Fig. 9(d)). Note that the two initial faces to be stuck are isomorphic. Otherwise, it would be impossible to get a well-behave object.

The Mathematical Definition. When a geometric operation is defined, graphics designers constructively define the resulting object either by a mathematical definition or an algorithm. With the n-G-map model, they enumerate all the darts and links of the new object. Actually, they give a definition closer to code than to the intuitive description. For instance, the mathematical definition of sewing is:

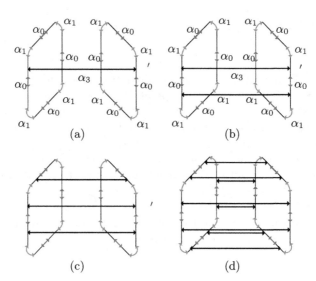

Fig. 9. How to sew two faces.

Definition 3. Let $G = (D, \alpha_0, ..., \alpha_n)$ a n-G-map, $d, d' \in D$ two darts and $i \leq n$ a natural number. Let φ an isomorphism from $< \alpha_0, \ldots, \alpha_{i-2}, \alpha_{i+2}, \ldots, \alpha_n > (d)$ to $< \alpha_0, \ldots, \alpha_{i-2}, \alpha_{i+2}, \ldots, \alpha_n > (d')$ such that $\varphi(d) = d'$. We define $G' = (D', \alpha'_0, ..., \alpha'_n)$ as follows:

1. $D' = D$
2. $\forall j, 0 \leq j \leq n, j \neq i, \alpha'_j = \alpha_j$
3. $\alpha'_i(e) = \begin{cases} \varphi(e) & \text{if } e \in < \alpha_0, \ldots, \alpha_{i-2}, \alpha_{i+2}, \ldots, \alpha_n > (d) \quad (3.1) \\ \varphi^{-1}(e) & \text{if } e \in < \alpha_0, \ldots, \alpha_{i-2}, \alpha_{i+2}, \ldots, \alpha_n > (d') \quad (3.2) \\ \alpha_i(e) & \text{else.} \quad (3.3) \end{cases}$

G' is the result of the i-sewing between d and d' from the n-G-map G. This definition asks for an explanation:

- The isomorphism φ from $< \alpha_0, \ldots, \alpha_{i-2}, \alpha_{i+2}, \ldots, \alpha_n > (d)$ to $< \alpha_0, \ldots, \alpha_{i-2}, \alpha_{i+2}, \ldots, \alpha_n > (d')$ represents the new links added by the sewing. The choice of these two orbits is understandable by remembering the constraints of the n-G-map model. In particular, $\alpha_i \circ \alpha_j$ is an involution for all $n \geq j > i+1 \geq 0$. As d and d' are sewed by α_i, it is necessary to sew the darts $\alpha_j(d)$ and $\alpha_j(d')$ by α_i for all $j \in \{0, 1, \ldots, i-1, i+1, \ldots, n\}$ to ensure this constraint. Step by step, all the darts of the two orbits and are linked by α_i.
- Points (2) and (3.3) give the unchanged links.
- Points (3.1) and (3.2) introduce the new links. The point (3.2) is necessary to ensure α_i to be an involution.

The CASL Specification. Our aim is to formalize the intuitive description of the sewing. We just want to explicitly give as few information as possible

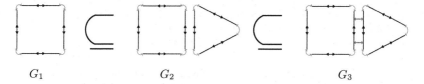

G_1 G_2 G_3

Fig. 10. The inclusion principle.

by taking advantage of the model properties. Our main idea is based on the fact that the expected resulting object is made up of the two original objects with the additional information that the two marked darts are linked. Any other properties should be clarified. The resulting object may be seen as the smallest object including the two initial objects such that the two darts are linked. To capture this description way, we have first introduced a notion of inclusion between n-G-maps, specified in Spec. 6. This predicate can be read as follows: a n-G-map G is included in a n-G-map G' if and only if the darts of G are included in the darts of G', and if whenever a dart is linked to another dart for any given dimension in G then it is linked to the same dart for the same dimension in G'. In particular, a dart d linked to itself by α_i in G , can be linked by α_i to another dart in G'. In Fig. 10, for instance, G_1 is included in G_2, itself included in G_3.

---

pred $\_ \subseteq \_$: $Gmap[n] \times Gmap[n]$
vars G, G' : $Gmap[n]$
- $G \subseteq G' \Leftrightarrow$
  $(G.D \subseteq G'.D \;\wedge$
  $\forall d, d' : Dart \bullet \forall i : Dim[n] \bullet$
  $(d \in G.D \;\wedge\; d' \in G.D \;\wedge\; \neg(d = d') \;\wedge\; G.\alpha(i)(d) = d') \Rightarrow G'.\alpha(i)(d) = d')$

Spec. 6. $\subseteq$ predicate, in CASL, specifying the n-G-map inclusion.

This is sufficient to specify the sewing between a dart d and a dart d' for the dimension i in a n-G-map G as the smallest n-G-map containing G such that $\alpha_i(d) = d'$. The existence of this n-G-map can be easily proven[6]. Now, to specify the sewing operation, we introduce a local predicate *predSew* (see Spec. 7) gathering all the criteria that the sewing n-G-map must verify:

- the sewing n-G-map contains the initial n-G-map and
- the darts d and d' are linked by α_i.

Thanks to this predicate, we specify the sewing operation as expected: the sewing between d and d' for the dimension i in G is the smallest n-G-map verifying the predicate (see Spec. 8).

---

[6] An easy way to prove the existence and the uniqueness of the n-G-map consists in showing that the n-G-map of the mathematical definition is the right n-G-map: it verifies the conditions and it is the smallest one.

---

pred $predSew$: $Gmap[n] \times Gmap[n] \times Dart \times Dart \times Dim[n]$
vars $G, Gafter$: $Gmap[n]$; d, d' : $Dart$; i : $Dim[n]$
- $predSew(Gafter, G, d, d', i) \Leftrightarrow (G \subseteq Gafter \wedge Gafter.\alpha(i)(d) = d')$

Spec. 7. Predicate for the sewing operation in CASL.

---

local
  pred $predSew$: $Gmap[n] \times Gmap[n] \times Dart \times Dart \times Dim[n]$
    ...
within
  op $\_.sew$: $Gmap[n] \times Dart \times Dart \times Dim[n] \to? Gmap[n]$
  vars G : $Gmap[n]$;
      d, d' : $Dart$;
      i : $Dim[n]$
  - $def\ G.sew(d, d', i) \Leftrightarrow ...$ %(sew_dom)%
  - $def\ G.sew(d, d', i) \Rightarrow predSew(G.sew(d, d', i), G, d, d', i)$ %(sew_def1)%
  - $def\ G.sew(d, d', i) \Rightarrow$
    $(\forall G' : Gmap[n] \bullet predSew(G', G, d, d', i) \Rightarrow G.sew(d, d', i) \subseteq G')$
                                                     %(sew_def2)%

Spec. 8. Specification of the sewing operation in CASL.

In the axiom *sew_def1* we specify that the n-G-map resulting from the sewing operation verifies the predicate *predSew*, i.e. the necessary criteria of the sewing operation, and the axiom *sew_def2* indicates it is the smallest n-G-map verifying this predicate. Note that this operation is partial (we put a question mark) to explicit that the operation is only defined if the two faces are isomorphic. These characteristics are defined in the axiom *sew_dom* which indicates when the sewing operation is defined (see [LA01]). In the axioms *sew_def1* and *sew_def2*, the precondition is necessary because the predicates are existential in CASL: whenever an argument is undefined, the predicates are false. Therefore if the *sew* operation is not defined, the predicate is false. Preconditions ensure the correctness of the axioms in undefined cases. Note that the CASL specification of the sewing operation and the mathematical definition have almost the same size. But we get a more abstract description, closer to the intuitive description of graphics designers.

4.2   Specification Methodology

The sewing operation is specified as the smallest n-G-map verifying a list of criteria[7]. In this way, only the necessary changes are expressed and the other ones are deduced from n-G-map properties. In the case of the sewing operation,

---

[7] From a mathematical point of view, it has been demonstrated that such a minimal n-G-map exists, is unique and exactly corresponds to the expected sewed n-G-map.

we just point out the two darts we want to link by a particular α_i, for example α_2. And as the expected object is represented by a n-G-map, some others links are naturally established to verify the n-G-map properties such that $\alpha_0 \circ \alpha_2$ is an involution. Thanks to this approach, we faithfully formalize the intuitive description of graphics designers. This point is very important since it allows one to increase the faith in the gap between informal requirements and formal one. So our methodology for specifying a geometric operation may be simply summarized: it consists in specifying the result of a geometric operation as the smallest/largest n-G-map verifying a list of criteria. Among these criteria, there always exists a criterion indicating the resulting n-G-map contains (or is contained in) a particular n-G-map. So we have the generic pattern of Spec. 9 to specify a geometric operation in CASL.

---

local
    pred $predOp$: ...
    vars ...
    • $predOp(...) \Leftrightarrow$ List of CRITERIA
within
    op $\_.op$: $Gmap[n] \times ... \rightarrow (?)\,Gmap[n]$
    vars G : $Gmap[n]; ...$
    • $predOp(G.op(...), ...)$
    • $\forall G' : Gmap[n]$ • $predOp(G', ...) \Rightarrow G.op(...) \subseteq G'$ or $G' \subseteq G.op(...)$

Spec. 9. Pattern to specify geometric operations in CASL.

Another real contribution of this methodology is to provide a new level of definition for geometric operations. This level can be considered as more abstract than the usual one. As we have already pointed out it for the sewing operation, a geometric operation concerning the n-G-map model is usually defined from a mathematical point of view by enumerating all the darts and describing all the links of the new n-G-map. Then our approach is less constructive and even closer from requirements.

4.3   Other Basic Operations on n-G-maps

The sewing operation is a necessary basic operation to build a n-G-map. The other ones are adding a dart, removing a dart, and unsewing a dart. In this part we just provide the specifications of these operations underlining the necessary criteria.

Dart Adding. Let us consider a n-G-map G. Following our specification methodology, the result of adding a dart in G is of course the smallest n-G-map containing G and a new dart (see Spec. 10).

---

local
>   **pred** $predAddDart$: $Gmap[n] \times Gmap[n] \times Dart$
>   **vars** $G, Gafter$: $Gmap[n]$;
> d : $Dart$
>   • $predAddDart(Gafter, G, d) \Leftrightarrow$
> $(\neg(d \in G.D) \;\wedge\; Gafter.D = G.D + d \;\wedge\; G \subseteq Gafter)$

within
>   **op** $\_\_.addDart$: $Gmap[n] \rightarrow Gmap[n]$
>   **var** $G : Gmap[n]$
>   • $\exists d : Dart \bullet predAddDart(G.addDart, G, d) \;\wedge$
> $(\forall G' : Gmap[n] \bullet predAddDart(G', G, d) \Rightarrow G.addDart \subseteq G')$

Spec. 10. Specification of the dart adding in CASL.

The operation $\_\_ + \_\_$: $FinSet[S] \times S \rightarrow FinSet[S]$ is provided by CASL libraries [RMS00] and simply specifies the adding of an element into a set. Like the sewing operation, the addition of a dart is specified as the smallest n-G-map verifying several criteria. Each operation adding new elements (darts and/or links) in the object is specified as being the smallest n-G-map verifying some well-chosen criteria.

Dart Removal. This operation does the reverse of the dart addition. In this case we must select a dart d to remove. The resulting n-G-map obviously does not contain the dart d. Moreover, it is the largest n-G-map contained in the initial one. Indeed, unlike the addition and sewing operations, an element is removed from the n-G-map. The resulting n-G-map, then, is smaller than the initial one and specified as the largest n-G-map verifying criteria (see Spec. 11).

The operation $\_\_ - \_\_$: $FinSet[S] \times S \rightarrow FinSet[S]$ is provided by CASL libraries [RMS00] and simply consists in removing an element from a set.

Dart Unsewing. This operation does the reverse of the sewing one. To unsew, a dart d and a dimension i are selected. Then the resulting n-G-map is the largest one contained in the initial one such that d is free for α_i (see Spec. 12).

---

local
>   **pred** $predRemove$: $Gmap[n] \times Gmap[n] \times Dart$
>   **vars** $G, Gafter$: $Gmap[n]$;
> d : $Dart$
>   • $predRemove(Gafter, G, d) \Leftrightarrow (Gafter \subseteq G \wedge Gafter.D = G.D - d)$

within
>   **op** $\_\_.remove$: $Gmap[n] \times Dart \rightarrow Gmap[n]$
>   **vars** G : $Gmap[n]$;
> d : $Dart$
>   • $predRemove(G.remove(d), G, d)$
>   • $\forall G' : Gmap[n] \bullet predRemove(G', G, d) \Rightarrow G' \subseteq G.remove(d)$

Spec. 11. Specification of the dart removal in CASL.

---

local

 pred $predUnsew$: $Gmap[n] \times Gmap[n] \times Dart \times Dim[n]$
 vars $G, Gafter$: $Gmap[n]$;
   d : $Dart$;
   i : $Dim[n]$
 • $predUnsew(Gafter, G, d, i) \Leftrightarrow (Gafter \subseteq G \land Gafter.\alpha(i)(d) = d)$

within

 op $\_\_.unsew$: $Gmap[n] \times Dart \times Dim[n] \to? \; Gmap[n]$
 vars G : $Gmap[n]$;
   d : $Dart$;
   i : $Dim[n]$
 • $def\, G.unsew(d, i) \Leftrightarrow d \in G.D$
 • $def\, G.unsew(d, i) \Rightarrow predUnsew(G.unsew(d, i), G, d, i)$
 • $def\, G.unsew(d, i) \Rightarrow$
   $(\forall G' : Gmap[n] \bullet predUnsew(G', G, d, i) \Rightarrow G' \subseteq G.unsew(d, i))$

Spec. 12. Specification of the unsewing operation in CASL.

Fig. 11. A cube whom we round a corner with the n-G-map model.

5 Application of our Methodology to the Rounding Operation

The rounding operation is one of the most complex and useful operations in geometric modelling. It is a central operation in CAD, and it is used for making finishing touches. The rounding operation has already been defined mathematically by enumerating all the darts and links of the resulting n-G-map [Elt94]. This definition is complex and difficult to read because it isn't intuitive and handles a lot of parameters. Thanks to our methodology, the rounding operation will be more abstractly defined. Then dealing with the usual particular cases is avoided [VMV94].

As in the previous part, we present the different necessary criteria for specifying the rounding operation. Due to the intrinsic complexity of the operation, these criteria are obviously more complex than those of basic operations. To introduce them, we consider the simple example of a cube whom we round a corner. In fact, rounding a vertex in a 3-dimensional object consists in replacing it by a little face as shown in Fig. 11.

The specification of the rounding operation is presented in Spec. 13. The rounding n-G-map is more complex than the initial one: some darts are added and new links appear. The resulting n-G-map is defined as the smallest n-G-map verifying three criteria. The first criterion (C_1) is that the rounding n-G-map

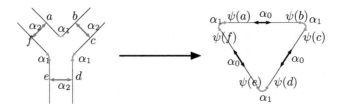

Fig. 12. The duality notion.

contains the n-G-map for which all the darts of the vertex to be rounded are no more linked by α_1 (see Fig. 13(b)). In other words, the orbit $< \alpha_1, \alpha_2, \ldots, \alpha_n >$ (d) has been broken for the dimension 1. The n-G-map of Fig. 13(f) contains the n-G-map of Fig. 13(b). Generally, to round an i-cell[8], we break it for the dimension $i + 1$.

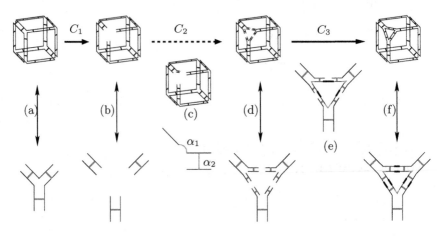

Fig. 13. The different steps of the rounding operation.

A piece of n-G-map to add is identified for each dart of the vertex that we round. Briefly, if we round an i-cell in a n-G-map, we add for each dart d of the i-cell, $n - i$ darts $\{d_{i+1}, \ldots, d_n\}$ such that:

- $\alpha_{i+1}(d) = d_{i+1}$ and,
- $\alpha_{j+1}(d_j) = d_{j+1}$ $(i + 1 < j < n)$ (see Fig. 13(c) and Fig. 13(d)).

This corresponds to our second criterion (C_2). At this step we almost recognize the expected result (see Fig. 13(d)). To get the n-G-map of Fig. 13(f) some links must be added. Among the missing links, some can be easily deduced

---

[8] The i-cell related to dart d is the orbit $< \alpha_0, \ldots, \alpha_{i-1}, \alpha_{i+1}, \ldots, \alpha_n > (d)$. Note that, a 0-cell is a vertex, a 1-cell is an edge, a 2-cell is a face and a 3-cell is a volume.

---

local

    op $\_.\psi^-$: $Gmap[n] \times Dim[n] \times Dart \to? \; Dart$

    vars G : $Gmap[n];$ d, d' : $Dart;$ i, j : $Dim[n]$

    • $def \; G.\psi^i(d) \Leftrightarrow d \in G.D$

    • $\neg (G.\psi^i(d) \in G.D)$

    • $G.\psi^i(d) = G.\psi^j(d') \Rightarrow (d = d' \wedge i = j)$

    pred $isRounded$: $Gmap[n] \times Gmap[n] \times Dart \times Dim[n]$

    vars $G, Gafter$: $Gmap[n];$ d : $Dart;$ i : $Dim[n]$

    • $isRounded(Gafter, G, d, i) \Leftrightarrow$

        $(G.unsewSet(G.cell(d,i), (i+1) \; as \; Dim[n]) \subseteq Gafter \; \wedge$ %%$\mathbf{C_1}$

        $\forall d' : Dart \bullet d' \in G.cell(d,i).D \Rightarrow$

        $(Gafter.\alpha(i+1 \; as \; Dim[n])(d') = G.\psi^{(i+1) \; as \; Dim[n]}(d') \wedge$

        $\forall k : Dim[n] \bullet ((i+1 \le k \wedge k \le n-?1) \Rightarrow$

        $Gafter.\alpha(k+1 \; as \; Dim[n])(G.\psi^k(d')) = G.\psi^{(k+1) \; as \; Dim[n]}(d')) \wedge$ %%$\mathbf{C_2}$

        $\forall d1, d2 : Dart \bullet (d1 \in G.cell(d,i).D \wedge d2 \in G.cell(d,i).D) \Rightarrow$

        $(\forall k : Dim[n] \bullet k \le n-?1 \Rightarrow$

        $((k < i \wedge G.\alpha(k)(d1) = d2) \Rightarrow Gafter.\alpha(k)(G.\psi^n(d1)) = G.\psi^n(d2)) \wedge$

        $((k > i \wedge G.\alpha(k)(d1) = d2) \Rightarrow$

        $Gafter.\alpha(k-?1 \; as \; Dim[n])(G.\psi^n(d1)) = G.\psi^n(d2)))))$ %%$\mathbf{C_3}$

within

    op $\_.rounding$: $Gmap[n] \times Dart \times Dim[n] \to? \; Gmap[n]$

    vars G : $Gmap[n];$ i, j : $Dim[n];$ d : $dart$

    • $def \; G.rounding(d, i) \Leftrightarrow (d \in G.D \wedge i < n)$

    • $def \; G.rounding(d, i) \Rightarrow$

        $isRounded(G.rounding(d,i), G, d, i)$

    • $def \; G.rounding(d, i) \Rightarrow$

        $(\forall G' : Gmap[n] \bullet isRounded(G', G, d, i) \Rightarrow G.rounding(d,i) \subseteq G')$

Spec. 13. Specification of the rounding operation in CASL.

from the initial i-cell being round. In fact, we exactly know the structure of the introduced face. This is our third criterion (C_3) which is the exact formalization of the duality notion[9]. In Fig. 12, the initial cell is represented on the left side and the introduced face on the right side. By considering the one-to-one bijection ψ in Fig. 12, we have the general property for the rounding of an i-cell C in a n-G-map:

- if $\alpha_j(d) = d'$ with $j < i$ then $(\alpha_j \circ \psi)(d) = \psi(d')$.
- if $\alpha_j(d) = d'$ with $j > i$ then $(\alpha_{j-1} \circ \psi)(d) = \psi(d')$.

This is sufficient enough to get the 2-G-map of Fig. 13(f) from the one of Fig. 13(d). The last links are ensured by the n-G-map properties. Indeed by using the duality notion, we have just explicitly given the result of Fig. 13(e) which is not a n-G-map as $\alpha_0 \circ \alpha_2$ is not an involution here. Since we specify that the result is a n-G-map, we actually get the result of Fig. 13(f).

---

[9] The duality notion corresponds to the one-to-one bijection between the original i-cell to be rounded and the new n-cell added by rounding.

We will not comment any more the specification of rouding. Let us just point out one can recognize the specification pattern introduced previously (introduction of a local predicate and use of the inclusion notion). In this specification, $G.unsew(G.cell(d, i).D, (i+1)$ as $Dim[n])$ provides the largest 3-G-map included in G and such as all the darts belonging in[10] $G.cell(d, i)$ are unsewed[11] for α_i.

6   Contributions of CASL

Using CASL and more generally axiomatic specifications has enabled us to differently tackle and design geometric operations. Previously geometric operations were essentially defined as they are coded by graphist designers. There are several different particular cases with many parameters. The design level was therefore very low until now. But sometimes, graphics designers succeeded in deducing mathematical definitions. These definitions were close to programming methods and written in a constructive style. If we are interested in geometric operations concerning n-G-maps, these definitions totally describe the resulting n-G-map by enumerating all its darts and links. This approach reveals a lack of abstraction which involves a lack of generality in definitions.

¿From a software design point of view, an abstract formalization of geometric operations allows one to achieve general and intuitive definitions. To get this formalization, we start from the intuitive and high-level explanations of geometric modelling experts. We straightforwardly specify their intuitive descriptions into CASL specifications. This methodology contributes to easily validate the specification of requirements.

The syntactic and semantic richness of CASL and its specification libraries allowed us to easily write an high-level abstract specification. The result appears to be clear, concise, structured and syntacticaly close to usual mathematical notations. More generally, our specifications, then, provide more abstract definitions of geometric operations than the existing mathematical ones. The enumeration of darts is avoided. We prefer to take care about the n-G-map properties and use them to avoid to totally construct the object. Besides, a new specification methodology for geometric operations has been established. It is based on a notion of inclusion between n-G-maps which faithfully reflects the informal descriptions made by geometric modelling specialists. This methodology led us to a new rouding definition, clearer, more abstract and more general than the existing definition [Elt94] by suppressing all the redundancies [LMA+00].

Note that our specifications are really readable and understandable by graphics designers thanks to the expressive and powerful syntax of CASL. This is a very important point to promote the use of formal methods for designing geometric

---

[10] $G.cell(d, i)$ provides the n-G-map of the orbit $< \alpha_0, ..., \alpha_{i-1}, \alpha_{i+1}, ..., \alpha_n > (d)$ itself included in G.

[11] $unsew : Gmap[n] \times FinSet[Dart] \times Dim[n] \rightarrow Gmap[n]$ is an operation which unsews a set of darts for a particular dimension. $G.unsew(C, i)$ unsews all the darts of C for the involution α_i. This operation is specified using the $unsew$ specification of a dart given beforehand.

modelling software. Moreover, by abstractly specifying the rounding operation with our methodogy, we have been able to only consider a local treatment to the darts. We only focus on the parts of the object which must be altered and the rest of the object remains unchanged. If one takes this line to extremes, there is a particular treatment for each dart included in the modified parts. And the darts not involved in the operation remain unchanged. This point is very important in the topology-based modellers as efficient implementations of operations are precisely based on the principle of traversing through the darts of the changed part of the n-G-map applying the same processing for each dart.

We are really satisfied by the CASL language and its main features as pretty-printing syntax, structuration, libraries, subsort and partiality management ... It has allowed us to easily specify geometric modelling notions and get a new description level for geometric modelling. Two additional features, though, would be particularly convenient to us if they were provided with CASL. The first one is high order. Such a feature would allow us to get a more powerful syntax. For instance, functions as α_i woud be more simply expressed. In fact, from our point of view, high order is especially missed with respect to our goal of being close to mathematical habit of graphics designers. Some works extend CASL with higher-order [MHK00], but we do not use them since first of all, we want to use CASL tools. The second missing feature according to our crew is weak predicates, namely predicates which are true as soon as one of its arguments is undefined. As mentionned in Section 4, most of our axioms include a precondition about the definedness of a geometric operation. With such weak predicates, we could have written our axioms without these systematic preconditions. Of course, introducing such weak predicates within CASL would involve two types of predicates with different semantics. Then writing specifications would become more subtle because of the choice between these two kinds of predicates.

7   Concluding Remarks

In this paper, we have presented our use of formal methods for geometric modelling. Some works about using formal methods for geometric modelling yet exist [PD98,PD00], but with a different aim. The main goal of these works is to formally certify mathematical definitions or properties of the underlying mathematical model, while our goal is to use formal methods to develop software. We have succeeded in providing abstract specifications, easily readable by graphics designers and a methodology aiming at guiding any new specification activity in the topology-based modelling area.

The work described in this paper is the result of a joint research effort between researchers in geometric modelling area and researchers in software engineering area. Contributions of CASL (see previous Section) have allowed us to get a satisfactory result from a geometric modelling point of view. Moreover, our description of the rounding operation is really used to develop within the framework of a French national project devoted to the design of a new generation of modellers. The geometric modelling researchers we work with are already satisfied. Still they are waiting for the sequel.

8   Future Directions

For the moment, a high-level specification of geometric operations has been provided. It consists of pure topologic operations and geometric operations. In this paper, we were interested in topology-based modelling, and more precisely we worked with the n-G-map model. So we have just presented the topologic part of the operations and not their geometric counterpart (for a first idea, look at [LAGB00]). For instance, in the specifications presented here we do not take care about the coordinates of vertex in the space. And each new face introduced by rounding operation could be represented by a parametric surface as a Bezier patch or a NURBS patch. To ensure a satisfactory realistic aspect, continuity constraints between surfaces should be managed. To continue our work, we have two distinct perspectives:

- *to refine our specification towards a more concrete one.* The key point for this refinement would be the use of a reference specification considered as a rather low level one, which would correspond to a kernel of basic operations that we will not refine, but whose we have an implementation considered both as reliable and efficient by researchers in geometric modelling. Indeed, we need optimized code in geometric modelling and it is well known that a complete refinement stepwise doesn't naturally provide an efficient code. Therefore we should just specify high-level operations as sewing or rounding and refine their requirement specifications towards more concrete ones based on our kernel specification. Note that this kernel is a generalisation of the software described in [BD94]. The refinement process has also to be provided with a methodology.
- *to extend our work to the design of a complete modeller.* This work involves developing dedicated specification libraries for geometric modelling by taking into account many other operations and continuity problems for particular surface types.

However, in the two cases, we need tools:

- a prover to validate our refinement steps. Currently we wait for the new version of HOL-CASL which will allow us to do large proofs with structured specifications.
- a code generator to prototype and quickly validate our abstract geometric definitions.
- at last, a test case generator to verify a wide-ranging project as a complete geometric modeller. Indeed it is generally too long to prove all the design steps of a complete software.

References

[BD94]   Y. Bertrand and J.-F. Dufourd. Algebraic Specification of a 3D-modeller
            Based on Hypermaps. *Computer vision, graphical model, and image pro-
            cessing*, 56(1):29–60, 1994.

[Duf97]   J.-F. Dufourd. Algebras and formal specifications in geometric modeling.
            The Visual Computer, 13:131–154, 1997. Springer-Verlag.

[Elt94]   H. Elter. *Etude de structures combinatoires pour la représentation de com-
            plexes cellulaires*. PhD thesis, Université de Strasbourg, 1994.

[FBBD96]   L. Fuchs, D. Bechmann, Y. Bertrand, and J.-F. Dufourd. Formal specifica-
            tion for free-form curves and surfaces. In *Spring Conference on Computer
            Graphics*, Bratislava, 1996.

[LA01]   F. Ledoux and A. Arnould. Geospec: specification libraries for geometric
            modelling, sept. 2001.
            `http://www.sic.sp2mi.univ-poitiers.fr/GL/GeoSpec`.

[LAGB00]   F. Ledoux, A. Arnould, P. Le Gall, and Y. Bertrand. A High-Level Opera-
            tion in 3D Modelling: a CASL Case Study. Technical Report 52, Université
            d'Évry, 2000.
            `ftp://ftp.lami.univ-evry.fr/pub/publications/reports/index.html`.

[Lie91]   P. Lienhardt. Topological models for boundary representations: a com-
            parison with n-dimensional generalized maps. *Computer-Aided Design*,
            23(1):59–82, 1991.

[Lie94]   P. Lienhardt. N-dimensional generalized combinatorial maps and cellular
            quasi-manifolds. *International Journal of Computational Geometry and
            Applications*, 1994.

[LMA⁺00]   F. Ledoux, J.-M. Mota, A. Arnould, C. Dubois, P. Le Gall, and Y.
            Bertrand. Formal specification for a mathematics-based application do-
            main: geometric modelling. Technical Report 51, Université d'Évry, 2000.
            `ftp://ftp.lami.univ-evry.fr/pub/publications/reports/index.html`.

[Man83]   M. Mantyla. Computational topology: A study of topological manipula-
            tions and interrogations in computer graphics and geometric modeling.
            Acta Polytech. Scand. Math. and Comput. Sci. Ser., MA37:1–49, 1983.

[MHK00]   T Mossakowski, Anne Haxthausen, and Bernd KriegBruckner. Subsorted
            partial higher-order logic as an extension of casl. In *Christine Choppy, Di-
            dier Bert, and Peter Mosses (eds.): Recent Developments in Algebraic De-
            velopment Techniques, Lecture Notes in Computer Science*, volume 1827,
            pages 126–145, Chateau de Bonas, France, 2000. Springer-Verlag.

[Mos99]   P. D. Mosses. CASL: A guided tour of its design. *Lecture Notes in Com-
            puter Science*, 1589:216–240, 1999.

[MRSS01a]   Till Mossakowski, Markus Roggenbach, Lutz Schroder, and Pascal
            Schmidt). CASL tool set (CATS) version 0.73, sept. 2001.
            `http://www.tzi.de/cofi/CASL/CATS/download.html`.

[MRSS01b]   Till Mossakowski, Markus Roggenbach, Lutz Schroder, and Pascal
            Schmidt). HOL-CASL system version 0.6, sept. 2001.
            `http://www.tzi.de/cofi/CASL/`.

[oLD00a]   CoFI (Common Framework Initiative) Task Group on Language Design.
            CASL case studies, 2000.
            `http://www.pst.informatik.uni-muenchen.de/baumeist/CoFI/case.html`.

[oLD00b]   CoFI (Common Framework Initiative) Task Group on Language Design.
            CASL the common algebraic specification language summary, June 2000.
            `ftp://ftp.brics.dk/Projects/CoFI`.

[PD98]   F. Puitg and J.-F. Dufourd. Formal specification and theorem proving breakthroughs in geometric modeling. In *Proc. 11th International Theorem Proving in Higher Order Logics Conference*, pages 401–422, 1998.

[PD00]   François Puitg and Jean-François Dufourd. Formalizing mathematics in higher-order logic: A case study in geometric modelling. *Theoretical Computer Science*, 234(1–2):1–57, 2000.

[RMS00]   M. Roggenbach, T. Mossakowski, and L. Schroder.
            Basic   datatypes   in   CASL.   CoFI   Note   L-12-Version   0.4.1,
            `http://www.brics.dk/Projects/CoFI/Notes/L-12/index.html`, May 2000.

[VMV94]   Tamás Várady, Ralph R. Martin, and Janos Vida. A survey of blending methods that use parametric surfaces. *Computer-Aided Design*, 26(5):341–365, 1994.

A Compositional Approach
to Connector Construction[*]

Antónia Lopes[1], Michel Wermelinger[2,3], and José Luiz Fiadeiro[1,3]

[1] Department of Informatics, Faculty of Sciences, University of Lisbon, Campo Grande,
1749-016 Lisboa, Portugal
mal@di.fc.ul.pt
[2] Department of Informatics, Faculty of Sciences and Technology, New University of
Lisbon, 2829-516 Caparica, Portugal
[3]ATX Software SA
Alameda António Sérgio 7, 1A, 2795-023 Linda-a-Velha, Portugal
mw@di.fct.unl.pt,jose@fiadeiro.org

Abstract. We develop a notion of higher-order connector towards supporting the systematic construction of architectural connectors for software design. The idea is that individual properties of connectors, such as security
and fault-tolerance, can be designed separately as higher-order connectors.
Complex connectors can then be described as suitable combinations of
higher-order connectors and basic connectors. We use CommUnity, a Unity-
like parallel program design language that we have been using for formalising aspects of architectural design, for illustrating our approach to the compositional construction of connectors and also to motivate the categorical
semantics of higher-order connectors that we propose.

1 Introduction

Although components have always been considered the fundamental building blocks of
software systems, the way the components of a system interact may also be determi-
nant on the system properties. Recently, component interactions were recognised also
to be first-class design entities, and architectural connectors have emerged as a powerful tool for supporting the description of these interactions. However, as argued in
[11], the current level of support for connectors is still far from the one components
have.

At an architectural level of design, component interactions can be very simple (for
instance a shared variable) but also very complex (for instance database-acessing protocols). Hence, it is of great interest to have mechanisms for designing connectors in an
incremental and compositional way as well as principled ways of extending existing
ones, promoting reuse. Furthermore, as argued in [3], modularising the different kinds

---

[*] This research was partially supported by Fundação para a Ciência e Tecnologia through
project POSI/32717/00 (FAST — Formal Approach to Software Architecture).

of services involved in interaction protocols makes it easier to evolve systems (possibly at run-time), because service modules may be added only when necessary, hence preventing performance penalties when such complex interactions are not required.

In this work we take a step towards this goal by proposing a specification mechanism that allows independent aspects such as compression, fault-tolerance, security, monitoring, *etc.*, to be specified separately, and then composed and integrated with existing connectors. In this way, it becomes possible to benefit from the multiple combinations of different services, ideally chosen *à la carte*. More concretely, we propose a notion of higher-order connector — a connector that takes a connector as parameter — through which is possible to describe the superposition of certain capabilities over the form of coordination that is handled by the connector that is passed as an actual argument.

We follow Garlan's proposal of considering higher-order connectors as operators with which new connectors can be built up from old connectors [8,9]. Concretely, we define a higher-order connector through a (formal) parameter declaration and a body connector that models the nature of the service that is superposed on instantiation of the formal parameter. For instance, the monitoring of messages in a unidirectional communication can be captured by a higher-order connector with a parameter *Unidirectional-comm* that specifies the kind of connectors to which the service can be applied, and a body connector that describes how an actual parameter is adapted in order to transmit certain messages to a monitoring component. A higher-order connector can be applied to any connector that instantiates the formal parameter, giving rise to a connector with the new capabilities. In the case of monitoring, the higher-order connector can be applied, for instance, to a connector that models asynchronous communication between a sender and a receiver.

This notion of higher-order connector extends our preliminary proposal presented in [12]. Furthermore, we not only develop this notion over a specific language — CommUnity, but also delineate a notion of higher-order connector which is not specific to any Architecture Description Language. Such generalisation capitalizes on our previous work [4,5,6,13] on the formal underpinning of connectors, namely on the categorical framework that establishes the semantics of architectural connectors independently of specific choices of design languages and behavioural models.   Due to space limitation, there are some important details of the categorical semantics of higher-order connectors that will not be discussed.

We start by presenting the program design language CommUnity, a Unity-like [2] parallel program design language that we have been using for formalising aspects of architectural design. In this paper, we will use CommUnity to illustrate the notion of higher-order connector we wish to put forward and also to show how categorical techniques can be used to formalise a design formalism. This is important because the formal semantics we shall present for connectors and higher-order connectors relies on the fact that category theory provides a convenient framework for describing a design formalism, namely designs, configurations and relationships between designs, such as refinement.

2 CommUnity

CommUnity is a program design language in the style of Unity [2], that also combines elements from IP (Interacting Processes) [7]. The language is independent of the actual data types used and, hence, we assume there are pre-defined sorts and functions given by a fixed algebraic signature. For the purposes of examples, we assume this algebraic signature contains sorts *bool, nat, list, queue* with the usual operations; ordered pairs with projection functions *fst* (first element) and *snd* (second element); a function *if(cond,then-expr,else-expr)* with the obvious meaning.

2.1 Component Designs

A CommUnity design is of the form

```
design P is
out        out(V)
in         in(V)
prv        prv(V)
```
$$\mathbf{do}\ \underset{g\in sh(\Gamma)}{[]}\quad g:\ L(g),\ U(g)\ \rightarrow\ \underset{v\in D(g)}{||}\quad v:\in F(g,v)$$

$$\underset{g\in prv(\Gamma)}{[]}\ \mathbf{prv}\quad g:\ L(g),\ U(g)\ \rightarrow\ \underset{v\in D(g)}{||}\quad v:\in F(g,v)$$

where

- *V* is the set of *variables* that can be declared as *input, output* or *private*. Input variables are read from the environment of the component but cannot be modified by the component. Output and private variables are local to the component, i.e., they cannot be modified by the environment. We use *loc(V)* to denote the set of local variables. Output variables can be read by the environment but private variables cannot. Each variable v is typed with a sort *sort(v)*.
- Γ is the set of *action names*. Actions can be declared either as *private* or *shared*. Private actions represent internal computations and their execution is uniquely under the control of the component. In contrast, shared actions represent interactions between the component and the environment and their execution is also under the control of the environment.
- For each action g, $D(g)$ consists of the local variables that action g can change – its write frame. For every local variable v, we also denote by $D(v)$ the set of actions that can change v.
- For each action g, $L(g)$ and $U(g)$ are two conditions such that $U(g) \supset L(g)$. They establish an interval in which the enabling condition of g must lie: $L(g)$ is the lower bound of the interval, i.e., it is implied by the enabling condition; $U(g)$ is the upper bound, i.e., it implies the enabling condition. Therefore, the negation of $L(g)$ establishes a *blocking* condition, whereas $U(g)$ establishes a *progress* condition. The enabling condition is fully determined when $L(g)$ and $U(g)$ are equivalent, in which case we write only one condition.
- For each action g and variable v in $D(g)$, $F(g,v)$ is a expression that denotes a set – the set of values that can be assigned to v. We abbreviate deterministic assignments

of the form $v:\in\{t\}$ as $v:=t$. When an action has empty domain, we use the expression `skip` instead.

When, for every $g\in\Gamma$, $L(g)$ and $U(g)$ coincide, and $F(g,v)$ is a singleton for every local variable v, then the design is called a *program*. The behaviour of a program without input variables is as follows. At each execution step, one of the actions whose enabling condition holds of the current state is selected, and its assignments are executed atomically in parallel. Furthermore, private actions that are infinitely often enabled are guaranteed to be selected infinitely often.

As an example consider the design presented below that models a box consisting of a button, a sensor and a light. Its purpose is to allow a patient to request help in case of medical emergency, with the transmission of the current value of the sensor (e.g., pulse). Pressing the button, which is modelled by the execution of *hreq*, turns on the light, which is modelled by variable *off* becoming false. The light is turned off when the help request is acknowledged. After the button is pressed, the current value of the sensor is read, which is modelled by the execution of the private action *read*, and made available for transmission in the output variable *data*. The private variable *rd* is used to distinguish between states in which the value in *data* is the value to be transmitted or not.

```
design help is
in      sensor:nat
out     data:nat, off: bool
prv     rd: bool
do      hreq: off → off:=false
[] prv read: ¬rd ∧ ¬off →data:=sensor‖ rd:=true
[]      hack: rd → rd:=false ‖ off:=true
```

2.2 Configurations

In CommUnity it is also possible to describe a system as the interconnection of a number of interacting component designs by defining a configuration. The model of interaction between components is based on action synchronisation and the interconnection of input variables of a component with output variables of other components. Although these are common forms of interaction, CommUnity requires interaction between components — name bindings — to be made explicit in configurations. Name bindings are established as relationships between the signatures of the corresponding components and are defined with the help of additional signatures (representing the interaction points) and signature maps.

Let us consider, for instance, that we want to describe a system in which the messages from the *help* component are sent (to a receiver) through a bounded channel. The design *buffer* parametrised by sort s described below models the bounded channel, more concretely it models a bounded buffer with a FIFO discipline.

```
design buffer[s] is
in      i:s
out     o:s
prv     rd:bool;  q:queue(K,s)
do      put: ¬full(q) → q:=enqueue(i,q)
[] prv  next: ¬empty(q)∧¬rd → o:=head(q) ‖ q:=tail(q) ‖ rd:=true
[]      get: rd → rd:=false
```

This buffer can store, through the action *put*, messages of sort *s*, received from the environment through the input variable *i*, as long there is space for them. It can also make stored messages available to the environment through the output variable *o* and the action *next*. Naturally, this activity is possible only when there are messages in store and the current message in *o* has already been read by the environment, which is modelled by the action *get* and the private variable *rd*.

In order to establish that messages from the *help* component are sent (to a receiver) through a bounded channel, we consider the following configuration

where we use *sig(P)* to denote the signature of *P*, and *channel(nat)* is a signature that consists of an input variable of sort *nat* and a shared action. The names of this variable and of this action are not relevant: they are only placeholders for the name bindings.

In this configuration, the input variable of *channel* is mapped to the output variable *data* of the *help* component and to the input variable *i* of *buffer*. This establishes an I/O interconnection between *help* and *buffer*. Moreover, the actions *hack* of *help* and *put* of *buffer* are mapped to the shared action of *channel*. This defines that *help* and *buffer* must synchronise each time either of them wants to perform the corresponding action.

Signatures and signature morphisms can be formally defined as follows.

A design signature is a tuple $<V,\Gamma,tv,ta,D>$ where
- *V is an S-indexed family of mutually disjoint finite sets,*
- *Γ is a finite set,*
- *$tv: V\rightarrow\{out,in,prv\}$ is a total function,*
- *$ta: \Gamma\rightarrow\{sh,prv\}$ is a total function,*
- *$D: \Gamma\rightarrow2^{loc(V)}$ is a total function.*

A morphism $\sigma: \theta_1\rightarrow\theta_2$ is a pair $<\sigma_{var},\sigma_{ac}>$ where
- *$\sigma_{var}: V_1\rightarrow V_2$ is a total function satisfying:*
  - *1. $sort_2(\sigma_{var}(v))=sort_1(v)$ for every $v\in V_1$;*
  - *2. $\sigma_{var}(o)\in out(V_2)$ for every $o\in out(V_1)$;*
  - *3. $\sigma_{var}(i)\in out(V_2)\cup in(V_2)$ for every $i\in in(V_1)$;*
  - *4. $\sigma_{var}(p)\in prv(V_2)$ for every $p\in prv(V_1)$.*
- *$\sigma_{ac}: \Gamma_2\rightarrow\Gamma_1$ is a partial mapping satisfying for every $g\in\Gamma_2$ s.t. $\sigma_{ac}(g)$ is defined:*
  - *5.   if $g\in sh(\Gamma_2)$ then $\sigma_{ac}(g)\in sh(\Gamma_1)$;*
  - *6.   if $g\in prv(\Gamma_2)$ then $\sigma_{ac}(g)\in prv(\Gamma_1)$;*

7. $\sigma_{var}(D_1(\sigma_{ac}(g))) \subseteq D_2(g)$;

8. $\sigma_{ac}(D_2(\sigma_{var}(v))) \subseteq D_1(v)$ *for every* $v \in loc(V_1)$.

*Design signatures and signature morphisms constitute a category **SIGN**.*

A morphism $\sigma: \theta_1 \rightarrow \theta_2$ identifies a way in which a component with signature θ_1 is embedded in a larger system with signature θ_2. More concretely, σ_{var} identifies for each variable of the component the corresponding variable of the system, and σ_{ac} identifies the action of the component that is involved in each action of the system, if ever. Notice that input variables of a component may become output variables of the system (condition 3) because we consider that the result of interconnecting an input variable of a component with an output variable of another component is an output variable of the system. Conditions 7 and 8 require that change within a component is completely encapsulated in the structure of actions defined for the component.

Not every diagram of signatures represents a meaningful configuration in the sense that there are restrictions on the way that we can interconnect components that are not captured by the notion of morphism alone but require the whole diagram. The two following rules express the restrictions on diagrams that make them well-formed configurations:

- An output variable of a component cannot be connected (directly or indirectly through input variables) with output variables of the same or other components.
- Private variables and private actions cannot be involved in the connections.

The second rule ensures that private variables cannot be read by the environment and that the execution of private actions is uniquely under the control of the component.

In the sequel, for simplicity, rather than using diagrams involving signatures and signature morphisms, we will mainly use a more user-friendly notation. For instance, the configuration presented previously could be described as follows.

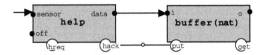

In this notation, the name bindings are still explicit but are expressed in terms of arcs that connect variables and actions directly. These configurations can be easily translated into categorical diagrams involving signatures and signature morphisms.

Configuration Semantics

The semantics of configuration diagrams relies on an extension of the notion of signature morphism that allows us to establish relationships between designs. Design morphisms capture relationships between components and the systems that they are part-of. They can be seen to provide a formalisation for a notion of superposition that is similar to those that have been used for parallel program design [2,10].

A superposition morphism $\sigma: P_1 \rightarrow P_2$ *consists of a signature morphism* $\sigma: \theta_1 \rightarrow \theta_2$ *s.t., for every* $g \in \Gamma_2$ *s.t.* $\sigma_{ac}(g)$ *is defined:*

1. *for every $v \in D_1(\sigma_{ac}(g))$, $\Phi \vDash (F_2(g, \sigma_{var}(v)) \subseteq \sigma(F_1(\sigma_{ac}(g), v)))$;*
2. *$\Phi \vDash (L_2(g) \supset \sigma(L_1(\sigma_{ac}(g))))$;*
3. *$\Phi \vDash (U_2(g) \supset \sigma(U_1(\sigma_{ac}(g))))$;*

where $\vDash$ *denotes validity in the first-order sense. Designs and superposition morphisms constitute the category* **c-DSGN**.

A morphism $\sigma : P_1 \to P_2$ identifies a way in which P_1 is "augmented" to become P_2 so that P_2 can be considered as having been obtained from P_1 through the superposition of additional behaviour. The conditions say that the effects of the actions have to be preserved or made more deterministic and that the bounds for the enabledeness of each action cannot be weakened. Strengthening of the lower bound is typical of superposition and reflects the fact that all the components that participate in the execution of a joint action have to give their permission for the action to be performed. On the other hand, it is clear that progress for a joint action can only be guaranteed when all the components involved can locally guarantee so.

Any diagram **D** in **SIGN** that establishes the interactions between a given set of components can be trivially lifted to a diagram **D'** of designs and superposition morphisms: the signature of each design is replaced by the design itself; every channel *ch* in **D** is replaced by **dsgn(ch)**— the design with signature *ch*, tautological bounds for the enabledeness of each action and the least deterministic assignment for the variables in the write frame of each action $(v : \in sort(v))$. Defined in this way, **dsgn(ch)** is a design that is "neutral" with respect to the establishment of superposition morphisms in the sense that every signature morphism $\sigma : ch \to sig(P)$ defines a superposition morphism $\sigma : dsgn(ch) \to P$. For simplicity, we will continue to use *ch* instead of **dsgn(ch)** in the configuration diagrams.

On the account of this transformation, every configuration can be transformed into a single design that represents the whole system by taking the colimit of the diagram **D'** in the category **c-DSGN**. Very roughly, the colimit "merges" the input variables identified with an output variable into that output variable, it establishes the synchronisation sets and it assigns to each of these sets an action whose bounds for enabledeness are the conjunction of the local bounds of each action in the set and whose assignments are the parallel composition of the assignments performed locally by each action in the set. In the next section we will present an example of the colimit construction.

2.3   Refinement

CommUnity supports several mechanisms for underspecification which can be reduced or eliminated through refinement. Concretely, in CommUnity designs, actions may be underspecified in the sense that their enabling conditions may not be fully determined and subject to refinement by reducing the interval established by *L* and *U*, and their effects on the variables may be undetermined and also subject to refinement by replacing the assignment sets by proper subsets.

Refinement of CommUnity designs can be modelled by the following morphisms.

A refinement morphism $\sigma: P_1 \rightarrow P_2$ consists of a signature morphism $\sigma: \theta_1 \rightarrow \theta_2$ s.t.:

1. $\sigma_{var}(inp(V_1)) \subseteq inp(V_2)$ *and* $\sigma_{var} \downarrow (out(V_1) \cup inp(V_1))$ *is injective;*
2. $\sigma_{ac}^{-1}(g) \neq \emptyset,\ g \in sh(\Gamma_1);$

For every $g \in \Gamma_2$ s.t. $\sigma_{ac}(g)$ is defined:

3. *for every* $v \in D_1(\sigma_{ac}(g)), \Phi \vDash (F_2(g, \sigma_{var}(v)) \subseteq \sigma(F_1(\sigma_{ac}(g), v)));$
4. $\Phi \vDash (L_2(g) \supset \sigma(L_1(\sigma_{ac}(g)))).$

For every $g_1 \in \Gamma_1$:

5. $\Phi \vDash (\sigma(U_1(g_1)) \supset \bigvee_{\sigma_{ac}(g_2) = g_1} U_2(g_2)).$

*Designs and refinement morphisms constitute the category **r-DSGN**.*

A refinement morphism supports the identification of a way in which a design P_1 is refined by another design P_2. Each variable of P_1 has a corresponding variable in P_2 and each action g of P_1 is implemented by the set of actions $\sigma_{ac}^{-1}(g)$ in the sense that $\sigma_{ac}^{-1}(g)$ is a menu of refinements for action g. The actions for which σ_{ac} is left undefined (the new actions) and the variables which are not in $\sigma_{var}(V_1)$ (the new variables) introduce more detail in the refined description of the component.

Condition 1 ensures that an input variable cannot be made local by refinement and that different variables of the interface cannot be collapsed into a single one (refinement does not alter the border between the system and its environment). Condition 2 ensures that those actions that model interaction between the design and its environment have to be implemented. Conditions 4 and 5 state that the "interval" of (allowed) non-determinism defined by the two bounds for enabledeness can only be preserved or reduced by refinement (refinement, pointing in the direction of an implementation, should reduce allowed non-determinism). The non-determinism of assignments must be preserved or decreased (condition 3).

For instance, the *help* design presented previously is a refinement of   a generic sender of messages.

```
design    sender[s] is
out       o:s
prv       rd:bool
do   prv  prod: ¬rd,false→o:∈s ‖ rd:=true
     []   send: rd,false→rd:=false
```

Notice that, in this design, it is left unspecified when and how many messages the sender will send as well as the discipline of production. The underlying refinement morphism is

$\eta: sender(nat) \rightarrow help \qquad \eta(o) = data, \eta(rd) = rd, \eta(read) = prod, \eta(send) = hack$

3   Architectural Connectors

Software Architecture has put forward connectors as first-class entities for modelling interactions between systems components. According to [1], a connector is defined by a set of *roles* and a *glue* specification. Each role describes the behaviour that is expected of each of the interacting parts, i.e., it determines the obligations that they have to fulfill to become instances of the roles. The glue describes how the activities of the role instances are coordinated.

Using the mechanisms that we have just described for configuration design in CommUnity, it is not difficult to come up with a formal notion of connector:

- *A connection consists of*
    - *two designs G and R, called the glue and the role of the connection, respectively;*
    - *a signature θ and two superposition morphisms $\sigma{:}dsgn(\theta){\rightarrow}G$, $\mu{:}dsgn(\theta){\rightarrow}R$ connecting the glue and the role.*
- *A connector is a finite set of connections with the same glue that, together, constitute a well-formed configuration.*

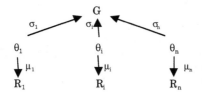

- *The semantics of a connector is the colimit of the diagram formed by its connections.*

For instance, asynchronous communication of values of type *s* through a bounded channel can be modelled by a connector *Async* with two roles — *sender* and *receiver*. These roles define the behaviour required of the components to which the connector can be applied. For the *sender* we require that it does not produce another message before the previous one has been processed. After producing a message, the *sender* should expect an acknowledgement to produce a new message. For the *receiver*, we simply require that it has an action that models the reception of a message. In CommUnity, the role *sender*   is modelled by the design *sender[s]* defined previously and the *receiver* can be designed as follows.

```
design receiver[s] is
in     i:s
do     rec: true,false→skip
```

In order to leave unspecified when and how many messages the receiver will receive the progress guard of *rec* is false (in this way, the enabling condition can be as strong as we wish).

The glue of *Async* is the design *buffer[s]* presented previously. It prevents the sender from sending a new message when there is no space and prevents the receiver from reading a new message when there are no messages.

Finally, the configuration below establishes how the roles and the glue of *Async* are connected.

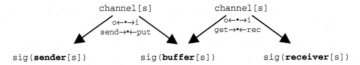

$$sig(\textbf{sender}[s])\qquad sig(\textbf{buffer}[s])\qquad sig(\textbf{receiver}[s])$$

The left-hand side morphisms define that *sender* and *buffer* must synchronise on actions *send* and *put*, and establish the interconnection of the output variable o of *sender* with the input variable i of *buffer*. On the other hand, the right-hand side morphisms define that *buffer* and *receiver* must synchronise on actions *get* and *rec* and establish the interconnection of the output variable o of *buffer* with the input variable i of *receiver*.

The semantics of connector *Async* is given by the colimit of its configuration diagram, which returns the design given, up to an isomorphism, by

```
design async[s] is
out      o_s,o_b:s
prv      rd_s,rd_r:bool; q:queue(K,s)
do prv   prod:¬rd_s, false→ o_s:∈s||rd_s:=true
[]       send|put:¬full(q)∧¬rd_s,false→q:=enqueue(o_s,q)||rd_s:=false
[] prv   next:¬empty(q)∧¬rd_r→ o_b:=head(q)|| q:=tail(q)||rd_r:=true
[]       rec|get: rd_r, false → rd_r:=false
```

This design provides the means for global properties of the protocol that *Async* defines to be derived. For instance, reasoning about this design, it is possible to conclude that no messages are lost and that the correctness of the transmission/ reception of data (in order message delivery) does not depend on the speed at which messages are produced and consumed.

The connectors we have described so far are connector *types* in the sense that they can be instantiated. More concretely, the roles of a connector type can be instantiated with specific designs. Role instantiation has to obey a compatibility requirement expressed via the refinement relationship and, hence, an instantiation of a connector can now be defined as follows:

– *An instantiation of a connection with role R consists of a design P together with a refinement morphism $\eta:R{\to}P$.*
– *An instantiation of a connector consists of an instantiation for each of its connections.*

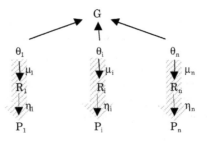

Figure 1

- *The semantics of a connector instantiation is the colimit of the diagram in c-DSGN formed as described above by composing the role morphism of each connection with its instantiation.*

For instance, the connector *Async* can be used to interconnect the *help* component with a given receiver. This can be achieved by instantiating the role *sender* of *Async* with *help,* through the refinement morphism η defined previously.

It is important to notice that the formalisation of architectural connectors we presented is not specific to any design formalism insofar as the involved categories are not fixed. However, there are some properties that a design formalism needs to satisfy to support such architectural concepts. For instance, the semantics of a connector instantiation relies on the possibility of composition of the role morphism of each connection with its instantiation.

This property and other properties that a design formalism need to satisfy to support such architectural concepts are discussed in detail in [4]. Herein, we shall simply explain why it is possible to compose a role morphism with its instantiation in the case of CommUnity.

In CommUnity, each instantiation $\eta:R{\rightarrow}P$ of a connection is composable with μ in order to define $\mu;\phi:\theta{\rightarrow}sig(P)$ because θ is according to the rules that define well-formed configurations and, hence, has no private variables, which means that at the level of signatures the refinement morphism has the same properties as a composition morphism over θ. As we have already argued in section 2.2, every such signature morphism can be lifted to a design morphism $\mu;\phi:dsgn(\theta){\rightarrow}P$. Hence, an instantiation of a connector defines a diagram in *c-DSGN* that connects the role instances to the glue. Moreover, because each connection is according to the rules set for well-formed configurations as detailed in the previous section, the diagram defined by the instantiation is, indeed, a configuration and, hence, has a colimit.

In the sequel, we will use $C+(\eta_i)$ to denote the configuration depicted in figure 1, obtained by composing the role morphism of each connection i of C with its instantiation η_i.

4 Higher-Order Architectural Connectors

As explained before, it is important that connectors can be designed compositionally, by combining different interaction capabilities. In particular, it is important to have principled forms of adapting connectors to new situations, for instance in order to incorporate compression, fault-tolerance, security, monitoring, etc.

Let us consider for instance *compression*. In this case, the goal is to adapt a connector that represents a unidirectional communication protocol in order to compress data for transmission in a transparent way.

A generic unidirectional communication protocol can be modelled by the binary connector *Uni-comm[s]*

where

```
design glue[s] is
in      i:s
out     o:s
do      put: true,false → skip
[]  prv prod: true,false → o:∈s
[]      get: true,false → skip
```

and *sender[s]* and *receiver[s]* are defined as before. Notice that this glue leaves completely unspecified the way in which messages are processed and transmitted.

Our aim is to install a compression/decompression service over *Uni-comm*. That is to say, our aim is to apply an operator to *Uni-comm* such that, in the resulting connector, a message sent by the sender is compressed before it is transmitted through *Uni-comm* and then decompressed before it is delivered to the receiver.

It is not difficult to realize that this form of coordination of the sender and receiver activities, embodied by the glue of the new connector, can be obtained by instantiating the sender role of *Uni-comm* with a component *comp* that compresses messages before it transmits them, and by instantiating the sender role of *Uni-comm* with a component *decomp* that decompresses the message it receives.

Such components can be designed in CommUnity as follows

```
design comp is
in      di:t
out     co:s
prv     v:t; rd,msg:bool
do      drec: ¬msg → v:=di‖ msg:=true
[] prv  comp:¬rd∧ msg → co:=comp(v)‖ rd:=true
[]      csend:rd → rd:=false‖ msg:=false

design decomp is
in      ci:s
out     do:t
prv     v:s; rd,msg:bool
do      crec: ¬msg → v:=ci‖ msg:=true
[] prv  dec:¬rd∧msg → do:=decomp(v)‖ rd:=true
[]      dsend: rd → rd:=false‖ msg:=false
```

and the instantiation of *Uni-comm* with *comp* and *decomp* can be described by the refinement morphisms

$$\eta^*_s\text{:}sender(s)\rightarrow comp \qquad \eta^*_s(o)=co, \eta^*_s(rd)=rd, \eta^*_s(comp)=prod, \eta^*_s(csend)=send$$
$$\eta^*_r\text{:}receiver(s)\rightarrow decomp \ \ \eta^*_r(i)=ci, \eta^*_r(crec)=rec$$

In this way, the glue of the new connector is given by

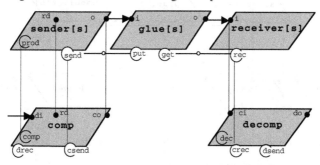

where hairlines were used to represent the refinement relationship.

It remains to define that, in the new connector, the messages sent by the sender are received by the *comp* component and that *decomp* delivers the decompressed messages to the receiver. The connector *Compression* depicted below models this form of coordination.

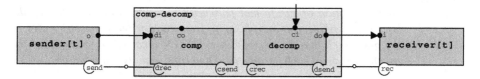

Notice that the glue, *comp-decomp*, is defined in terms of a configuration where *comp* and *decomp* do not interact.

In summary, the procedure we described for installing the compress/decompress service over *Uni-comm*, that models a generic unidirectional communication protocol, is described by

- the connector *Compression;*
- the refinement morphisms η_s:*sender(s)*→*comp-decomp* and η_r:*receiver(s)* → *comp-decomp* induced, respectively, by η^*_s and η^*_r (because *comp* and *decomp* do not interact, any component refined by one of them is also refined by their composition).

It is not difficult to realise that this procedure can be applied to more specific connectors provided they model unidirectional communication protocols. In fact, we may regard *Uni-comm* as a formal parameter of this description, that can be instantiated with different connectors. The connector *Compression* can be regarded as the body of the definition, that models the nature of the service that is superposed on instantiation of the formal parameter. Finally, the refinement morphisms establish a relationship between the parameter and the body connectors.

Putting the three previous pictures together we get a graphical representation of this parameterised entity—*Compression(Uni-comm)*, and is an example of what we call a higher-order connector.

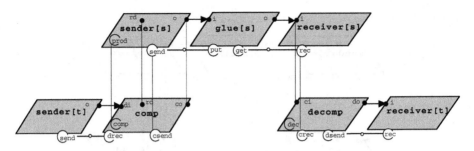

Higher-order connectors can be formally defined as follows.

— *A higher-order connector (hoc) consists of*

- *a connector pC, called the formal parameter of the hoc; its roles, glue and connections are called, respectively, the parametric roles, the parametric glue and the parametric connections of the hoc;*

- *a connector C – its roles and glue are also called the roles and the glue of the hoc;*

- *an instantiation of the formal parameter connector with the glue of the hoc, i.e., a refinement morphism η_i from each of the parametric roles to the glue, such that the diagram in **c-DSGN** obtained by composing the role morphism of each parametric connection with its instantiation*

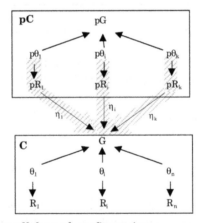

constitutes a well-formed configuration.

— *The semantics of a higher-order connector is the connector depicted below. Its roles are the roles of C and its glue is G', a design returned by the colimit of the configuration pC+(η_i), obtained by composing the role morphism of each parametric connection with its instantiation.*

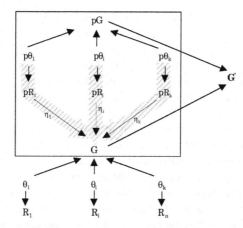

Now it remains to explain the procedure of parameter passing, i.e., how a higher-order connector can be applied to a specific connector and how the resulting connector is obtained.

Let us suppose that we want to install the compression service over the *Async* connector. In this case, it is not difficult to realize that we may replace the formal parameter of *Compression(Uni-comm)* by *Async* because this connector does model a unidirectional communication protocol. More concretely, *Async* has exactly the same roles that *Uni-comm* and its glue is a refinement of *Uni-comm*'s glue.

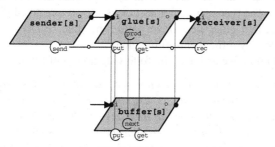

The construction of a new connector from the given higher-order connector and the actual parameter connector is straightforward. We only need to compose the interconnections of the *buffer* to *sender* and *receiver* with the refinements η_s and η_r that define the instantiation of *Uni-comm* with *comp* and *decomp*, respectively. For example, variable *co* of *comp* becomes connected to the input variable *i* of *buffer* because *co* corresponds to the variable *o* of *sender* which in turn is, in *Async*, connected to *i*. The resulting configuration fully defines the connector *Compression(Async)*. Its roles are *sender* and *receiver* and its glue — *c-buffer-d*, is defined in terms of a configuration involving *comp*, *decomp* and *buffer* as shown below.

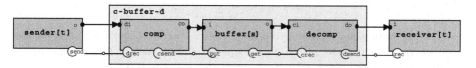

In this example, the formal and the actual connectors have exactly the same roles and the instantiation of the higher-order connector is established merely by a refinement morphism from the glue of the former to the glue of the later. However, the instantiation of a higher-order connector can be defined in a more general way, through the definition of a notion of fitting morphism between connectors. The instantiation of a higher-order connector is then established by a fitting morphism from the formal to the actual connector.

We shall now discuss the suitable notion of fitting morphism between connectors. Because the formal parameter of a higher-order connector defines the kind of connectors to which the higher-order connector can be applied, its instantiation has to obey a compatibility requirement. Intuitively, a fitting morphism from a connector C_1 to a connector C_2 must express that it is possible to use, in the design of a given system, C_2 in place of C_1 in the sense that the functionality of the system is preserved.

We first notice that, for this be possible, the two connectors must have the same number of roles. Furthermore, C_2 has to admit to be instantiated with the same components than C_1. That is to say, every restriction on the components to which C_2 can be applied must also be a restriction imposed by C_1. In this way, fitting morphisms must establish a correspondence between the roles of C_1 and C_2 and must require that each of the roles of C_2 is refined by the corresponding role of C_1.

We have seen that connectors may be based on glues that are not fully developed as designs (may be underspecified) and, nevertheless, the concrete commitments that have already been made determine in some extent the type of interconnection that the connector will ensure. The type of interconnection is clearly preserved if we simply consider a less unspecified glue, i.e., if we refine the glue. Hence, fitting morphisms must allow for arbitrary refinements of the glue.

Having this in mind, we arrive at the following notion of fitting morphism:

– *A fitting morphism ϕ from a connection $<\sigma_1: \textbf{dsgn}(\theta_1) \rightarrow G_1, \mu_1:\textbf{dsgn}(\theta_1) \rightarrow R_1>$ to a connection $<\sigma_2:\textbf{dsgn}(\theta_2) \rightarrow G_2, \mu_2:\textbf{dsgn}(\theta_2) \rightarrow R_2>$ consists of a pair $<\phi_G: G_1 \rightarrow G_2, \phi_R:R_2 \rightarrow R_1>$ of refinement morphisms s.t. the interconnection $<\sigma_1, \mu_1>+\phi_G$ of R_1 with G_2 is refined by the interconnection $<\sigma_2, \mu_2>+\phi_R$ (that is to say, there exists a refinement morphism from the colimit $<\sigma_1, \mu_1>+\phi_G$ of to the colimit of $<\sigma_2, \mu_2>+\phi_R$).*

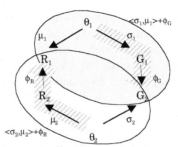

– *A fitting morphism ϕ from a connector C_1 to a connector C_2 with the same number of connections consists of a fitting morphism ϕ from each of C_1's connections to each of C_2's connections, all with the same glue refinement ϕ_G.*

Based on these fitting morphisms between connectors, we may finally define the instantiation of a higher-order connector.

– *An instantiation of a higher-order connector with formal parameter pC (figure 2) consists of a connector C^A (the actual parameter) together with a fitting morphism $\phi:pC \rightarrow C^A$, such that the diagram in c-DSGN obtained by composing the role morphisms of each actual connection with the corresponding fitting component and then with the role instantiation (figure 3) constitutes a well-formed configuration.*

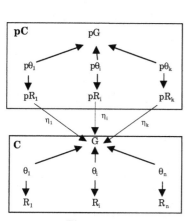

Fig. 2

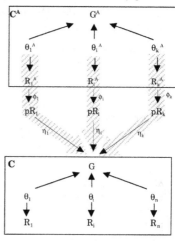

Fig. 3

– *The semantics of a higher-order connector instantiation is the connector with the same roles as C and its glue is a design returned by the colimit of the configuration $C^A+(\phi_i;\eta_i)$.*

For simplicity, we defined higher-order connectors with one parameter only. However, the definition can be extended to the case of several parameters in a straightforward way. We end this section by presenting an example of a higher-order connector with two parameters that models a fault-tolerance service. Specifically, we aim to describe a way of combining two unidirectional communication protocols in order to obtain a unidirectional communication protocol that provides reliable and in order message delivery, in the presence of messages loss and duplication faults.

The idea behind this fault-tolerance service is that messages must be numbered before they are transmitted and then must be de-numbered before they are delivered to the receiver. Moreover, the messages that arrive out-of-order must be ignored (not transmitted to the receiver) and in response to a message an acknowledgment with its number must be transmitted.

This service can be modelled by a higher-order connector with two parameters. One represents the connector used to transmit the messages (after being numbered) and the other represents the connector used to transmit the acknowlegment messages (which in fact will be natural numbers) in the opposite direction. Clearly, each of these connectors should model a unidirectional communication protocol.

The higher-order connector itself, *Ft(Uni-comm[t],Uni-comm[t*nat])*, consists of

– the connector *Numbering* defined by

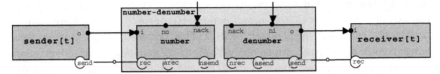

where the glue, *Number-Denumber*, is defined in terms of a configuration with the following two components:

```
design number is
in      i:t,nack:nat
out     no:t*nat
prv     b: list(t), current:t+null, k:nat, rd: bool
do      rec: ¬full(b)→b:=b.i
[]  prv take:¬empty(b)∧current=null→current:=hd(b)||b:=tl(b)||k:=k+1
[]  prv nprod:¬rd∧current≠null → rd:=true||no:=<current,k>
[]      nsend: rd → rd:=false
[]      arec: true → current:=if(nack=k,null,current)

design denumber is
in      ni:t*nat
out     nack:nat, o:t
prv     b: list(t), n,k:nat, rdₐ,rd₈: bool
do      nrec: ¬rdₐ∧¬full(b) → b:=if(snd(ni)=k+1,b.fst(ni),b)
                              || k:=if(snd(ni)=k+1,k+1,k)||n:=snd(ni)
[]  prv aprod: ¬rdₐ → nack:=n ||rdₐ:=true
[]      asend: rdₐ → rdₐ:=false
[]      prod: ¬empty(b)∧¬rd₈ → o:=hd(b)||b:=tl(b)||rd₈:=true
[]      send: rd₈ → rd₈:=false
```

Component *number* numbers the messages to be transmitted, sending repeatedly the same message until an acknowledgement with its number is received. Meanwhile, messages to be transmitted are stored in a buffer. Component *denumber* receives numbered messages through *ni* and transmits through *nack* an acknowlegment with the message number. Meanwhile, it stores the messages that arrive in order and delivers them to the receiver.

– the connectors *Uni-comm[t*nat]* and *Uni-comm[nat]* — the two formal parameters;

– the refinement morphisms

η_s:*sender(t*nat)→number-denumber*, η_r:*receiver(t*nat)→number-denumber*
κ_s:*sender(nat)→number-denumber*, κ_r:*receiver(nat)→number-denumber*

induced respectively by

η^*_s:*sender(t*nat)→number*
$\qquad \eta^*_s(o)=no, \eta^*_s(rd)=rd, \eta^*_s(nprod)=prod, \eta^*_s(nsend)=send$
η^*_r:*receiver(t*nat)→denumber*
$\qquad \eta^*_r(i)=ni, \eta^*_r(nrec)=rec$
κ^*_s:*sender(nat)→denumber*
$\qquad \kappa^*_s(o)=nack, \kappa^*_s(rd)=rd_a, \kappa^*_s(aprod)=prod, \kappa^*_s(asend)=send$
κ^*_r:*receiver(nat)→number*

$\kappa^*_r(i)=nack, \kappa^*_r(arec)=rec$

It is important to notice that the two parameters of the higher-order connector *Ft(Uni-comm[t],Uni-comm[t*nat])* can be instantiated with two different connectors but they can also be both instantiated with the same connector, provided this connector encompasses both capabilities — transmission of numbered messages in one direction and transmission of natural number in the other direction.

5   Conclusions

In this paper we proposed a notion of higher-order connector and showed that this abstraction indeed facilitates the separation of concerns in the development of complex connectors and their compositional construction. In particular, we have shown how a compression service and fault-tolerance service can be modelled separately as higher-order connectors. Furthermore, we continued our previous work on the use of Category Theory to formalise key notions of software architecture. Building on the categorical semantics for the notion of architectural connector, we formalised higher-order connectors and established their categorical semantics.

Our definitions agree with Garlan's original proposal [8] of a hoc as an operator over connectors for supporting connector construction through incremental transformation, hence allowing one to define more complex interactions in a more systematic way. More concretely, Garlan and Spitznagel [9] propose that a connector transformation be modelled as a function — from one or more connectors to a new connector — defined in terms of its inputs, preconditions on its application and postconditions on its result. They formalise these ideas in the context of a particular ADL, namely Wright, relying on the specific language and semantics of CSP.

We showed that a transformation of a connector can be modelled by a parameterised entity that is essentially constituted by two connectors. One of these connectors is the formal parameter that defines the kind of connectors the transformation can be applied to. The other connector — the body of the hoc — concerns the transformation itself. Owing to the formal semantics of hocs, the transformation can be understood and analysed. Although we have used CommUnity to motivate and illustrate our ideas, by adopting a categorical framework for formalising these ideas, we achieved independence relatively to the ADL.

Due to space limitation, we did not address the composition of higher-order connectors but the categorical framework leads naturally to a notion of composition of higher-order connectors that is useful for combining orthogonal properties. Such composition is naturally defined by considering a more general form of instantiation — *parameterised instantiation*, more specifically the instantiation of hocs with hocs. This composition supports the combination of different kinds of functionality, modelled separately by different higher-order connectors, giving rise also to a higher-order connector. In this way, it is possible to analyse the properties that such compositions exhibit, namely to investigate whether undesirable properties emerge and desirable properties are preserved.

As mentioned before, the individual specification of independent aspects such as compression and fault-tolerance as higher-order connectors makes it easier to evolve systems at run-time. Through run-time reconfiguration of the system architecture, namely through the replacement of connectors, such services may be added only when necessary, hence preventing performance penalties when such complex interactions are not required. Work reported in [15] addresses the support that is required for an architectural-driven process of reconfiguration in which connectors, as well as components, can be replaced, added or deleted.

References

1. R.Allen and D.Garlan, "A Formal Basis for Architectural Connectors", *ACM* TOSEM, 6(3):213-249, July 1997.
2. K.Chandy and J.Misra, *Parallel Program Design - A Foundation*, Addison-Wesley 1988.
3. G.Denker, J.Meseguer and C.Talcott, "Rewriting semantics of meta-objects and composable distributed services", Internal report, Computer Science Laboratory, SRI International, 1999.
4. J.L.Fiadeiro, A.Lopes and M.Wermelinger, "A Mathematical Semantics for Architectural Connectors". Submitted for publication (available at http://www.fiadeiro.org /jose/papers)
5. J.L.Fiadeiro and A.Lopes, "Algebraic Semantics of Coordination, or what is in a signature?", in *AMAST'98*, A.Haeberer (ed), LNCS 1548, Springer-Verlag 1999.
6. J.L.Fiadeiro and A.Lopes, "Semantics of Architectural Connectors", in TAPSOFT'97, LNCS 1214, Springer-Verlag 1997, 505-519.
7. N.Francez and I.Forman, Interacting Processes, Addison-Wesley 1996.
8. D.Garlan, "Higher-Order Connectors", Presented at the Workshop on Compositional Software Architectures, Monterey, CA, January 6-7, 1998.
9. D.Garlan and B.Spitznagel, "Toward compositional construction of complex connectors", Proceedings of the Eighth International Symposium on the Foundations of Software Engineering (FSE-8), November 2000.
10. S.Katz, "A Superimposition Control Construct for Distributed Systems", *ACM TOPLAS* 15(2):337-356, 1993.
11. N.Mehta, N.Medvidovic and S.Phadke, "Towards a taxonomy of software connectors", Proc. of 22nd International Conference on Software Engineering, ACM Press, 2000, 178-187.
12. M.Wermelinger, A.Lopes and J.L.Fiadeiro, "Superposing Connectors", in *Proc. 10th International Workshop on Software Specification and Design*, IEEE Computer Society Press 2000, 87-94. (also available at http://ctp.di.fct.unl.pt/~mw/proj/fast /index.html)
13. M.Wermelinger and J. L. Fiadeiro, "Connectors for mobile programs", IEEE Trans. on Software Eng., 24(5):331--341, May 1998.
14. M.Wermelinger and J.L.Fiadeiro, "Algebraic Software Architecture Reconfiguration", in *Software Engineering – ESEC/FSE'99*, LNCS 1687, pp. 393-409, Springer-Verlag 1999.
15. M.Wermelinger, A.Lopes and J.L.Fiadeiro, "A Graph Based Architectural (Re)configuration Language", Proc. ESEC/FSE'01, ACM Press, 2001. In print..

Institution Independent Static Analysis for CASL[*]

Till Mossakowski[1] and Bartek Klin[2]

[1] BISS, Department of Computer Science, Bremen University
[2] BRICS, Århus University

Abstract. We describe a way to make the static analysis for the in-the-large part of the Common Algebraic Specification Language (CASL) independent of the underlying logic that is used for specification in-the-small. The logic here is formalized as an institution with some extra components. Following the institution independent semantics of CASL in-the-large, we thus get an institution independent static analysis for CASL in-the-large. With this, it is possible to re-use the CASL static analysis for extensions of CASL, or even completely different logics. One only has to provide a static analysis for specifications in-the-small for the given logic. This then can be plugged into the generic static analysis for CASL in-the-large.

1 Introduction

The specification language CASL [Mos97,CoFa,CoF01], based on subsorted partial first-order logic, is the central language of a whole family of languages. Current research includes the development of *extensions* of CASL that also consider the specification of higher-order functions [MHKB00,SM] and of reactive [RR00], [BZ00] and object-oriented [ACZ00] behaviour. Several *restrictions* of CASL to sublanguages [Mosc] make it possible to use specialized tool support. Connected with this, CASL has a clean separation between specification in-the-small and specification in-the-large. Specification in-the-small here means specification of individual software modules using signatures and axioms, while specification in-the-large concerns the combination of both specifications and software modules. Following this design, also the semantics of CASL exhibits this separation of concerns: The semantics of specification in-the-small is based on a particular institution [GB92], while the semantics of specification in-the-large is defined over an arbitrary but fixed institution [CoFb,Mos00a]. (Strictly speaking, institutions here are replaced by so-called institutions with qualified symbols [Mos00a] in order to admit symbol sets and symbol maps as basic primitives, instead of signature morphisms.) This separation of levels in the semantics makes it possible to re-use the semantics of CASL in-the-large also for the extensions and sublanguages of CASL. Only the semantics of CASL in-the-small has to be adapted individually for each extension (and restricted appropriately for the sublanguages).

[*] This research was supported by the ESPRIT-funded CoFI Working Group 29432 and by the DFG project MULTIPLE.

M. Cerioli and G. Reggio (Eds.): WADT/CoFI 2001, LNCS 2267, pp. 221–237, 2002.

Based on the CASL semantics, we have implemented a static analyser for CASL, which is part of the CASL tool set CATS ([Mosa,Mos00b], see also the architecture of CATS in Fig. 1). Now for tool development it is desirable to have a similar separation of levels as in the semantics.

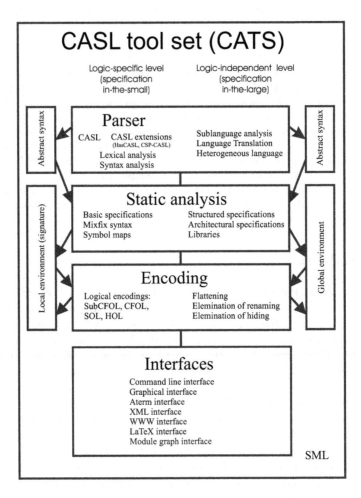

Fig. 1. Architecture of the CASL tool set (CATS)

In this work, we are therefore making our static analysis of CASL in-the-large institution independent, based on the institution independent semantics. This permits a similar economy as for the semantics: the static analysis tools for extensions and sublanguages of CASL need only re-implement the institution-specific part, while the static analysis of CASL in-the-large remains the same for all institutions. As a side-effect, the modular design of the CASL tool set now becomes a good case study for a CASL architectural specification. Architectural specifications [BST98] are a novel feature of CASL; they allow to describe branching points in system development by indicating units (modules) to be independently developed and showing how these units, once developed, are to be put together to produce the overall result. In the context of the present work, this decomposition roughly will be one into institution-specific and institution independent parts.

The main contribution of this work is not the introduction of new concepts or proof of new results. Rather, it is a bridge between theory and practice. We feel that this is quite an important topic, since it is important both to know whether and how the theoretical concepts really work in practice, and to give a solid theoretical basis for practical implementation work. Along these lines, this paper can also be seen as a step towards bootstrapping CASL by using CASL to develop CASL tools. Of course, much has to be done to achieve this goal, but we feel that such a successful bootstrap would be a very convincing argument in favour of CASL and the methodologies behind it.

The paper is organized as follows: Section 2 recalls the notion of institution with qualified symbols, while section 3 recalls the division of the CASL design into different layers (cleanly separating specification in-the-small from specification in-the-large). Based on this, in section 4, we informally describe how to separate these layers also for the static analysis, thus obtaining a generic static analysis for CASL in-the-large. In section 5, we specify this as a CASL architectural specification. Section 6 contains the conclusions.

2   Preliminaries: Institutions with Qualified Symbols

The notion of *institution* [GB92] formalizes what a logical system is. The theory of institutions takes a predominantly model-theoretic view of logic, with the satisfaction relation between models and logical sentences adopted as a primary notion. Somewhat unlike in the classical model-theory though, a family of such relations is considered at once, indexed by a category of signatures.

Definition 1. An *institution* **I** consists of:

- a category **Sign** of *signatures*,
- a functor **Sen**: **Sign** $\rightarrow$ **Set**, giving a set **Sen**(Σ) of Σ-*sentences* for each signature $\Sigma \in |\textbf{Sign}|$,
- a functor **Mod**: **Sign**^{op} $\rightarrow$ **Cat**[1], giving a class **Mod**(Σ) of Σ-*models* for each signature $\Sigma \in |\textbf{Sign}|$, and
- for $\Sigma \in |\textbf{Sign}|$, a *satisfaction relation* $\models_\Sigma \subseteq \textbf{Mod}(\Sigma) \times \textbf{Sen}(\Sigma)$,

such that for any signature morphism $\sigma \colon \Sigma \rightarrow \Sigma'$, Σ-sentence $\varphi \in \textbf{Sen}(\Sigma)$ and Σ'-model $M' \in \textbf{Mod}(\Sigma')$ the following *satisfaction condition* holds:

$$M' \models_{\Sigma'} \textbf{Sen}(\sigma)(\varphi) \iff \textbf{Mod}(\sigma)(M') \models_\Sigma \varphi.$$

---

[1] **Cat** is the (quasi-)category of all categories.

We write $M'|_\sigma$ for $\mathbf{Mod}(\sigma)(M')$, where $\sigma\colon \Sigma \longrightarrow \Sigma' \in \mathbf{Sign}$ and $M' \in \mathbf{Mod}(\Sigma')$.

In the CASL institution ([CoFb]), signatures are the usual many- and sub-sorted signatures with partial and total operations and predicates, and sentences are formulas of the subsorted partial first order logic with equality and sort generation constraints.

Institutions with qualified symbols add further structure to institutions, mainly to deal with names and qualifications of symbols and with generation of signature morphisms from concise and user-friendly symbol maps. Therefore, an institution with qualified symbols comes along equipped with an underlying set of (fully qualified) symbols, for each signature, and an underlying symbol translation map, for each signature morphism. Moreover, there also is a notion of raw symbols, which includes besides the fully qualified symbols also unqualified or partially qualified symbols as they may be input in specifications.

In order to formalize this, consider an institution $(\mathbf{Sign}, \mathbf{Sen}, \mathbf{Mod}, \models)$ additionally equipped with a faithful functor $|_-|\colon \mathbf{Sign} \longrightarrow \mathbf{Set}$ that extracts from each signature the set of (fully qualified) symbols that occur in it (i.e. $(\mathbf{Sign}, |_-|)$ is a concrete category [AHS90]). We assume that there is some fixed 'universe' of Sym of symbols that may be used by the specifier when writing symbols (and symbol mappings). In the CASL institution, the natural choice for $|\Sigma|$ is the set of fully qualified symbols of Σ (if we omit qualifications, $|_-|$ is no longer faithful, because symbols may be overloaded with different profiles).

Now in CASL symbol mappings, one may (either partially or completely) omit qualifications of symbols. This leads to the notion of *raw symbol*, which in the case of the CASL institution can be a qualified symbol, an unqualified symbol or a partially qualified symbol. The link between symbols and raw symbols is given by a *matching relation* specifying which symbols correspond to which raw symbols. Finally, in CASL, fitting maps for instantiations of parameterized specifications are automatically extended to compound identifiers, such that they also act on the components. In order to mimick this behaviour within an arbitrary institution, we further assume that there is a set ID of *compound identifiers*. This leads to the following definition:

Definition 2 ([Mos00a]). An *institution with qualified symbols* $(\mathbf{Sign}, \mathbf{Sen}, \mathbf{Mod}, \models, Sym, |_-|, \mathrm{ID}, RawSym, IDAsRawSym, SymAsRawSym, matches)$ consists of

- an institution $(\mathbf{Sign}, \mathbf{Sen}, \mathbf{Mod}, \models)$,
- a set of (fully qualified) symbols Sym,
- a faithful functor $|_-|\colon \mathbf{Sign} \longrightarrow \mathbf{Set}$,
- a set ID of compound identifiers,
- a set of raw symbols $RawSym$ with two injections $IDAsRawSym\colon \mathrm{ID} \longrightarrow RawSym$ and $SymAsRawSym\colon Sym \longrightarrow RawSym$,
- a matching relation $matches \subseteq Sym \times RawSym$ specifying which qualified symbols match which raw symbols,

such that

- $|\Sigma| \subseteq Sym$ for each $\Sigma \in |\mathbf{Sign}|$,

- for $id, id_1, \ldots, id_n \in$ ID, also $id[id_1, \ldots, id_n] \in$ ID (i.e. we can form compound identifiers),
- SY matches $SymAsRawSym(SY')$ iff $SY = SY'$ for $SY, SY' \in Sym^2$ and
- for each $SY \in Sym$, there is a unique $Ident \in$ ID with SY matching $IDAsRawSym(Ident)$, called the *name* of SY.

3   The Different Layers of Casl

The design of CASL has been structured in four different layers, which are largely orthogonal to each other. The first layer, CASL *basic specifications*, allows to formalize axiomatic requirements for a single software module in a specific logic. This corresponds to specification in-the-small. The other three layers are devoted to specification in-the-large: *Structured specifications* allow to combine specifications in a structured way, still refering to specification of *single* software modules. *Architectural specifications* allow to prescribe how to decompose the task of implementing a specification into smaller sub-tasks which can be implemented independently. Finally, basic, structured and architectural specifications can be collected into *libraries*.

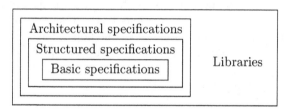

The CASL semantic concepts are structured in layers in a similar way:

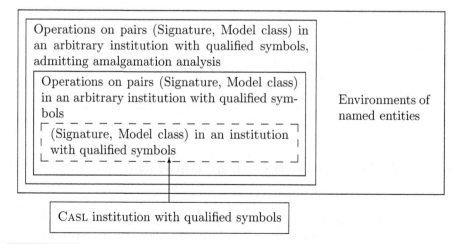

---

[2] This property is not technically needed in the semantics, but it is desirable since it means that for each ambiguity there is a qualification to resolve it.

The semantics of a CASL basic specification is a signature together with a model class within the CASL *institution with qualified symbols* (cf. Definition 2 above). Now, insofar as the other layers are concerned, the layer of CASL basic specifications can be replaced by any other institution with qualified symbols. More precisely, the semantics of structured specifications can be formalized over an *arbitrary* institution with qualified symbols. To define semantics of architectural specifications, the underlying institution needs to satisfy some additional conditions, connected to the amalgamation property. The CASL institution does not have the amalgamation property, but this problem can be circumvented using an embedding into an enriched institution that has the property, as described in [SMH+01].

4   Generic Static Analysis

Now the structure of static analysis follows this layer structuring as well. This enables us to turn the analysis of structured and architectural specifications into a generic program, with the interface consisting of sorts and operations capturing the static part of an institution with qualified symbols (i.e., the signature category and the symbol functor) and, moreover, the static analysis for basic specifications and for symbol maps. Note that symbol maps are usually considered to be part of structured specifications in the CASL documents. However, since they are institution-specific and there is no hope to make them institution-independent, we here count them to the layer of basic specifications.

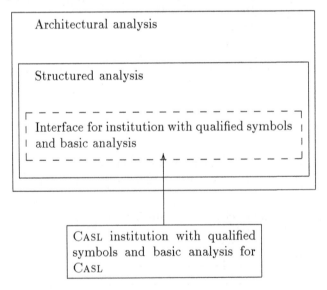

The generic structured and architectural analysis can then be instantiated with a program module providing the ingredients of the CASL institution with qualified symbols, and a basic analysis for it. Of course, it is now possible to supply any other institution with symbols here.

Note that we have omitted the layer of libraries so far. This has been deliberate, since the analysis of libraries is entirely orthogonal to the rest: we only need to assume that we have a notion of library item and a static analysis for library items. The library analysis then just performs the item analysis for each library item in a given library, and takes care of downloads from other libraries.

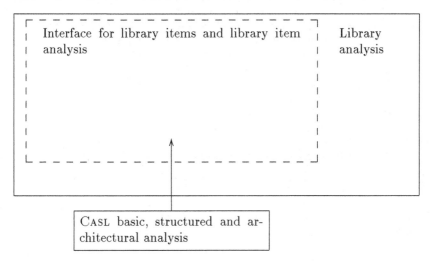

The generic library analysis can then be applied to the library item analysis for CASL basic, structured and architectural specifications. Note that we presently do not have in mind to apply the library analysis within other contexts (although it would be easy to do so). The use of a generic program here mainly has the reason of a clean separation of concerns.

5   Generic Static Analysis Specified in Casl

Our task is now to formalize the graphical visualization of the separation of layers in some way. This is exactly the task CASL architectural specifications have been designed for. Hence, an architectural specification of the overall structure of the static analysis is given in Fig. 2.

The unit I, satisfying the specification INSTITUTION (the specification is given in Fig. 3 and Fig. 4 below) contains implementation of all the institution-specific components needed in static analysis of structured and architectural specifications:

- Abstract syntax of basic specifications, symbol maps and symbol lists conforming to a simple interface specified in the specification BASICABSTRACT-SYNTAX (given below),
- a signature category together with a symbol functor and other components of an institution with qualified symbols,
- operations performing basic analysis for the institution, i.e., analysis of basic specifications, symbol maps and symbols lists (note that the analysis of

arch spec LIBRARYANALYSIS =
units I : INSTITUTION;
         SAS : INSTITUTION $\rightarrow$ STRUCTUREDABSTRACTSYNTAX;
         AAS : STRUCTUREDABSTRACTSYNTAX $\rightarrow$ ARCHABSTRACTSYNTAX;
         $GEnv$: INSTITUTION $\rightarrow$ GLOBALENV;
         $StrAna$: INSTITUTION $\times$ STRUCTUREDABSTRACTSYNTAX $\times$ GLOBALENV
                    $\rightarrow$ STRUCTUREDANALYSIS;
         $ArchAna$: INSTITUTION $\times$ ARCHABSTRACTSYNTAX $\times$ GLOBALENV $\times$
                    STRUCTUREDANALYSIS $\rightarrow$ ARCHANALYSIS;
         $LibAna$: ABSTRACTLIBITEMANALYSIS $\rightarrow$ LIBANALYSIS;
         SS $=$ $SAS[I]$;
         AS $=$ $AAS[SS]$;
         G $=$ $GEnv[I]$;
         Str $=$ $StrAna[I][SS][G]$;
         A $=$ $ArchAna[I][AS][G][Str]$;
result
         $LibAna$ [A **fit** $Entry \mapsto global\_entry$, $Env \mapsto global\_env$,
                    $analysis \mapsto arch\_analysis$]
end

Fig. 2. Specification of the static analysis of the CASL tool set as a CASL architectural specification

basic specifications returns a basic specification again, since mixfix grouping analysis might change the initial abstract syntax tree),
- some additional operations, which in principle can be defined over an arbitrary institution with qualified symbols, but only in a very inefficient to compute (if computable at all) way. Hence, to achieve good performance, the task of implementing these operations is assigned to the author of the unit I. These operations include computation of signature union, subsignature relation, induction of signature morphisms from raw symbol maps and a few others,
- an additional operation performing so-called sharing analysis, needed in the static analysis of architectural specifications. In the recently developed diagram semantics of architectural specifications [SMH+01,SMH+], each part of an architectural specification is evaluated to a diagram $D : \mathbf{I} \rightarrow \mathbf{Sign}$, where $\mathbf{I}$ is a finite category, and $\mathbf{Sign}$ is the signature category of the underlying institution. A family of models $\langle M_i \rangle_{i \in \mathbf{Ob(I)}}$ is *compatible* with D if for all $i \in \mathbf{Ob(I)}$, $M_i \in \mathbf{Ob(Mod}(D(i)))$, and for all $m : i \rightarrow j$ in $\mathbf{I}$, $M_i = \mathbf{Mod}(D(m))(M_j)$. For two diagrams $D : \mathbf{I} \rightarrow \mathbf{Sign}$, $D' : \mathbf{I}' \rightarrow \mathbf{Sign}$, where D' extends D, we say that D *ensures amalgamability for* D', if any model family compatible with D can be uniquely extended to a family compatible with D'. We can reduce this to the case where D' extends D by just one object and one morphism into this object. We now need a specification of diagrams in the category of signatures and an operation ensures_amalgamability checking whether the extension of a diagram with a new node and a new edge (corresponding to some architectural unit term)

ensures amalgamability. Further motivation for introducing this operation, together with a description of some difficulties in implementing it in CASL institution, is given in [KHT+,Kli00].

Based on this, several units containing institution independent operations are built:

- SS and AS, based on the syntactic part of I and containing definitions of abstract syntax of structural and architectural specifications,
- G, implementing data structures for global environments needed in the static analysis of structural and architectural specifications,
- Str and A, implementing static analysis of structural and architectural specifications.

The institution independence of these units is expressed by providing methods of constructing them (i.e. parameterized units SAS, AAS, $Genv$, $StrAna$ and $ArchAna$, respectively) that can be applied to any unit satisfying the specification INSTITUTION. Finally, the resulting unit A is given as a parameter to a parameterized unit $LibAna$, which, given a unit implementing static analysis of some abstract library items (here instantiated with structured and architectural specifications), implements static analysis of libraries of such items. We now specify an arbitrary institution with qualified symbols in CASL itself. This specification is used as interface specification for the described above (specification of the) generic structured and architectural analysis in Fig. 2.

Since our target implementation language is Standard ML [Pau91], a higher-order functional programming language, we use HasCASL [SM,MHKB00], a polymorphic higher-order extension of CASL designed for the specification of Haskell programs, as specification language. In particular, in the examples below, we will use polymorphic type constructors such as $Map\ a$, $Set\ a$, $Table\ a\ b$ and $[a]$ (for maps, sets, index tables and lists). However, note that the use of HasCASL is not really essential here and it is done only to get better readability. The specifications can be easily rewritten in plain CASL; but then one has to explicitly include a separate instantiation of lists, maps etc., for each required element type. Also, product sorts have to be introduced explicitly, and operations of predicate type have to be replaced by predicates.

Axioms in this specification are omitted as irrelevant for the general presentation given in this paper. All the needed axioms can be derived from the semantics of CASL.

The specification INSTITUTION, shown in Fig. 3, is essential for extending the CASL tool set CATS with static analysis for future extensions and/or restrictions of CASL. To reuse the static analysis of structured and architectural specifications in an enriched language, one should implement all the specified operations for the new language. If architectural specifications are omitted in the modified language, the implementation of the operation `ensures_amalgamability` can be omitted. For restrictions of CASL, one can either just use the ordinary CASL tool set, combined with a sublanguage analyzer that we have developed, or one can re-implement the analysis of a sublanguage (e.g. to obtain better efficiency) and use the generic analysis for CASL in-the-large as described above.

The specification STRUCTUREDABSTRACTSYNTAX of the abstract syntax for structured specifications is given in Fig. 5. It is based on the abstract syntax for basic specifications (including sorts *SYMB_ITEMS* and *SYMB_MAP_ITEMS* as a syntax for raw symbols and raw symbol maps). It follows the CASL abstract syntax [CoF01], except that the productions for *SYMB_ITEMS* and *SYMB_MAP_ITEMS* are omitted – they belong to the institution-specific part. We also provide a specification STRUCTUREDANALYSIS with the profiles of the analysis functions for structured specifications. Moreover, the corresponding specifications for the architectural level are given in Fig. 6.

The specification GLOBALENV, given in Fig 7, provides data structures for the information that is extracted by the static analysis. For structured specifications, the structure is roughly kept (mainly, symbol maps are replaced by

spec INSTITUTION =
    CATEGORY **with** *object* $\mapsto$ *sign*
and BASICABSTRACTSYNTAX **and** LIST **and** FINITESET **and** FINITEMAP
and DIAGRAM[CATEGORY **with** *object* $\mapsto$ *sign*]
then %% Basic analysis
    ops *basic_analysis* : *BASIC_SPEC* $\times$ *sign* $\to$
                                    BASIC_SPEC $\times$ *sign* $\times$ [*FORMULA*];
        stat_symb_map_items : [*SYMB_MAP_ITEMS*] $\to$ Map *raw_symbol*;
        stat_symb_items : [*SYMB_ITEMS*] $\to$ [*raw_symbol*]
    %% Structured specifications: Symbols and symbol maps
    sorts *symbol* < *raw_symbol*;
        ID < *raw_symbol*
    ops ||__|| : *sign* $\to$ Set *symbol*;
        ||__|| : *morphism* $\to$ Map *symbol*;
        __ $\leq$ __ : *pred*(*symbol* $\times$ *symbol*);
        %% Ordering needed for efficient symbol tables
        __*matches*__ : *pred*(*symbol* $\times$ *raw_symbol*);
        empty_signature : *sign*;
        __[__] : *ID* $\times$ [*ID*] $\to$ *ID*;
        name : *symbol* $\to$ *ID*
    %% Architectural specifications: sharing analysis
    op *ensures_amalgamability* :
            diagram $\times$ *node* $\times$ *sign* $\times$ *edge* $\times$ *morphism* $\to$? *diagram*
    %% Derived operations
    ops *signature_union* : *sign* $\times$ *sign* $\to$? *sign*;
        final_union : *sign* $\times$ *sign* $\to$? *sign*;
        is_subsig : *pred*(*sign* $\times$ *sign*);
        generated_sign, *cogenerated_sign* : [*raw_symbol*] $\times$ *sign* $\to$ *morphism*;
        induced_from_morphism : Map *raw_symbol* $\times$ *sign* $\to$ *morphism*;
        induced_from_to_morphism : Map *raw_symbol* $\times$ *sign* $\times$ *sign* $\to$ *morphism*
    %% Axioms omitted
end

Fig. 3. Specification of institutions with qualified symbols

spec CATEGORY =
    sorts *object, morphism*
    ops   *id* : *object* → *morphism*;
          __o__ : *morphism* × *morphism* →? *morphism*;
          dom, cod : *morphism* → *object*
    %% Axioms omitted
end

spec DIAGRAM[CATEGORY] =
    LIST
then **sorts** *node, edge, diagram*
    ops   *dom, cod* : *edge* → *node*;
          fresh_node : *diagram* → *node*;
          fresh_edge : *diagram* × *node* × *node* → *edge*;
          add_node : *diagram* × *node* × *object* →? *diagram*;
          add_edge : *diagram* × *edge* × *morphism* →? *diagram*;
          empty_diagram : *diagram*;
          object_at_node : *diagram* × *node* →? *object*;
          morphism_at_edge : *diagram* × *edge* →? *morphism*;
          diagram_nodes : *diagram* → [*node*];
          diagram_edges : *diagram* → [*edge*];
          node_in_diagram : *pred*(*node* × *diagram*)
    %% Axioms omitted
end

spec BASICABSTRACTSYNTAX =
    %% Minimum needed for interface to structured analysis
    sorts *FORMULA, SYMB_ITEMS, SYMB_MAP_ITEMS, BASIC_SPEC*
end

Fig. 4. Auxiliary specifications used to specify institutions with qualified symbols

the corresponding signature morphisms), while basic specifications are fully expanded to signature and set of axioms that they denote. The data structures for architectural specifications closely follow their semantics, as given in [CoFb].

The specification ABSTRACTLIBITEMANALYSIS is given in Fig. 8. It contains a sort *LIB_ITEM* for the syntax of abstract library items (the constituents of a library), a sort *Env* for the semantics of these, and a sort *ITEM_NAME* to name them. Environments (sort *Env*) are then just tables of entries indexed by item names. The function *analysis* takes a *LIB_ITEM* and analysis it in the given environment.

The specification LIBANALYSIS is built on top of ABSTRACTLIBITEMANAL-YSIS. It provides data structures for both libraries with possible downloads from other libraries (this follows the CASL abstract syntax, except that an intermediate non-terminal *LIB_ITEM'* is needed here to separate the library item-specific from the library item-independent parts), and a function *check* that statically analyses a library, using the given function *analysis* from ABSTRACTLIBITEM-

spec STRUCTUREDABSTRACTSYNTAX =
       INSTITUTION **and** LIST **and** FINITEMAP **and** STRING
then **sorts** *SPEC_NAME* = *string*;
              VIEW_NAME = *string*

    free types *FIT_ARG*   ::= *fit_spec*(*SPEC*; [*SYMB_MAP_ITEMS*])
                                   | *fit_view*(*VIEW_NAME*; [*FIT_ARG*]);

                           RENAMING   ::= *renaming*([*SYMB_MAP_ITEMS*]);
                           RESTRICTION ::= *hide_spec*([*SYMB_ITEMS*])
                                   | *reveal_spec*([*SYMB_MAP_ITEMS*]);

                           GENERICITY ::= *genericity*(*PARAMS*; *IMPORTS*);
                           PARAMS   ::= *params*([*SPEC*]);
                           IMPORTS   ::= *imports*([*SPEC*]);

                           SPEC   ::= *basic*(*BASIC_SPEC*)
                                   | *translation*(*SPEC*; *RENAMING*)
                                 | *reduction*(*SPEC*; *RESTRICTION*)
                                 | *union_spec*([*SPEC*])
                                 | *extension*([*SPEC*])
                                 | *free_spec*(*SPEC*)
                                 | *local_spec*(*SPEC*; *SPEC*)
                                 | *closed_spec*(*SPEC*)
                                 | *spec_inst*(*SPEC_NAME*; [*FIT_ARG*]);

                   VIEW_TYPE ::= *view_type*(*SPEC*; *SPEC*);

                 PRE_LIB_ITEM ::= *spec_defn*(*SPEC_NAME*;
                                      GENERICITY;
                                      SPEC)
                               | *view_defn*(*VIEW_NAME*;
                                      GENERICITY;
                                      VIEW_TYPE)
end

spec STRUCTUREDANALYSIS =
       STRUCTUREDABSTRACTSYNTAX **and** GLOBALENV
then
ops *spec_analysis* : *sign* × *global_env* × *SPEC* →? *spec_lenv* × *SPEC*;
       structured_analysis :
              global_env × *PRE_LIB_ITEM* →? *global_env* × *PRE_LIB_ITEM*
       %% Axioms and hidden operations omitted
end

Fig. 5. Specification of structured abstract syntax and analysis

ANALYSIS. It returns an environment corresponding to the library, and also takes and returns a library environment, which is a table associating library names with environments.

spec ARCHABSTRACTSYNTAX =
STRUCTUREDABSTRACTSYNTAX
then
sorts $UNIT\_NAME = string$;
$ARCH\_SPEC\_NAME = string$;
$UNIT\_TYPE\_NAME = string$

free   types

$ARCH\_SPEC$	$::= basic\_arch\_spec([UNIT\_DECL\_DEFN];$
	$\qquad\qquad\qquad RESULT\_UNIT)$
	$\mid\quad named\_arch\_spec(ARCH\_SPEC\_NAME);$
$UNIT\_DECL\_DEFN$	$::= unit\_decl\_case(UNIT\_DECL)$
	$\mid\quad unit\_defn\_case(UNIT\_DEFN);$
$UNIT\_DECL$	$::= unit\_decl(UNIT\_NAME;\ UNIT\_SPEC;$
	$\qquad\qquad\qquad UNIT\_IMPORTED);$
$UNIT\_IMPORTED$	$::= unit\_imported([UNIT\_TERM]);$
$UNIT\_DEFN$	$::= unit\_defn(UNIT\_NAME;\ UNIT\_EXPRESSION);$
$UNIT\_SPEC$	$::= unit\_type\_case(UNIT\_TYPE)$
	$\mid\quad spec\_name\_case(SPEC\_NAME)$
	$\mid\quad arch\_spec\_case(ARCH\_SPEC)$
	$\mid\quad closed\_unit\_spec(UNIT\_SPEC);$
$UNIT\_TYPE$	$::= unit\_type([SPEC];\ SPEC);$
$RESULT\_UNIT$	$::= result\_unit(UNIT\_EXPRESSION);$
$UNIT\_EXPRESSION$	$::= unit\_expression([UNIT\_BINDING];\ UNIT\_TERM);$
$UNIT\_BINDING$	$::= unit\_binding(UNIT\_NAME;\ UNIT\_SPEC);$
$UNIT\_TERM$	$::= unit\_translation(UNIT\_TERM;\ RENAMING)$
	$\mid\quad unit\_reduction(UNIT\_TERM;\ RESTRICTION)$
	$\mid\quad amalgamation([UNIT\_TERM])$
	$\mid\quad local\_unit([UNIT\_DEFN];\ UNIT\_TERM)$
	$\mid\quad unit\_appl(UNIT\_NAME;\ [FIT\_ARG\_UNIT]);$
FIT_ARG_UNIT	$::= fit\_arg\_unit(UNIT\_TERM;\ [SYMB\_MAP\_ITEMS]);$
LIB_ITEM	$::= \textbf{sort}\ PRE\_LIB\_ITEM$
	$\mid\quad arch\_spec\_defn(ARCH\_SPEC\_NAME;$
	$\qquad\qquad\qquad ARCH\_SPEC)$
	$\mid\quad unit\_spec\_defn(SPEC\_NAME;\ UNIT\_SPEC)$

end

spec ARCHANALYSIS =
ARCHABSTRACTSYNTAX **and** STRUCTUREDANALYSIS
then
op $arch\_analysis : global\_env \times LIB\_ITEM \to? global\_env \times LIB\_ITEM$
%% Axioms and hidden operations omitted
end

Fig. 6. Specification of architectural abstract syntax and analysis

spec GLOBALENV =
        INSTITUTION **and** LIST **and** TABLE **and** STRING
then
sorts *SPEC_NAME* = *string*;
        UNIT_NAME = *string*;
        ITEM_NAME = *string*
free types
        spec_env ::= *basic_env*(*sign*; [*FORMULA*])
                | *translate_env*(*spec_env*; *morphism*)
                | *derive_env*(*spec_env*; *morphism*)
                | *union_env*([*spec_env*])
                | *extension_env*([*spec_env*])
                | *free_spec_env*(*spec_env*)
                | *closed_spec_env*(*spec_env*)
                | *spec_inst_env*(*SPEC_NAME*; *spec_env*; *morphism*; [*spec_env*]);
        %% intended use: (name, body, fitting morphism, actual args)
        spec_lenv ::= *SPEC_ENV*(*sign*; *sign*; *spec_env*)
        %% intended use: (flattened sign, flattened hidden sign, env)
op   *empty_spec_lenv* : *spec_lenv*
type *genericity_env* = (*spec_lenv* × [*spec_lenv*] × *sign*)
        %% intended use: (union of envs for all imports,
        %%   list of envs for formal parameters,
        %%   signature union of all imports and all formal parameters)
type *comp_sigs* = [*sign*];
        unit_sig = *comp_sigs* × *sign*;
        st_based_unit_ctx = *Table UNIT_NAME node*;
        based_par_unit_sig = *node* × *par_unit_sig*;
        st_par_unit_ctx = *Table UNIT_NAME based_par_unit_sig*;
        ext_st_unit_ctx = *st_par_unit_ctx* × *st_based_unit_ctx* × *diagram*;
        arch_sig = *ext_st_unit_ctx* × *unit_sig*
        %% architectural signatures
free type *global_entry* ::= *spec_defn_env*(*genericity_env*; *spec_lenv*)
                      | *view_defn_env*(*genericity_env*; *spec_lenv*; *morphism*; *spec_lenv*)
                      | *arch_spec_defn_env*(*arch_sig*)
                      | *unit_spec_defn_env*(*unit_sig*)
type *global_env* = *Table ITEM_NAME global_entry*
op   *empty_global_env* : *global_env*
end

Fig. 7. Specification of global environments

6   Conclusion and Future Work

We have shown how to structure the static analysis of CASL within the CASL
tool set (CATS) in a modular way. We therefore have followed the structuring
of the CASL design and semantics into several layers, and outlined how the
static analysis can be turned into a generic program which is parameterized over
an arbitrary institution (plus some extra components). We also have specified

spec ABSTRACTLIBITEMANALYSIS =
STRING **and** TABLE
then **sorts** $LIB\_ITEM, Entry$;
$\quad\quad\quad\quad$ $ITEM\_NAME = string$;
$\quad\quad\quad\quad$ $Env = Table\ ITEM\_NAME\ Entry$
$\quad\quad$ **ops** $analysis : Env \times LIB\_ITEM \to?\ Env \times LIB\_ITEM$
end

spec LIBANALYSIS =
ABSTRACTLIBITEMANALYSIS
then
sorts $LIB\_NAME = string$
free **types**
$\quad\quad$ $ITEM\_NAME\_OR\_MAP ::= item\_name(ITEM\_NAME)$
$\quad\quad\quad\quad\quad\quad\quad\quad\quad$ | $\quad item\_name\_map(ITEM\_NAME;\ ITEM\_NAME)$;
$\quad\quad$ LIB_ITEM' $\quad\quad\quad\quad\quad$ ::= **sort** LIB_ITEM
$\quad\quad\quad\quad\quad\quad\quad\quad\quad$ | $download\_items(LIB\_NAME;$
$\quad\quad\quad\quad\quad\quad\quad\quad\quad\quad\quad\quad\quad\quad$ $[ITEM\_NAME\_OR\_MAP])$;
$\quad\quad$ LIB_DEFN $\quad\quad\quad\quad$ ::= $lib\_defn(LIB\_NAME \times [LIB\_ITEM])$;
$\quad\quad$ Lib_Env $\quad\quad\quad\quad\quad$ ::= $lib\_env(Table\ LIB\_NAME\ (Env \times LIB\_DEFN))$
op $\quad$ $check : Lib\_Env \times LIB\_DEFN \to Lib\_Env \times Env \times LIB\_DEFN$
$\quad\quad$ %% Axioms and hidden operations omitted
end

Fig. 8. Specification of abstract library analysis

this modular structure of the CASL static analysis in CASL itself, using a CASL architectural specification with parameterized units.

At the implementation level, these parameterized units are realized as Standard ML functors (SML is the implementation language of the CASL tool set). Thus, this work can also be seen as a case study of relating CASL architectural specifications and Standard ML implementations. The size of the different components of the CASL tool set is as follows:

Parsing and printing	7.000 lines of code
Institution-specific analysis	9.000 lines of code
Institution-independent analysis	9.000 lines of code
Total	25.000 lines of code

This means that for replacing the CASL institution with some other institution like higher-order CASL, one has to re-program both the institution-specific analysis and the parsing and printing, while the generic institution-independent analysis can be re-used.

In the future, this should be applied to several extensions of CASL, like HO-CASL [MHKB00], HasCASL [SM], CASL-LTL [RR00], SB-CASL [BZ00], LB-CASL [ACZ00], etc.

We hope that also large parts of the institution-specific analysis of CASL basic specifications can be re-used, since these extensions are built on top of CASL. However, it is not clear whether this re-use can be turned into a generic program. At the moment, this kind of re-use seems to be more a "copy and paste" re-use. The same also holds for parsing and printing, since generally it seems to be difficult to build parsers for "grammars with holes" and to instantiate the holes with different (institution-specific) grammars later on.

A related direction of future work is the construction of a static analysis for heterogeneous CASL [Mosb]. Heterogeneous CASL combines several logics for basic specifications within one language, while the structured and architectural specifications basically are those of CASL.

References

[ACZ00]   D. Ancona, M. Cerioli, and E. Zucca. Extending CASL by late binding. In C. Choppy, D. Bert, and P. Mosses, editors, *Recent Trends in Algebraic Development Techniques, 14th International Workshop, WADT'99, Bonas, France*, volume 1827 of *Lecture Notes in Computer Science*. Springer-Verlag, 2000.

[AHS90]   J. Adámek, H. Herrlich, and G. Strecker. *Abstract and Concrete Categories*. Wiley, New York, 1990.

[BST98]   Michel Bidoit, Donald Sannella, and Andrzej Tarlecki. Architectural specifications in CASL. In *AMAST '98, Proc. 7th Intl. Conference on Algebraic Methodology and Software Technology, Manaus*, volume 1548 of *LNCS*, pages 341–357. Springer-Verlag, 1998.

[BZ00]   H. Baumeister and A. Zamulin. State-based extension of CASL. In *Proceedings IFM 2000*, volume 1945 of *Lecture Notes in Computer Science*. Springer-Verlag, 2000.

[CoFa]   CoFI. The Common Framework Initiative for algebraic specification and development, electronic archives. Notes and Documents accessible from http://www.brics.dk/Projects/CoFI/.

[CoFb]   CoFI Semantics Task Group. CASL – The CoFI Algebraic Specification Language – Semantics. Note S-9 (Documents/CASL/Semantics, version 1.0), in [CoFa], forthcoming.

[CoF01]   CoFI Language Design Task Group. CASL – The CoFI Algebraic Specification Language – Summary. Documents/CASL/Summary, in [CoFa], March 2001.

[GB92]   J. A. Goguen and R. M. Burstall. Institutions: Abstract model theory for specification and programming. *Journal of the Association for Computing Machinery*, 39:95–146, 1992. Predecessor in: LNCS 164, 221–256, 1984.

[KHT⁺]   B. Klin, P. Hoffman, A. Tarlecki, T. Mossakowski, and L. Schröder. Checking amalgamability conditions for CASL architectural specifications. In J. Sgall, A. Pultr, P. Kolman, editors, *Mathematical Foundations of Computer Science*, volume 2136 of *Lecture Notes in Computer Science*, pages 451–463. Springer-Verlag, 2001.

[Kli00]   B. Klin. An implementation of static semantics for architectural specifications in CASL (in Polish). Master's thesis, Warsaw University, 2000.

[MHKB00]   T. Mossakowski, A. Haxthausen, and B. Krieg-Brückner. Subsorted partial higher-order logic as an extension of CASL. In C. Choppy, D. Bert, and P. Mosses, editors, *Recent Trends in Algebraic Development Techniques, 14th International Workshop, WADT'99, Bonas, France*, volume 1827 of *Lecture Notes in Computer Science*, pages 126–145. Springer-Verlag, 2000.

[Mosa]   T. Mossakowski.   The CASL   tool   set.   Available   at `http://www.tzi.de/cofi/CATS`.

[Mosb]   T. Mossakowski. Heterogeneous development graphs and heterogeneous borrowing. Submitted.

[Mosc]   Till Mossakowski. Relating CASL with other specification languages: the institution level. *Theoretical Computer Science*. To appear.

[Mos97]   Peter D. Mosses. CoFI: The Common Framework Initiative for Algebraic Specification and Development. In *TAPSOFT '97, Proc. Intl. Symp. on Theory and Practice of Software Development*, volume 1214 of *LNCS*, pages 115–137. Springer-Verlag, 1997.

[Mos00a]   T. Mossakowski. Specification in an arbitrary institution with symbols. In C. Choppy, D. Bert, and P. Mosses, editors, *Recent Trends in Algebraic Development Techniques, 14th International Workshop, WADT'99, Bonas, France*, volume 1827 of *Lecture Notes in Computer Science*, pages 252–270. Springer-Verlag, 2000.

[Mos00b]   Till Mossakowski. CASL: From semantics to tools. In S. Graf and M. Schwartzbach, editors, *TACAS 2000*, volume 1785 of *Lecture Notes in Computer Science*, pages 93–108. Springer-Verlag, 2000.

[Pau91]   L. C. Paulson. *ML for the Working Programmer*. Cambridge University Press, 1991.

[RR00]   G. Reggio and L. Repetto. CASL-CHART: a combination of statecharts and of the algebraic specification language CASL. In *Proc. AMAST 2000*, volume 1816 of *Lecture Notes in Computer Science*. Springer Verlag, 2000.

[SM]   L. Schröder and T. Mossakowski. HasCASL: Towards integrated specification and development of Haskell programs. Submitted.

[SMH+01]   L. Schröder, T. Mossakowski, P. Hoffman, B. Klin, and A. Tarlecki. Semantics of architectural specifications in CASL. In H. Hußmann, editor, *Fundamental Approaches to Software Engineering*, volume 2029 of *Lecture Notes in Computer Science*, pages 253–268. Springer-Verlag, 2001.

[SMH+]   L. Schröder, T. Mossakowski, P. Hoffman, B. Klin, and A. Tarlecki. Amalgamation in the semantics of CASL. Submitted to Theoretical Computer Science.

Tight and Loose Semantics
for Transformation Systems

Fernando Orejas[2], Hartmut Ehrig[1], and Elvira Pino[2]

[1] FB 13 Informatik, Tech. Univ. Berlin,
Franklinstrasse 28/29, 10587 Berlin, Germany
`ehrig@cs.tu-berlin.de`
[2] Dpto de L.S.I., Universidad Politècnica de Catalunya, Campus Nord, Mòdul C6,
Jordi Girona 1-3, 08034 Barcelona, Spain
`{pino,orejas}@lsi.upc.es`

Abstract. When defining the requirements of a system, specification units typically are partial or incomplete descriptions of a system component. In this context, providing a complete description of a component means integrating all the existing partial views for that component. However, in many cases defining the semantics of this integration operation is not an easy task. In particular, this is the case when the framework used at the specification level is, in some sense, an "operational" one (e.g. a Petri net or a statechart). Moreover, this problem may also apply to the definition of compositional semantics for modular constructs for this kind of frameworks.

In this paper, we study this problem, at a general level. First, we define a general notion of framework whose semantics is defined in terms of transformations over states represented as algebras and characterize axiomatically the standard *tight* semantics. Then, inspired in the *double-pullback approach* defined for graph transformation, we axiomatically present a loose semantics for this class of transformation systems, exploring their compositional properties. In addition, we see how this approach may be applied to a number of formalisms.

1 Introduction

When defining the requirements of a system, specification units typically are partial or incomplete descriptions of a system component. For example, a *viewpoint* [8] describes a component of a system from a certain perspective (e.g., an external user's point of view), or a *use case* [18] describes (perhaps partially) one possible behaviour of a component or system. In this context, providing a complete description of a component means integrating all the existing partial views for that component. However, in many cases defining the semantics of this integration operation is not an easy task. In particular, this is the case when the framework used at the specification level is, in some sense, an "operational" one (e.g. a Petri net or a statechart). Moreover, this problem may also apply to the definition of compositional semantics for modular constructs for this kind of frameworks.

M. Cerioli and G. Reggio (Eds.): WADT/CoFI 2001, LNCS 2267, pp. 238–255, 2002.

In this paper, we study this problem at a general level. First, we define a general notion of framework whose semantics is defined in terms of transformations over states represented as algebras, and characterize axiomatically the standard *tight semantics*. The approach is inspired in the *double-pullback approach* [12,13] defined for graph transformation and, especially in [7]. In that paper, the double pullback approach is, in a way, extended in a conceptual way to deal with other frameworks. In this paper, the approach is different. In particular, we approach the problem in an axiomatic way characterizing tight and loose semantics for a general class of transformation systems, studying some compositionality properties for the loose case. The basic idea in the case of the loose semantics consist in considering that the transformations defined by a given specification should be interpreted as the minimal changes that should be performed over a given state. In contrast to the case of the tight semantics where the transformations define exactly the changes that should be performed over a given state. This is equivalent to eliminating the so-called frame assumption.

We believe that our approach may also be applicable for the definition of compositional semantics of modular constructs for certain classes of programming or specification languages. Actually, another source of inspiration was the definition of compositional semantics for logic programs with negation [15]. This was considered a difficult task due to the non-monotonic nature of negation as failure. In that paper, we avoided that problem by defining a loose semantics for this class of logic programs and, then, by defining adequate algebraic constructions to define a least model semantics.

In order to show the application of our approach, we use as running examples the cases of graph transformation using the double pushout approach, the single pushout approach and the double pullback approach, and the case of place/transition nets. Moreover, at the end of the paper we present the essential ideas underlying a loose semantics for logic programs with negation.

The paper is organized as follows. In section 2, we present a general formulation of tight transformation frameworks, showing how graph transformation by double and single pushouts and place/transition nets fit in our approach. In section three, we modify the previous definitions to characterize loose transformation frameworks, showing that graph transformation by double pullbacks is a loose framework, and we show how these ideas can be used to provide a loose semantics to tight transformation frameworks. Moreover, we define an operation for combining transformation rules, that may be considered complementary, and we provide an informal example of how our approach can be used to give semantics to viewpoints specifications. Finally, in section four, we show how one can restrict the loose interpretation of transformation rules by means of constraints and we apply these ideas to the definition of semantics of logic programs.

In this paper, we assume that the reader has a certain knowledge of the most standard algebraic concepts, together with the most usual notation (for details, for example, one may consult [1]). We also assume that the reader has some very basic knowledge of algebraic graph transformation (see e.g. [17]), place/transition nets [16] and logic programming [2].

2   Transformation Systems

In this section we present the basic ideas underlying our approach. First, we define a general notion of transformation framework and then we axiomatize the standard tight semantics that these formalisms usually have. As running examples to discuss our definitions we study graph transformation (double and single pushout approaches) and place/transition nets.

2.1   Basic Notions

We consider that a transformation system is some kind of specification that defines a relation (a transformation relation) over a well-defined class of states. In this sense, a transformation framework is a formalism for specifying transformation systems.

Some examples of transformation formalisms that we will consider along this paper are graph transformation systems (using the double pushout approach and the single pushout approach) and place/transition nets. Other obvious candidates are all kinds of approaches based on the transformation of algebras, such as Abstract State Machines (formerly called evolving algebras) [4,11], D-oids [3], algebraic approaches to object specification [6] or Algebra Transformation Systems, [10]. Moreover, in the last section we will deal with the special case of logic programs. In the case of graph transformation systems, states are graphs and transformation systems are sets of graph productions. In the case of the double pushout approach, productions are pairs of injective graph morphisms $L \leftarrow H \to R$, while in the case of the single pushout approach, productions are injective partial graph morphisms $L \to R$. In the case, of place/transition nets, states are net markings and the transformation systems are the nets themselves. The transformations denoted by a net are the possible firings.

In our case, to achieve a reasonable degree of generality, we assume that states are partial Σ-algebras. This covers most reasonable cases. For instance, in the case of graph transformation, graphs can be seen as total algebras over a signature including two sorts (one for vertices and one for edges) and two operations providing, respectively, the source and target of each edge. Similarly, in the case of place/transition nets, markings can be seen as algebras over a signature including the signature of the natural numbers and, in addition, having as many constants of sort natural as places in the net. However, we may consider that not all the algebras over this signature are valid states. In particular, we may consider that the only valid states are the algebras whose carrier associated to the sort *natural* corresponds to the set of natural numbers and where the interpretation of the natural number operators coincide with the corresponding natural numbers operations. On the other hand, we consider that a state transformation consists of two Σ-algebras (the source and target states) together with a tracking map, which is an injective partial map relating the elements of the source and target states that remain unchanged by the transformation. However, again, we may consider that not all the possible tracking maps are valid for a given formalism. For instance, in the case of place/transition nets only the identity should be a valid tracking map. The reason is that this is the only reasonable relation between the elements (the natural numbers) of the source and target states.

Following these intuitions, we may define a transformation framework $\mathcal{T}$ as a 5-tuple consisting of: a class of signatures $Sig_{\mathcal{T}}$; a mapping, $States : Sig_{\mathcal{T}} \rightarrow \mathcal{P}(PAlg)$, that associates to each signature Σ the class of partial Σ-algebras that can be considered allowable states in the given framework; another mapping, $Tmaps : Sig_{\mathcal{T}} \rightarrow PInjMaps$, that associates to each signature Σ the class of partial injective $Sorts(\Sigma)$-maps that can be used as tracking maps in transformations, a set of the transformation rules, $Rules$, allowed in the given framework; and a mapping, $Trafos : Rules \rightarrow StateTrafos$, that associates to each transformation rule the set of transformations defined by that rule. In particular, $StateTrafos$ is the set of all possible transformations, i.e. the set of all 4-tuples consisting of a signature Σ, two Σ-states which are the source and target of the transformation, and a tracking map t. We do not assume that the set of rules of a given transformation framework has any specific form.

Definition 1. *A transformation framework $\mathcal{T}$ is a 5-tuple ($Sig_{\mathcal{T}}$,$States$:$Sig_{\mathcal{T}} \rightarrow \mathcal{P}(PAlg)$,$Tmaps$:$Sig_{\mathcal{T}} \rightarrow PInjMaps$, $Rules$, $Trafos : Rules \rightarrow StateTrafos$), where $Sig_{\mathcal{T}}$ is included in Sig, the class of all signatures, and $StateTrafos$ is the set of all possible state transformations in $\mathcal{T}$. In particular, a state transformation in $StateTrafos$ is a 4-tuple $(\Sigma, A1, A2, t)$, where $A1$ and $A2 \in States(\Sigma)$ are called the source and target states, respectively, and $t=\{t_s:A1 \rightarrow A2\}_{s \in Sorts(\Sigma)}$ is the tracking map of the transformation, which is a family of partial injective functions, written $t : A1 \rightarrow A2$. In addition, we assume that, for every rule r, $Trafos(r)$ is closed up to isomorphism and tracking maps, i.e. if $(\Sigma, A1, A2, t)$ is in $Trafos(r)$, $h1 : A1 \rightarrow A1'$ and $h2 : A2 \rightarrow A2'$ are Σ -isomorphisms and $(t' : A1' \rightarrow A2') \in Tmaps(\Sigma)$, such that $dom(t') = h1(dom(t))$ and $t' \circ h1 = h2 \circ t$, then $(\Sigma, A1', A2', t')$ is also in $Trafos(r)$.*

Example 1. The double pushout approach to graph transformation can be defined formally as follows. $Sig_{\mathcal{T}}$, contains only the graphs signature:

$GRAPHS =$ **sorts** $vertex, edge$
$\qquad\qquad\qquad$ **opns** $source, target : edge \rightarrow vertex$

States are total algebras over the above signature. Rules are spans of injective homomorphisms, $L \leftarrow K \rightarrow R$. Tracking maps are pairs of arbitrary injective partial maps, $t_{vertex} : G_{vertex} \rightarrow H_{vertex}$ and $t_{edge} : G_{edge} \rightarrow H_{edge}$, defined as follows:

- For every n in D_{vertex}, $t_{vertex}(l^*(n)) = r^*(n)$
- For every n in $G_{vertex} \backslash l^*(D)$, $t_{vertex}(n)$ is undefined
- For every e in D_{edge}, $t_{edge}(l^*(e)) = r^*(e)$
- For every e in $G_{edge} \backslash l^*(D)$, $t_{edge}(e)$ is undefined

Finally, (Σ, G, H, t), is a transformation associated to the previous rule if there is a graph D and morphisms m, d, m^*, l^* and r^* such that diagrams (1) and (2) in figure 1 are pushouts. Then, for every rule r, $Trafos(r)$ is closed under isomorphism and tracking maps by the definition of the associated tracking maps and because pushouts are closed under isomorphism.

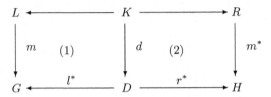

Fig. 1. Double pushout

Example 2. The single pushout approach to graph transformation can be defined formally as follows. Sig_T, *States* and *Tmaps* are defined as for the double pushout approach. *Rules* are partial injective homomorphisms, $p : L \rightarrow R$. Finally, (Σ, G, D, t), is a transformation associated to the previous rule if there are morphisms m, m^* and p^* such that the diagram in figure 2 is a pushout in the category of graphs and partial graph morphisms, and where t is the restriction of p^* to the category of sets and partial mappings.

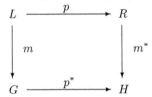

Fig. 2. Single pushout

Again, $Trafos(r)$ is closed under isomorphism and tracking maps because pushouts are closed under isomorphisms.

Example 3. Place/transition nets can be defined formally as follows. As said above, Sig_T consists of signatures extending the signature of natural numbers with a number of constants of sort *nat*, a constant representing each place in the net. *States* are total algebras extending the natural numbers. The only tracking maps considered are the identity functions. As said above, the reason is that, in this case, the states only contain value sorts (the natural numbers) and they should remain unchanged under transformation. *Rules* are the sets of transitions in a net. Finally, $(\Sigma, A1, A2, t)$ is a transformation associated to a net N if we can pass from the marking $A1$ to the marking $A2$ by the firing of some transition in N.

2.2   Tight Transformations Frameworks

Two basic ideas underlie standard (tight) transformation frameworks. The first one is what we may call the locality assumption. This principle says that, if a given rule defines a transformation on a state A, producing as result a state

B, then this rule may also be applied to a larger state A', including A in some well-defined sense, obtaining a state B', including B. The second basic idea is the so-called frame assumption. This assumption implies that, when applying a local transformation on part of a state A, the rest of the state should remain unchanged in the result.

In order to formalize these two assumptions, we must, first, define what we mean by saying that a state A' includes a state A. On the one hand, we may consider that if A is a subalgebra of A' then A' includes A. For instance, dealing with graphs as algebras, the subgraph relation would correspond to the subalgebra relation. On the other hand, we may also consider that if a signature Σ is included in Σ', and the states A and A' are, respectively a Σ-algebra and a Σ'-algebra, such that $A'|_{\Sigma} = A$, then A' also includes A. For instance, if A denotes a marking of a certain net, then an extension of A with respect to a larger signature would correspond to a marking of a net including more places. Mixing these two conditions, we may say that a Σ-state A is included in a Σ'-state A' if A is a subalgebra of $A'|_{\Sigma}$.

According to these ideas, we may say that a transformation framework is a tight transformation framework if the locality and the frame assumptions hold. This may be formalized as follows:

Definition 2. $\mathcal{T} = (Sig_{\mathcal{T}}, States, Tmaps, Rules, Trafos)$ *is a tight transformation framework if the following two properties hold:*

(1) *If* $\tau = (\Sigma, A, B, t) \in Trafos(r)$, *then for every* $\Sigma' \in Sig_{\mathcal{T}}$, $A', B' \in States(\Sigma')$, *such that* $\Sigma \subseteq \Sigma'$, $A'|_{\Sigma} = A$ *and* $B'|_{\Sigma} = B$, *and every tracking map* $t : A' \to B'$ *such that:*

- *For every* s *in* $Sorts(\Sigma), t'_s = t_s$
- *For every* s *in* $Sorts(\Sigma')\backslash Sorts(S), t'_s = id$
- *For every* σ *in* $\Sigma_{s1...sn,s}, \sigma_{B'} = \sigma_B$
- *For every* σ *in* $\Sigma'_{s1...sn,s}\backslash\Sigma$ *and for all* $(a1',\ldots,an')$ *in* $B'_{s1} \times \ldots \times B'_{sn}$

$$\sigma_{B'}(a1',\ldots,an') = t'_s(\sigma_{A'}(t_{s1}'^{-1}(a1'),\ldots,t_{sn}'^{-1}(an')))$$

where = *should be interpreted as strong equality (i.e.* $e = e'$ *if both expressions are defined and equal or both are undefined), and where if one subexpression in undefined (e.g.,* $t_{s1}'^{-1}(a1')$ *) then the result of the whole expression is considered undefined.*

we have that the transformation $(\Sigma', A', B', t') \in Trafos(r)$.

(2) *If* $\tau = (\Sigma, A, B, t) \in Trafos(r)$, *then for every* $A', B' \in States(\Sigma)$, *such that* $A \subseteq A'$, *and* $B \subseteq B'$, *and every tracking map* $t : A' \to B'$ *such that:*

- *For every* s *in* $Sorts(\Sigma)$,
  - $B'_s = (A'_s \backslash A_s) + B_s$, *where, as usual,* $+$ *denotes disjoint union.*
  - *For every* $a \in A_s, t'_s(a) = t_s(a)$.
  - *For every* $a \in A'_s \backslash A_s, t'_s(a) = a$.
- *For every* σ *in* $\Sigma_{s1...sn,s}$ *and for all* $(a1',\ldots,an')$ *in* $B'_{s1} \times \ldots \times B'_{sn}$

$$\sigma_{B'}(a1',\ldots,an') = t'_s(\sigma_{A'}(t_{s1}'^{-1}(a1'),\ldots,t_{sn}'^{-1}(an')))$$

we have that the transformation $(\Sigma', A', B', t') \in Trafos(r)$.

Properties (1) and (2) take care of the locality and frame assumptions for both kinds of state inclusion described above, and the combination of both properties takes care of the general notion of state inclusion. We could have provided a single property combining (1) and (2) in an obvious way. However we think that the splitting into two properties is simpler to deal with.

Proposition 1.   *1. The double pushout approach for graph transformation is a tight transformation framework.*
   2. The single pushout approach for graph transformation is a tight transformation framework.
   3. Place/transition nets are a tight transformation framework.

Proof.   1. Property (1) holds trivially for graph transformation since there is just one signature in Sig_T. With respect to property (2), let us suppose that $(\Sigma, G, H, t) \in Trafos(r)$ by means of the double pushout diagram in figure 1, and suppose that G', H' and t' are as in property (2) above. Then, we can build the double pushout diagram in figure 3, where all the vertical arrows are inclusions, and where the pushout complement D' is $D' = D + (G' \backslash G)$.

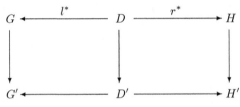

Fig. 3.

To show that D' is indeed a pushout complement we must show that the topmost left corner satisfies the so-called gluing conditions. In particular, we must show that the "dangling condition" holds (the so-called identification condition holds trivially in our case), i.e. that there cannot be an edge e in $G' \backslash G$ which is incident to a node in $G \backslash D$. Suppose that $n = source_{G'}(e)$ (or, similarly, $n = target_{G'}(e)$) is in $G \backslash D$ and e is in $G' \backslash G$. Then we would have that $t_{node}(n)$ is undefined, by the definition of the tracking map associated to the first double pushout, and, as a consequence, so is that $t'_{node}(n)$. However, according to the conditions stated in property (2), we would have that $t'_{edge}(e) = e$ and $source_{H'}(e) = t'_{edge}(source_{G'}(e))$, which means that H' would be ill-formed.
Now, the gluing of the two double pushout diagrams provides us with the desired transformation and the corresponding tracking map coincides with the definition in property (2)
   2. Again, it is enough to prove that property (2) holds. The proof is similar. Suppose that $(\Sigma, G, H, t) \in Trafos(r)$ by means of the pushout diagram in figure 2, and suppose that G', H' and t' are as in property (2) above. Then, the diagram in figure 4, where the vertical arrows are inclusions, is a pushout.

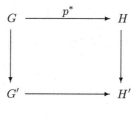

Fig. 4.

Finally, the gluing of the two pushout diagrams provides us with the desired transformation and the corresponding tracking map coincides with the definition in property (2).

3. In this case it is property (2) that holds trivially, since the only states allowed differ only on the number and the value of the constants representing the places of the nets. This means that there do not exist two states A and B such that A is a strict subalgebra of B. With respect to property (1), it should be clear that, independently of the places in a net the effect of a firing only modifies the values of the constants associated to the positions involved in the transition.

■

3   Loose Transformation Frameworks

As discussed in the introduction, requirements specifications are typically incomplete specifications, i.e. they usually describe some partial aspects of the behaviour of a system. In the case of specifications using some kind of transformation framework, this means that the transformation rules associated to a given event only describe, partially, the state transformations that should happen when this event occurs. According to this intuition, we may consider that, in this context, an event may cause some additional transformations not specified by the given transformation system. This is equivalent to eliminating the frame assumption when defining the semantics of a transformation system. This can be obtained by relaxing properties (1) and (2) of tight transformation systems, just asking for the satisfaction of the locality assumption. In particular, the below properties, (3) and (4), state that if a certain transformation can be applied to a state A, yielding a state B, then this transformation should be also applicable to any larger state A'. In this case, any state B' including B, in some well-defined sense, could be a possible result of the transformation. The idea is that the concrete state B' obtained depends on the additional changes caused by the "environment".

In our opinion, this kind of loose transformation frameworks are adequate for certain forms of requirements specifications. This is the case, for instance, of the so-called viewpoints approach. The idea is that, when specifying a given system, one describes its functionality by means of several complementary points of view (e.g. the external user view, the system manager view, etc.) Then, the complete specification of the system is given by the integration or combination of all the views. In what follows, we informally describe an example of the specification of a system using this approach, to motivate the rest of the section.

Example 4. Let us consider as a simple example the specification of a logistics information system for a certain company. We assume that this company has a certain number of stores, where the items produced by the company are kept in stock, and certain number of shops that sell these items to the end customer. Moreover the company has also a number of trucks, whose location is not fixed a priori, for delivering the items. From time to time, a certain shop may ask the system for a supply of N units of a certain item. Then the system must decide from which store are these items supplied and register the delivery on the company accounts. In what follows, we briefly describe this supply operation from different viewpoints.

- A shop manager viewpoint: From the point of view of the manager of a shop, the effect of asking for the supply of N units of a certain item could be that the stock of one store (maybe more than one as possible choices) should be decreased by N units and the stock in the shop of that item should be increased by N units.
- The logistics manager viewpoint: From the point of view of the person in charge of the logistics of the company, the effect of asking for the supply of N units of a certain item to a given shop could be that the stock of that item on a certain store should be decreased by N units and the stock in the shop of that item should be increased by N units. In addition, that store is selected among the stores having enough stock for that delivery, on the basis of the transportation cost, considering the distance of the store to the shop and to the closer free truck. Moreover, the selected truck should be marked as busy.
- The accounting manager viewpoint: From the point of view of the person in charge of the accounting of the company, the effect of asking for the supply of N units of a certain item to a given shop could be that the global value of the stocked items should be decreased by the value of the items delivered and the income of the company should be increased by that value.

Now we could formalize these viewpoints by means of some kind of algebra transformation approach. Each viewpoint would be described by a transformation rule that describes only part of the transformations associated to the supply operation. It may be noted that, in this example, the second and the third (or the first and the third) viewpoints are, in a way, orthogonal. Their associated transformations would deal with different parts of the "global" state of the system. However, in the case of the shop manager viewpoint and the logistics manager viewpoint, their associated transformations partially overlap. In particular, with respect to that overlapping the logistics manager viewpoint can be seen as a refinement of the shop manager viewpoint, since in the latter case the choice of the store is nondeterministic.

Then, the complete specification of the supply operation would be the combination, using the operation described at the end of the section, of the transformation rules associated to these three viewpoints.

Definition 3. $\mathcal{T} = (Sig_\mathcal{T}, States, Tmaps, Rules, Trafos)$ *is a loose transformation framework if the following two properties hold:*

(3) *If* $\tau = (\Sigma, A, B, t) \in Trafos(r)$, *then for every* $\Sigma' \in Sig_\mathcal{T}$ *and every* $A', B' \in States(\Sigma')$, *such that* $\Sigma \subseteq \Sigma'$, $A'|_\Sigma = A$ *and* $B'|_\Sigma = B$, *and for every* t', *such that* $t'|_{Sorts(\Sigma)} = t$, *we have that the transformation* $(\Sigma', A', B', t') \in Trafos(r)$.
(4) *If* $\tau = (\Sigma, A, B, t) \in Trafos(r)$, *then for every* $A', B' \in States(\Sigma')$, *such that* $A \subseteq A'$ *and* $B \subseteq B'$, *and for every* t', *such that* $t'|_A = t$, *we have that the transformation* $(\Sigma', A', B', t') \in Trafos(r)$.

where $t'|_{Sorts(\Sigma)}$ *and* $t'|_A$ *denote, respectively, the restriction of* t' *to the sorts of the signature* Σ *and the restriction of* t' *to the elements of the subalgebra* A.

Definition and Proposition 1 *The double pullback approach for graph transformation is a loose transformation framework*

Proof. The double pullback approach to graph transformation can be defined formally as follows. $Sig_\mathcal{T}$, *States*, *Rules* and tracking maps are as for the double pushout approach. Then, (Σ, G, D, t), is a transformation associated to the rule $L \leftarrow K \rightarrow R$, if there is a graph D and morphisms m, d, m^*, l^* and r^* such that l^* and r^* are injective homomorphisms and diagrams (1) and (2) in figure 5 are pullbacks, and where t is defined as follows:

- For every n in D_{vertex}, $t_{vertex}(l^*(n)) = r^*(n)$
- For every n in $G_{vertex} \backslash l^*(D)$, $t_{vertex}(n)$ is undefined
- For every e in D_{edge}, $t_{edge}(l^*(e)) = r^*(e)$
- For every e in $G_{edge} \backslash l^*(D)$, $t_{edge}(e)$ is undefined

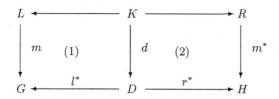

Fig. 5. Double pullback

Moreover, for every rule r, $Trafos(r)$ is closed under isomorphism and tracking maps by the definition of the associated tracking maps and because pullbacks are closed under isomorphism.

In this case, we just have to prove property (4). Suppose that $(\Sigma, G, H, t) \in Trafos(r)$ by means of the double pullback diagram in figure 5, and suppose that $G \subseteq G'$, $H \subseteq H'$ and $t'|_{G'} = t$. Then, we can build the double pullback diagram in figure 6, where $D' = D + (dom(t') \backslash dom(t))$ and where all the vertical arrows are inclusions. The gluing of the two double pullback diagrams provides us with the desired transformation.

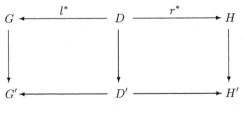

Fig. 6.

Additionally, the tracking map associated to this double pullback coincides, by construction, with t'. ∎

Obviously, every loose transformation framework is also a tight transformation framework, but the converse is not true in general.

Fact 2 *If $\mathcal{T}$ is a loose transformation framework then $\mathcal{T}$ is a tight transformation framework.*

This fact may seem counterintuitive, since, together with the previous proposition, it states that the double pullback approach is a tight transformation framework. An alternative definition of tight and loose transformation frameworks could consist in, first, considering that, in a transformation framework, every rule r defines a set of minimal transformations, $MinTrafo(r)$. Then, we would say that a transformation framework is tight (resp. loose) if, for every rule r, $Trafo(r)$ is the closure of $MinTrafo(r)$ with respect to properties (1) and (2) (resp. with respect to properties (3) and (4)). According to that alternative definition, in general, loose transformation frameworks would not be tight. Instead, we could consider tight transformations as special cases of loose transformations, as double pushout transformations can be seen as special cases of double pullback transformations, as we will see below. This would probably fit better our intuition.

On the other hand, the previous fact gives us a way of defining a loose transformation semantics for tight transformation frameworks. In particular, it is enough to make a closure, with respect to the properties (3) and (4) above, of the transformations associated to the given rules:

Definition 4. *Let $\mathcal{T} = (Sig_{\mathcal{T}}, States, Tmaps, Rules, Trafos)$ be a transformation framework and r be a rule in Rules, we define the loose extension of $Trafos(r)$, denoted $Loose(Trafos(r))$ as the least set of transformations, satisfying properties (3) and (4) that includes $Trafos(r)$. We also define the loose transformation framework associated with $\mathcal{T}$, $Loose(\mathcal{T})$ as:*

$$Loose(\mathcal{T}) = (Sig_{\mathcal{T}}, States, Tmaps, Rules, LTrafos)$$

where, for every r, $LTrafos(r) = Loose(Trafos(r))$

One may wonder whether the double pullback approach is the loose version of the double pushout approach. The answer is negative. There are some double pullback transformations which can not be obtained by the loose closure of the double pushout transformations. The reason is that, given a production $L \leftarrow K \rightarrow R$ and a morphism from L to a graph G we can not always build a double pushout to define a transformation. It is required that certain conditions (the so-called gluing conditions) hold. However, for building a double pullback the required conditions are looser. In this context, a double pullback transformation, where the gluing conditions are not satisfied, would not correspond to any loose extension of a double pushout transformation.

Definition and Proposition 3 *The faithful double pullback approach for graph transformation is the loose transformation framework associated with the double pushout approach.*

Proof. The double pullback transformation in figure 7 is faithful if m satisfies the identification condition. This means that if $x1$ and $x2$ are elements (vertices or edges) of L such that $m(x1) = m(x2)$ then, either $x1 = x2$, or there exist $y1$ and $y2$ in K such that $l(y1) = x1$ and $l(y2) = x2$. Then the faithful dou-

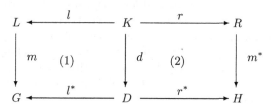

Fig. 7.

ble pullback approach is defined as the double pullback approach, but where only faithful transformations are allowed. The proof that this framework is a loose transformation framework is like the one for the general double pullback approach. The proof that every faithful double pullback transformation is an extension, with respect to property (4), of a double pushout transformation can be found in [13]. ∎

The loose versions of the single pushout approach and of place/transition nets are quite obvious: we can loosely transform graph G into graph H by means of the graph production $L \rightarrow R$, if there are graphs G' and H' such that $G' \subseteq G$, $H' \subseteq H$ and the diagram in figure 8 is a pushout.

By analogy with the case of double pushouts/pullbacks, one may wonder whether this can be characterized by means of a simple pullback, but the answer is no. The problem is that a pushout diagram in the category of graphs and

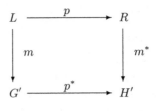

Fig. 8.

partial morphisms, such as the one in figure 8 may be not a pullback in that category as the following counter-example shows. If L is a graph with just two nodes, say $n1$ and $n2$, R and G' are graphs with just one node, say $n1$, p is a partial morphism binding $n1$ in L to $n1$ in R and m is a morphism binding $n1$ and $n2$ in L to $n1$ in G'. Then the result of the pushout, H' would be the empty graph. However, the pullback of p^* and m^* would be the empty graph and not L.

With respect to the case of place/transition nets, the result of a loose transformation associated to a given transition for a certain marking M would be any other marking such that the value of the positions associated to the transition are the ones specified by the transition, but the values of the rest of the positions may have changed in any arbitrary way.

As said above, the intuition behind loose transformation frameworks is that rules specify incompletely a certain class of transformations. In this context, we define an operation for combining the effects of two (or more rules). The idea is that, using this operation, one can complete or refine an incomplete transformation specification. The definition of this operation is quite simple: one can transform a state A into a state B if a "parallel" application of the combined rules provide this effect:

Definition 5. *Let $\mathcal{T} = (Sig_T, States, Tmaps, Rules, Trafos)$ be a loose transformation framework, the combination of two rules r and r', $r|r'$ is a rule that defines the following set of transformations:*

$(\Sigma, A1, A2, t) \in Trafos(r|r')$ if and only if there exist $\Sigma 0, \Sigma 0' \in Sig_T$, $B0, C0 \in States(\Sigma 0)$, $B0', C0' \in States(\Sigma 0')$, and tracking maps $t0 : B0 \to C0$ and $t0' : B0' \to C0'$, such that:

- *$(\Sigma 0, B0, C0, t0) \in Trafos(r)$*
- *$(\Sigma 0', B0', C0', t0') \in Trafos(r')$*
- *$B0 \subseteq A1|_{\Sigma 0}$ and $B0' \subseteq A1|_{\Sigma 0'}$*
- *$C0 \subseteq A2|_{\Sigma 0}$ and $C0' \subseteq A2|_{\Sigma 0'}$*
- *$t0 = t|_{B0}$ and $t0' = t|_{B0'}$ where $t|_{B0}$ and $t'|_A$ denotes, respectively, the restriction of t' to the sorts of the signature $\Sigma 0$ and to the elements of the subalgebra $B0$.*

As one can expect, the combination of two rules has a very simple compositional semantics:

Proposition 2. $Trafos(r|r') = Trafos(r) \cap Trafos(r')$

Proof. If $(\Sigma, A1, A2, t) \in Trafos(r|r')$ then, according to the previous definition, there exist $\Sigma0 \in Sig_T$, $B0 \in States(\Sigma0)$, $B0' \in States(\Sigma0')$, and a tracking map $t0 : B0 \to C0$, such $(\Sigma0, B0, C0, t0) \in Trafos(r)$, $B0 \subseteq A1|_{\Sigma0}$ and $t0 = t|_{B0}$.

This means that $(\Sigma, A1|_{\Sigma0}, A2|_{\Sigma0}, t|_{\Sigma0}) \in Trafos(r)$, since the framework is closed under (4). But this implies that $(\Sigma, A1, A2, t) \in Trafos(r)$ since the framework is closed under (3). The proof for r' is similar.

Conversely, if $(\Sigma, A1, A2, t) \in Trafos(r)$ and $(\Sigma, A1, A2, t) \in Trafos(r')$ then we have that $(\Sigma, A1, A2, t) \in Trafos(r|r')$: it is enough to take $\Sigma0 = \Sigma = \Sigma0'$, $B0 = A1 = B0'$, $C0 = A2 = C0'$, and $t0 = t = t0'$. ■

4   Loose Transformation Frameworks with Constraints

One may consider that loose transformation frameworks are too loose for many practical applications. In particular, we may know that state transformations defined by a given rule should not produce any effect on a certain part of the state, i.e. we may want to have some restricted form of the frame assumption. This is handled in the double pullback approach by defining open transformations over typed graphs [14,12], allowing us to express that part of the given graph should not change after a transformation. We could have defined something similar by assuming that, given a certain signature $\Sigma0$, when applying a transformation over a Σ-state A, with Σ including $\Sigma0$, this transformation should not produce any change on the $\Sigma0$ part of A. This means that, if B is the result of the transformation, then $B|_{\Sigma0} = A|_{\Sigma0}$. However, this may be not enough. We may also want that the resulting state satisfies certain conditions with respect to the source state. For instance, if the given signature Σ includes a constant c of sort integer, we may want to restrict the transformations over Σ-states so that the value of c in the target state is always greater that the value of c in the source state. A simple way of doing this is to assume that transformation frameworks may be equipped with a set of constraints that may be imposed on the rules. We may assume that these constraints are some kind of logic formula. However, in order to be general enough, we will not assume that constraints have any specific form. Instead, we will define constraints in a kind of institution-independent way [9], similarly to the related notion of logic of constraints in [5]

Definition 6. *Let* $\mathcal{T} = (Sig_T, States, Tmaps, Rules, Trafos)$ *be a transformation framework, a* logic of constraints *for* $\mathcal{T}$ *is a pair* $(Constraints, \models)$, *where* $Constraints : Sig_T \to Set$ *is a functor associating to every signature* Σ *in* Sig_T *a set of constraints over state transformations on that signature, and* $\models$ *is a* Σ-*indexed family of relations between* $Constraints(\Sigma)$ *and* Σ-*transformations,* $\{\models_\Sigma \subseteq StateTrafos(\Sigma) \times Constraints(\Sigma)\}_{\Sigma \in Sig_T}$, *where* $StateTrafos(\Sigma)$ *is the set of all possible* Σ-*transformations* (Σ, A, B, t). *In addition we require the following satisfaction condition:*

For every $\Sigma 1, \Sigma 2 \in Sig_T$, *every signature morphism* $h : \Sigma 1 \to \Sigma 2$, *every* $\Sigma 1$-*constraint* $c1$, *every* $\Sigma 2$-*states* $A2, B2$ *and every tracking map* $t : A2 \to B2$, *we have that:*

$$(\Sigma 2, A2, B2, t) \models_{\Sigma 2} c2 \ iff \ (\Sigma 1, A2|_h, B2|_h, t|_h \models_{\Sigma 1} c1)$$

where $c2 = Constraints(h)(c1)$, *and where* $t|_h$ *denotes the reduction of* t *along* h, *i.e. for every* $s \in \Sigma 1$, $(t|_h)_s = t|_{h(s)}$.

As usual, the satisfaction condition ensures that constraints over a given signature are uniformly translated over signature morphisms. This implies, in particular, that if a constraint is satisfied by a transformation over a given state, then this constraint (or, rather, its translation) would also be satisfied by any extension of this transformation over a state having a larger signature.

Now, we can extend the previous definition of loose transformation frameworks to include constraints. The idea, on the one hand, is to consider that each rule considered is equipped with a set of constraints. To this aim, we will first define the class of all constraints associated to a given logic and, then, extend the constraints satisfaction relation to sets of constraints over different signatures.

Definition 7. *Let* $L = (Constraints, \models)$ *be a logic of constraints over a transformation framework* T, *we define* $Constr(L)$, *the class of all constraints associated to* L, *as:*

$$Constr(L) = \bigcup_{\Sigma \in Sig_T} \{(\Sigma, c)/ \ c \in Constraints(\Sigma)\}$$

If C *is a set of constraints* $C \subseteq Constr(L)$, *we say that a transformation* (Σ, A, B, t) *satisfies* C, *denoted* $(\Sigma, A, B, t) \models C$ *if and only if* $\forall (\Sigma, c) \in C$, $(\Sigma, A, B, t) \models_{\Sigma} c$.

On the other hand, it is assumed that properties 3) and 4) should apply only to transformations satisfying the given constraint:

Definition 8. *Let* $T = (Sig_T, States, Tmaps, Rules, Trafos)$ *be a transformation framework and* $L = (Constraints, \models)$ *be a logic of constraints over* T, *then* T *is a loose transformation framework with respect to the set of constrained rules* $CRules \subseteq Rules \times \mathcal{P}(Constr(L))$ *if and only if for every* $(r, C) \in CRules$ *the following two properties hold:*

(5) *If* $\tau = (\Sigma, A, B, t) \in Trafos(r)$ *and* $(\Sigma, A, B, t) \models C$, *then for every* $\Sigma' \in Sig_T$ *and every* $A', B' \in States(\Sigma')$, *such that* $\Sigma \subseteq \Sigma'$, $A'|_{\Sigma} = A$ *and* $B'|_{\Sigma} = B$, *and for every* t', *such that* $t'|_{Sorts(\Sigma)} = t$, *if* $(\Sigma', A', B', t') \models C$ *then* $(\Sigma', A', B', t') \in Trafos(r)$.

(6) *If* $\tau = (\Sigma, A, B, t) \in Trafos(r)$ *and* $(\Sigma, A, B, t) \models C$, *then for every* $A', B' \in States(\Sigma)$, *such that* $A \subseteq A'$ *and* $B \subseteq B'$, *and for every* t', *such that* $t'|_A = t$, *if* $(\Sigma', A', B', t') \models C$ *then* $(\Sigma', A', B', t') \in Trafos(r)$.

Now, we can extend the composition operation defined in the previous section to the composition of constrained rules:

Definition 9. *Let $\mathcal{T} = (Sig_{\mathcal{T}}, States, Tmaps, Rules, Trafos)$ be a loose transformation framework, the combination of two constrained rules (r, C) and (r', C'), $(r, C)|(r', C')$ is the constrained rule $(r|r', C \cup C')$.*

Proposition 3. *If $\mathcal{T}$ is a loose transformation with respect to the set of constrained rules $CRules$, then for all rules (r, C) and (r', C') in $CRules$:*

$$Trafos((r, C)|(r', C')) = Trafos((r, C)) \cap Trafos((r', C'))$$

Proof. If $(\Sigma, A1, A2, t) \in Trafos(r|r')$ and $(\Sigma, A1, A2, t) \models C \cup C'$ then we have that $(\Sigma, A1, A2, t) \in Trafos(r)$ and, obviously, $(\Sigma, A1, A2, t) \models C$. The converse is trivial. ∎

Now, using the approach of loose transformation frameworks with constraints, we will see how we can define a compositional semantics for logic programs with negation. For simplicity, we will just consider the propositional case. The approach presented in the example is a (very) simplified reformulation of part of the work presented in [15], where the general first-order case is considered.

Example 5. A propositional logic program with negation is a set of ground clauses having the form:

$$a : -\ell_1, .., \ell_n$$

where a is an atom and each ℓ_i is a positive or a negative literal, i.e. a_i or $\neg a_i$, where a_i is an atom. There are several ways to define the semantics of a logic program P. Here we will just consider the semantics in terms of a (continuous) immediate consequence operator T_P. This semantics is used to define a least model semantics by means of the least fixpoint of T_P. This operator is defined as a mapping transforming logical structures into logical structures, where a logical structure, in this context, is a 3-valued Herbrand structure that is, a pair $\mathcal{A} = (A^+, A^-)$ of sets of atoms (representing the positive and the negative information in the structure) satisfying $A_i^+ \cap A_i^- =$ (consistency condition). The idea of the immediate consequence operator is that $T_P(\mathcal{A})$ should contain all the atoms that are one-step consequences of the program P and the information in $\mathcal{A}$. This means that $T_P(\mathcal{A}) = (B^+, B^-)$, where:

$B^+ = A^+ \cup \{a | \exists a : -\ell_1, .., \ell_n \in P$ such that $\ell_1, .., \ell_n \in \mathcal{A}\}$ (a)
$B^- = A^- \cup \{a | \forall a : -\ell_1, .., \ell_n \in P, \exists i$ such that $\neg\ell_i \in \mathcal{A}\}$ (b)

where $\ell \in \mathcal{A}$ means that $\ell \in A^+$, in case ℓ is positive, or $\ell \in A^-$, in case ℓ is negative. Definition (a) is the standard one when dealing with positive programs (without negation). On the other hand, Definition (b) is a very weak version of negation-as-failure.

Now, we can reformulate the above definitions as a tight transformation framework. States are sets of atoms (which can easily be formulated in several ways as algebras over an appropriate signature Σ), rules are programs, states are

3-valued Herbrand structures, and the transformation defined by a program P on a state $\mathcal{A}$ is $T_P(\mathcal{A})$. Finally, since the transformations associated to a program always add atoms to the given structure, we may consider that the tracking maps are always the identity.

In order to define a compositional semantics, we will now present a loose semantics for this class of programs. However, this definition is not a direct application of definition 11. The key idea is to consider that, if we think that a program P is incomplete, then the transformation associated to P, when applied to a state $\mathcal{A} = (A^+, A^-)$, should *at least* add to A^+ all the immediate positive consequences (i.e. $\{a | \exists a : -\ell_1, .., \ell_n \in P$ such that $\ell_1, .., \ell_n \in \mathcal{A}\}$) and should *at most* add to A^- all the immediate negative consequences (i.e. $\{a | \forall a : -\ell_1, .., \ell_n \in P, \exists i,$ such that $\neg \ell_i \in \mathcal{A}\}$). The reason is that adding more clauses to P would provide additional positive consequences but fewer negative consequences.

Now, we may obtain the desired semantics as follows. First, we reformulate the tight semantics considering programs as if they were positive. This means that $(\Sigma, \mathcal{A}, \mathcal{B}, t)$ is a transformation associated to P if $B^+ = A^+ \cup \{a | \exists a : -\ell_1, .., \ell_n \in P$ such that $\ell_1, .., \ell_n \in \mathcal{A}\}$, $B^- = A^-$ and t is the identity.

Then, we consider the loose extension of this framework. In this case, we can see that this loose extension $LTrafo(P)$ would include all transformations $(\Sigma, \mathcal{A}, \mathcal{B}, t)$ such that there are structures $\mathcal{A}' \subseteq \mathcal{A}$ and $\mathcal{B}' \subseteq \mathcal{B}$, such that $B'^+ = A'^+ \cup \{a | \exists a : -\ell_1, .., \ell_n \in P$ such that $\ell_1, .., \ell_n \in \mathcal{A}\}$ and $B'^- = A'^-$. However, this loose extension is not yet what we want. We have to make sure that the negative atoms that can be added in $\mathcal{B}$ with respect to $\mathcal{B}'$ are consistent with the negation as failure rule. This can be done by imposing a constraint on rules. In particular, this constraint C would be:

$\forall a \in B^- \backslash A'^-, \forall a : -\ell_1, .., \ell_n \in P, \exists i,$ such that $\neg \ell_i \in \mathcal{A}$

It is not difficult to prove that the following compositionality property holds:

$$LTrafos((P1 \cup P2, C)) = LTrafos((P1, C)) \cap LTrafos((P1, C))$$

5   Conclusions

The work presented in this paper has been motivated by the need of providing a loose semantics to transformation systems, in order to deal with requirements specifications and to provide compositional semantics to modular units. In this sense, we have introduced a general axiomatic approach to deal with transformation systems, showing how this approach can be applied to a number of cases. In particular, the double and single pushout approaches and the double-pullback approach to graph transformation, place/transition nets and logic programs with negation have been considered.

First, we have defined a notion of tight transformation framework, characterizing the locality and frame assumptions that underlie most operational approaches. Then, we have seen how, by eliminating the frame assumption, one can define loose transformation frameworks that can be used to give loose semantics to tight transformation approaches. Finally, we have seen how one can impose some constraints to restrict, as needed, the semantics of loose frameworks.

Acknowledgements

This work has been partially supported by the CICYT project HEMOSS (ref. TIC98-0949-C02-01) and CIRIT GRC 1999SGR-150.

References

1. E. Astesiano. H.-J. Kreowski, B. Krieg-Brueckner, *Algebraic Foundations of System Specification*, IFIP State of the Art Reports, Springer, 1999.
2. K.R. Apt,, Logic Programming, in *Handbook of Theoretical Computer Science, Vol B: Formal Models and Semantics*, Chapter 10, Elsevier, 1990.
3. Astesiano, E., Zucca, E.: D-oids: A Model for Dynamic Data Types. *Mathematical Structures in Computer Science* 5 (1995), pp. 257–282
4. Boerger, E., Huggins, J.K.: Abstract State Machines 1988-1998; Commented ASM Bibliography, *Bull. EATCS* 64 (1998), pp. 105–127
5. H. Ehrig, A Categorical Concept of Constraints for Algebraic Specifications; in: *Categorical Methods in Computer Science - with Aspects from Topology*, (H. Ehrig, H. Herrlich, H.-J. Kreowski, G. Preuss, eds.), Springer LNCS 393 (1989).
6. H.-D. Ehrich, Object specification, in [AKK 99], 1999.
7. H. Ehrig, R. Heckel, M. Llabres, F. Orejas, J. Padberg, G. Rozenberg A Rule-Based Framework with Incomplete Information, in *Theory and Applications of Graph Transformations* (H. Ehrig, G. Engels, H.-J. Kreowski, G. Rozenberg eds.), Springer LNCS 1764 (2000) 85–102.
8. A. Finkelstein, J. Kramer, B. Nuseibeh, M. Goedicke, L. Finkelstein, Viewpoints: A framework for integrating multiple perspectives in system development, *Int. J. of Software and Knowledge Engineering* 2,1 (1992) 31–58.
9. J.A. Goguen, R.M. Burstall. Institutions: Abstract model theory for specification and programming, *J. of the ACM* 39(1), 95–146, (1992)
10. Grosse-Rhode, M.: Algebra Transformation Systems and their Composition, Springer LNCS 1382 (1998), pp. 107–122
11. Gurevitch, Y.: Evolving Algebras, A Tutorial Introduction. *Bull. EATCS* 43 (1991), pp. 264-284
12. R. Heckel, *Open Graph Transformation Systems: A New Approach to the Compositional Modelling of Concurrent and Reactive Systems*, PhD Thesis, TU Berlin 1998.
13. R. Heckel, H. Ehrig, U. Wolter, A. Corradini, Double-Pullback Transitions and Coalgebraic Loose Semantics for Graph Transformation Systems, *Applied Categorical Structures* 9 (2001) 83–110.
14. R. Heckel, A. Corradini, H. Ehrig, M. Loewe, Horizontal and Vertical Structuring of Typed Graph Transformation Systems, *Mathematical Structures in Computer Science* 6(6) (1996) 613–648.
15. P. Lucio, F. Orejas, E. Pino. An algebraic framework for the definition of compositional semantics of Normal Logic Programs. *Journal of Logic Programming* 40(1) (1999) 89–124.
16. Reisig, W.: *Petri Nets*. EATCS Monographs in TCS 4, Springer Verlag, 1985
17. Rozenberg, G. (ed.): *Handbook of Graph Grammars and Computing by Graph Transformation, Vol 1 Foundations*, World Scientific, 1997
18. G. Schneider, J. Winters, *Applying use cases*, Addison Wesley, 1998.

Presentations for Abstract Context Institutions[*]

Wiesław Pawłowski

Institute of Computer Science, Polish Academy of Sciences
ul. Abrahama 18, 81-825 Sopot, Poland
w.pawlowski@ipipan.gda.pl

Abstract. The paper discusses a generalization of the notion of *context institution* and introduces a suitable notion of *presentation* for it. With an appropriate notion of morphism presentations constitute a category, whose structural properties can be used to systematically construct logical systems (i.e. abstract context institutions).

1 Introduction

Context institutions, as introduced in [12], enrich the inner structure of institutions (cf. [6,19]) by adding notions such as *contexts* and *substitutions*.

Context institutions are "concrete" in a sense similar to *concrete institutions* of Bidoit and Tarlecki [1] – for every signature, the category of models is concrete over the category of *indexed sets*, and consequently, there is a notion of a *carrier* for each model. Similarly contexts "contain" *sorted sets* of variables. The *satisfaction relation* relates *(open) formulae* over a given context and *valuations*. The latter are just functions from variables to carriers.

In the present paper we shall generalize context institutions by dropping the "concreteness" assumption. The resulting notion of *abstract context institution* has a simpler definition, while retaining the structural power of the original notion, and allowing for a wider range of examples.

We shall also introduce a notion of *presentation* for abstract context institutions. Presentations, also known as *parchments*, have been originally invented as a tool for proving the *satisfaction condition* for institutions (cf. [5]). In [18], parchments have been proposed as a framework for systematic construction of logical systems. This idea has been further developed in a series of papers [9,10,11].

Structurally, presentations for abstract context institutions are similar to the *model-theoretic parchments* of [11]. Internally however, they are based on a suitable notion of *metastructure* (see also [13]), rather than many-sorted algebras. The added "flexibility" of metastructures allows one to define certain useful constructions, which are not possible at the level of (model-theoretic) parchments.

After presenting some preliminaries, in Section 3 we shall define the notion of abstract context institution, giving also an informal motivation for its various

---

[*] This research has been partially supported by ESPRIT working group 29432 (CoFI WG), and KBN grant No. 8 T11C 037 16.

M. Cerioli and G. Reggio (Eds.): WADT/CoFI 2001, LNCS 2267, pp. 256–279, 2002.

components. We shall illustrate the introduced notion by means of an example. Then, specializing the notion of *institution morphism* (cf. [6]), we shall define the category of abstract context institutions and briefly discuss its properties.

In Section 4 we shall introduce *presentations* for abstract context institutions and give several examples as an illustration. In Section 5 we shall define a notion of *presentation morphism*, describe properties of categories of presentations and show how they can be used for structural construction of logics (abstract context institutions). Finally, in Section 6 we shall present some concluding remarks.

2   Preliminaries

In what follows we assume that the Reader is familiar with the basic notions of category theory (cf. [7,2]) and universal algebra (cf. [3,8]). Therefore the purpose of this section is mainly to fix notation being used throughout the rest of the paper and to introduce some less standard constructions.

Indexed and Sorted Sets. Let S be a set. The category of S-*indexed sets* $\mathbf{ISet}[S]$ is a category having S-indexed families of sets $\langle X_s \rangle_{s \in S}$ as objects, and S-indexed functions, i.e., families $\langle h_s : X_s \to Y_s \rangle_{s \in S}$ as morphisms. In fact, indexed sets form an *indexed category* $\mathbf{ISet} : \mathbf{Set}^{op} \to \mathbf{Cat}$ (cf. [20]).

For every $S \in |\mathbf{Set}|$, the full subcategory of $\mathbf{ISet}[S]$ determined by the following condition:

$$\forall X \in |\mathbf{SSet}[S]| \quad \forall s_1, s_2 \in S \quad s_1 \neq s_2 \ \Rightarrow \ X_{s_1} \cap X_{s_2} = \emptyset$$

will be called the category of S-*sorted sets*, and denoted by $\mathbf{SSet}[S]$.

The correspondence $S \mapsto \mathbf{SSet}[S]$, in an obvious way, extends to a functor $\mathbf{SSet} : \mathbf{Set} \to \mathbf{Cat}$.

For every function $f : S \to S'$, every S-sorted set X and every S'-indexed set V, there is a bijection:

$$(\_)_f^{X,V} : \mathbf{ISet}[S](X, \mathbf{ISet}_f(V)) \to \mathbf{ISet}[S'](\mathbf{SSet}_f(X), V).$$

This bijection is actually natural in both X and V.[1]

Composition in Categories. In most cases the composition of morphisms f and g in a category will be denoted by $f \, ; g$ (we shall always use the diagrammatic order). In the case of natural transformations, their *vertical* composition will be denoted by $\alpha \, ; \beta$, and the *horizontal* one by $\gamma * \delta$.

Variables and Categories of Substitutions. In what follows, the category of (many-sorted) algebraic signatures will be denoted by $\mathbf{AlgSig}$. Let *Var* be an infinite (countable) *vocabulary of variable symbols*. We shall assume, that *Var*

---

[1] In other words $(\_)_f^{-,-} : \mathbf{ISet}[S](\_, \mathbf{ISet}_f(\_)) \Rightarrow \mathbf{ISet}[S'](\mathbf{SSet}_f(\_), \_)$ is a natural isomorphism between the appropriately defined "generalized hom-functors".

comes with a fixed *choice function* i.e., a function $choice : (\mathcal{P}(\mathit{Var}) \setminus \{\emptyset\}) \to \mathit{Var}$, s.t. $choice(V) \in V$, for every non-empty set of variables V.

For every algebraic signature $\Sigma = \langle S, \Omega \rangle$, by the *category of Σ-substitutions*, we shall mean a category $\mathcal{T}_{\Sigma}$ defined as follows:

- objects: finite S-sorted sets of elements of Var, where "X is finite" means, that $\bigcup \{ X_s \mid s \in S \}$ is finite;
- morphisms: a morphism $f : X \to Y$ is an arbitrary S-sorted function $f : X \to |T_{\Sigma}(Y)|$, where $T_{\Sigma}(Y)$ denotes the algebra of Σ-terms with variables from Y.

For morphisms $f : X \to Y$ and $g : Y \to Z$ in $\mathcal{T}_{\Sigma}$, their composition is defined by $f ; g \,\hat{=}\, f ; |g^{\sharp}|$, where $g^{\sharp}$ is the free extension of g to a Σ-homomorphism.

The construction of the category $\mathcal{T}_{\Sigma}$ can be extended to a functor $\mathcal{T}_{\_}$: **AlgSig** $\to$ **SCat**, where **SCat** denotes the category of *small categories*. The construction of $\mathcal{T}_{\_}$ is pretty standard. Perhaps the only nontrivial part consists of showing that $\mathcal{T}_{\_}$ preserves composition (cf. [14]).

Categories of Diagrams. Let **K** be an arbitrary category. The category of *small diagrams in* **K** is a category **sDgm(K)**, whose objects are pairs $\langle \mathbf{A}, F \rangle$, where **A** is a *small* category and $F : \mathbf{A} \to \mathbf{K}$ is a functor. A morphism from $\langle \mathbf{A}, F \rangle$ to $\langle \mathbf{B}, G \rangle$ is a pair $\langle H, \alpha \rangle$, such that $H : \mathbf{A} \to \mathbf{B}$ is a functor and $\alpha : F \Rightarrow H ; G$ is a natural transformation. Composition of morphisms $\langle H_1, \alpha \rangle : \langle \mathbf{A}_1, F_1 \rangle \to \langle \mathbf{A}_1', F_1' \rangle$ and $\langle H_2, \beta \rangle : \langle \mathbf{A}_2, F_2 \rangle \to \langle \mathbf{A}_2', F_2' \rangle$ is defined as $\langle H_1 ; H_2, \ \alpha ; (H_1 * \beta) \rangle$.

For every category **K**, there is a "projection" $\mathbf{Pr_K} : \mathbf{sDgm(K)} \to \mathbf{SCat}$. This projection is a functor defined by: $\mathbf{Pr_K}(\langle \mathbf{A}, F \rangle) \,\hat{=}\, \mathbf{A}$ and $\mathbf{Pr_K}(\langle H, \alpha \rangle) \,\hat{=}\, H$. Using it, for any category **A** and any functor $G : \mathbf{A} \to \mathbf{sDgm(K)}$, we obtain a functor $\mathbf{bas}(G) : \mathbf{A} \to \mathbf{SCat}$ given by the composition $G ; \mathbf{Pr_K}$. Similarly, for any natural transformation $\alpha : G_1 \Rightarrow G_2$ we obtain a natural transformation $\mathbf{bas}(\alpha) : G_1 ; \mathbf{Pr_K} \Rightarrow G_2 ; \mathbf{Pr_K}$ given by the composition $\alpha * \mathbf{Pr_K}$.

We shall call $\mathbf{bas}(G)$ and $\mathbf{bas}(\alpha)$ the *base for F* and the *base for α* respectively. The *base construction* actually defines a functor between appropriate functor categories.

Elements Construction. Let **CoFun(Class)** denotes the category whose objects are *contravariant functors* $F : \mathbf{C}^{op} \to \mathbf{Class}$, where **C** is an arbitrary (not necessarily small) category, and **Class** is the category of classes (i.e. "potentially large" sets). A morphism from $F : \mathbf{C}^{op} \to \mathbf{Class}$ to $G : \mathbf{D}^{op} \to \mathbf{Class}$ is a pair $\langle H, \alpha \rangle$, such that $H : \mathbf{C} \to \mathbf{D}$ is a functor and $\alpha : F \Rightarrow H^{op} ; G$ is a natural transformation.

Then, there is a functor $\mathbf{Elts} : \mathbf{CoFun(Class)} \to \mathbf{Cat}$ given by:

- *action on objects:* $\mathbf{Elts}(F : \mathbf{C}^{op} \to \mathbf{Class})$ is a category whose objects are pairs of the form $\langle C, c \rangle$, where $C \in |\mathbf{C}|$ and $c \in F(C)$. $f : \langle C_1, c_1 \rangle \to \langle C_2, c_2 \rangle$ is a morphism in $\mathbf{Elts}(F)$ if $f : C_1 \to C_2$ in **C** and $F(f)(c_2) = c_1$,

- *action on morphisms*: for any morphism $\langle H, \alpha \rangle : F \to G$ in the category **CoFun(Class)**, the functor **Elts**($\langle H, \alpha \rangle$) is given by: $\langle C, c \rangle \mapsto \langle H(C), \alpha_C(c) \rangle$ and $f \mapsto H(f)$.

For any functor $F : \mathbf{C}^{op} \to \mathbf{Class}$, the category **Elts**($F$) is usually called the *category of elements for* F (cf. [2]). By **ind**(F) we shall denote the obvious "projection" functor from **Elts**(F) to **C**.

3 Abstract Context Institutions

3.1 Definition

Signatures. Similarly as in the case of ordinary institutions, abstract context institutions are built around the notion of *signature*. Signatures provide a *vocabulary* for constructing formulae. Since in computer science applications it is often desirable to be able to change the notation being used, signatures are required to form a category. This category will usually be denoted by **Sig** (possibly with a suitable superscript).

Contexts, Substitutions and Formulae. For every signature Σ, we want to consider (possibly *open*) *formulae* over Σ. Therefore we shall assume that for every Σ, there is a (small) category of Σ-*contexts* **Ctxt**$_\Sigma$. Context morphisms are meant to model *substitutions*.

The fact that for every Σ-context we have a corresponding set of formulae (built "over" that context), will be modeled by a functor **Frm**$_\Sigma$: **Ctxt**$_\Sigma \to$ **Set**. For any context morphism (substitution), its image under **Frm**$_\Sigma$, is a function "performing" the substitution.

To take the change of notation into account, we shall eventually define the *formula functor* as a functor **Frm** : **Sig** $\to$ **sDgm(Set)**.

Models and Valuations. The model structure of abstract context institutions will be given, as in the case of institutions, by a functor **Mod** : **Sig**$^{op} \to$ **Class**. The "semantical part" of abstract context institutions will also contain the notion of *valuations*.

Let us take an arbitrary signature Σ and model $M \in \mathbf{Mod}_\Sigma$. We shall assume, that for every Σ-context Γ there is a set **Val**$_{\Sigma,M}(\Gamma)$, whose elements will be called *valuations of* Γ *in* M. In typical cases valuations are just (suitably indexed) functions from (similarly sorted) sets of variables into the "carrier" of the model.

To better motivate the remaining parts of the valuation structure let us look at very simple example. Let us assume that our contexts are just sets of variables, models are sets of natural numbers and valuations are total functions between them. Let us take two contexts $\{x\}$ and $\{y\}$ and a context morphism (i.e., substitution) t defined by $t(x) = 2 * y$. For an arbitrary valuation of the context $\{y\}$ – e.g., $[y \mapsto 5]$, we can "evaluate" the substitution t to get the "corresponding" valuation of the context $\{x\}$, which in this case will be $[x \mapsto 10]$.

In a general case it means, that for every context morphism $t : \Gamma \to \Delta$ there is a function $\mathbf{Val}_{\Sigma,M}(t) : \mathbf{Val}_{\Sigma,M}(\Delta) \to \mathbf{Val}_{\Sigma,M}(\Gamma)$ between the corresponding sets of valuations. It seems natural to require that "evaluation" of a composition of two substitutions has the same effect as a composition of their evaluations. We shall ensure this by requiring the functoriality of $\mathbf{Val}_{\Sigma,M} : \mathbf{Ctxt}^{op} \to \mathbf{Set}$.

As in the case of formulae, every signature morphism "generates" a corresponding "translation" of valuations. To illustrate this, let us assume that our category of signatures is the category of sets $\mathbf{Set}$. As contexts lets us take sorted sets of variables (elements of a fixed vocabulary of variable symbols), and as models – indexed sets of natural numbers.

Let $S = \{s_1, s_2\}$ and $S' = \{s'\}$ be signatures. Let us take a signature morphism $\sigma : S \to S'$ given by $\sigma(s_1) = \sigma(s_2) = s'$, and an arbitrary S'-model (i.e., an S'-indexed set of natural numbers) – for example $M' = \{3, 7\}$. The *reduct* of M' under $\mathbf{Mod}_\sigma$ is an S-indexed set M defined by: $M_{s_1} = M_{s_2} = \{3, 7\}$.

Let X given by $X_{s_1} = \{x\}$, $X_{s_2} = \{y\}$ be a S-context. Let us consider a valuation $v \in \mathbf{Val}_{S,M}(X)$ s.t. $v = \langle [x \mapsto 3]_{s_1}, [y \mapsto 7]_{s_2} \rangle$. In an obvious way we can translate v "along" σ, obtaining a valuation $v' \in \mathbf{Val}_{S',M'}(\{x,y\})$ $(\{x,y\} = \mathbf{Ctxt}_\sigma(X))$ s.t. $v' = \langle [x \mapsto 3, y \mapsto 7]_{s'} \rangle$. The valuation v' was obtained from v by suitable "reindexing" corresponding to the signature morphism σ.

Similarly, in a general case, we shall assume that for every: signature morphism $\sigma : \Sigma \to \Sigma'$, Σ-context Γ and Σ'-model M' there is a function, denoted by $\mathbf{val}_\sigma(\Gamma)$ which for every Σ-valuation from the set $\mathbf{Val}_{\Sigma,\mathbf{Mod}_\sigma(M')}(\Gamma)$ assigns a Σ'-valuation from $\mathbf{Val}_{\Sigma',M'}(\mathbf{Ctxt}_\sigma(\Gamma))$. Using the notion of the *category of elements* (see Sect. 2) we can formalize it requiring that the *valuation structure* of an abstract context institution is given by a functor $\mathbf{Val} : \mathbf{Elts}(\mathbf{Mod}) \to \mathbf{sDgm}(\mathbf{Set})$, such that for every object $\langle \Sigma, M \rangle$ in $\mathbf{Elts}(\mathbf{Mod})$ the domain of $\mathbf{Val}_{\Sigma,M}$ is the category $\mathbf{Ctxt}_\Sigma^{op}$ (see the *coherence condition* below).

Satisfaction Relation. For every signature Σ, model $M \in \mathbf{Mod}_\Sigma$ and Σ-context Γ, the *satisfaction relation* $M[\_] \models_{\Sigma,\Gamma} \_$ will be a binary relation between the set of valuations $\mathbf{Val}_{\Sigma,M}(\Gamma)$ and the set of formulae $\mathbf{Frm}_\Sigma(\Gamma)$.

The structural components of an abstract context institution are tight together by two "semantical" conditions: the *satisfaction condition* and the *substitution condition*. The former expresses that fact that the satisfiability of a formula is invariant under the change of notation (i.e., signature morphisms), and the latter – that it is also invariant under substitutions (i.e., context morphisms).

Definition 1. *A context institution $\mathfrak{C}$ consists of:*

- *a category $\mathbf{Sig}^\mathfrak{C}$ of signatures,*
- *a formula functor $\mathbf{Frm}^\mathfrak{C} : \mathbf{Sig}^\mathfrak{C} \to \mathbf{sDgm}(\mathbf{Set})$, whose base (see Sect. 2) will be called the context functor for $\mathfrak{C}$ and denoted by $\mathbf{Ctxt}^\mathfrak{C}$ (hence $\mathbf{Ctxt}^\mathfrak{C} : \mathbf{Sig}^\mathfrak{C} \to \mathbf{SCat}$),*
- *a model functor $\mathbf{Mod}^\mathfrak{C} : (\mathbf{Sig}^\mathfrak{C})^{op} \to \mathbf{Class}$,*
- *a valuation functor $\mathbf{Val}^\mathfrak{C} : \mathbf{Elts}(\mathbf{Mod}^\mathfrak{C}) \to \mathbf{sDgm}(\mathbf{Set})$ and*

- *for every: signature Σ, model $M \in \mathbf{Mod}^{\mathfrak{C}}_\Sigma$, context $\Gamma \in |\mathbf{Ctxt}^{\mathfrak{C}}_\Sigma|$, a binary satisfaction relation:*

$$M[\_] \models^{\mathfrak{C}}_{\Sigma,\Gamma} \_ \subseteq \mathbf{Val}^{\mathfrak{C}}_{\Sigma,M}(\Gamma) \times \mathbf{Frm}^{\mathfrak{C}}_\Sigma(\Gamma)$$

such that the following three conditions are satisfied:

- Coherence Condition: $\mathbf{bas}(\mathbf{ind}(\mathbf{Mod}^{\mathfrak{C}}) ; \mathbf{Frm}^{\mathfrak{C}}) = \mathbf{bas}(\mathbf{Val}^{\mathfrak{C}}) ; (\_)^{op}$ *this condition says that for every signature Σ and model $M \in \mathbf{Mod}^{\mathfrak{C}}_\Sigma$, $\mathbf{Val}^{\mathfrak{C}}_{\Sigma,M}$: $(\mathbf{Ctxt}^{\mathfrak{C}}(\Sigma))^{op} \to \mathbf{Set}$, and that for every signature morphism $\sigma : \Sigma \to \Sigma'$, the first element of the pair $\mathbf{Val}^{\mathfrak{C}}_\sigma$ is a functor $(\mathbf{Ctxt}^{\mathfrak{C}}_\sigma)^{op} : (\mathbf{Ctxt}^{\mathfrak{C}}_\Sigma)^{op} \to (\mathbf{Ctxt}^{\mathfrak{C}}_{\Sigma'})^{op}$,*
- Satisfaction Condition: *for every: morphism $\sigma : \langle \Sigma, M \rangle \to \langle \Sigma', M' \rangle$ in $\mathbf{Elts}(\mathbf{Mod}^{\mathfrak{C}})$ (i.e., $\sigma : \Sigma \to \Sigma'$ in $\mathbf{Sig}^{\mathfrak{C}}$ s.t. $M = \mathbf{Mod}^{\mathfrak{C}}_\sigma(M')$), context $\Gamma \in |\mathbf{Ctxt}^{\mathfrak{C}}_\Sigma|$, valuation $v \in \mathbf{Val}^{\mathfrak{C}}_{\Sigma,M}(\Gamma)$ and formula $\phi \in \mathbf{Frm}^{\mathfrak{C}}_\Sigma(\Gamma)$:*

$$M[v] \models^{\mathfrak{C}}_{\Sigma,\Gamma} \phi \quad \text{iff} \quad M'[\mathbf{val}^{\mathfrak{C}}_\sigma(\Gamma)(v)] \models^{\mathfrak{C}}_{\Sigma',\mathbf{Ctxt}^{\mathfrak{C}}_\sigma(\Gamma)} \mathbf{frm}^{\mathfrak{C}}_\sigma(\Gamma)(\phi),$$

where the natural transformations $\mathbf{val}^{\mathfrak{C}}_\sigma : \mathbf{Val}^{\mathfrak{C}}_{\Sigma,M} \Rightarrow (\mathbf{Ctxt}^{\mathfrak{C}}_\sigma)^{op} ; \mathbf{Val}^{\mathfrak{C}}_{\Sigma',M'}$ and $\mathbf{frm}^{\mathfrak{C}}_\sigma : \mathbf{Frm}^{\mathfrak{C}}_\Sigma \Rightarrow \mathbf{Ctxt}^{\mathfrak{C}}_\sigma ; \mathbf{Frm}^{\mathfrak{C}}_{\Sigma'}$ denote the second element of $\mathbf{Val}^{\mathfrak{C}}_\sigma$ and the second element of $\mathbf{Frm}^{\mathfrak{C}}_\sigma$ respectively.

- Substitution Condition: *for every: $\langle \Sigma, M \rangle \in |\mathbf{Elts}(\mathbf{Mod}^{\mathfrak{C}})|$, context morphism $f : \Gamma \to \Delta$, formula $\phi \in \mathbf{Frm}^{\mathfrak{C}}_\Sigma(\Gamma)$ and valuation $v \in \mathbf{Val}^{\mathfrak{C}}_{\Sigma,M}(\Delta)$:*

$$M[\mathbf{Val}^{\mathfrak{C}}_{\Sigma,M}(f)(v)] \models^{\mathfrak{C}}_{\Sigma,\Gamma} \phi \quad \text{iff} \quad M[v] \models^{\mathfrak{C}}_{\Sigma,\Delta} \mathbf{Frm}^{\mathfrak{C}}_\Sigma(f)(\phi).$$

Due to the lack of space, we shall illustrate the introduced notion by giving only one simple example – the abstract context institution of (many-sorted) *equational logic*. Many other examples, among them of partial, modal, and higher-order logics, can be found in [14]. Also examples of context institutions from [12] can be viewed as abstract context institutions.

Informally speaking, one can claim that every logical system with a "reasonable" notion of *substitution*, satisfying a *substitution lemma* can be represented by an abstract context institution. By *substitution lemma* we mean a property saying that for every substitution f, model M and formula ϕ of the given logic:

$$M \models \phi \quad \text{iff} \quad M \models f(\phi)$$

where $f(\phi)$ denotes a formula obtained from ϕ by performing the substitution f. The property expressef by the substitution lemma is closely related to the *substitution condition*.

3.2 Example: Many-Sorted Equational Logic

The equational logic example will not be presented in full detail, but we hope that the Reader will be able to easily fill the gaps wherever necessary.

The category of signatures $\mathbf{Sig}^{EL}$ of the abstract context institution EL for the many-sorted equational logic is the category $\mathbf{AlgSig}$ of (many-sorted) algebraic signatures and their morphisms. The *context functor* $\mathbf{Ctxt}^{EL}$ is the functor $\mathcal{T}_- : \mathbf{AlgSig} \to \mathbf{SCat}$, defined in Sect. 2.

For a given Σ-context X the set of formulae $\mathbf{Frm}_\Sigma^{EL}(X)$ is the set of Σ-equations with variables from X. Both the Σ-formula functor $\mathbf{Frm}_\Sigma^{EL} : \mathbf{Ctxt}_\Sigma^{EL} \to \mathbf{Set}$ and the formula functor $\mathbf{Frm}^{EL} : \mathbf{AlgSig} \to \mathbf{sDgm}$ are defined in the usual way (cf. [12]). In particular, for any algebraic signature morphism $\sigma : \Sigma \to \Sigma'$, $\mathbf{Frm}_\sigma^{EL}$ is a pair $\langle \mathbf{Ctxt}_\sigma^{EL}, \mathbf{frm}_\sigma^{EL} \rangle$, where $\mathbf{frm}_\sigma^{EL} : \mathbf{Frm}_\Sigma^{EL} \Rightarrow \mathbf{Ctxt}_\sigma^{EL} ; \mathbf{Frm}_{\Sigma'}^{EL}$ is a natural transformation given by:

$$\mathbf{frm}_\sigma^{EL}(X)(t_1 \equiv t_2) \mathrel{\widehat{=}} |(\iota_X^\sigma)^\sharp|(t_1) \equiv |(\iota_X^\sigma)^\sharp|(t_2),$$

where $\iota_X^\sigma : X \to \mathbf{ISet}_\sigma(|T_{\Sigma'}(\mathbf{SSet}_\sigma(X))|)^2$ is the obvious inclusion.

The model functor $\mathbf{Mod}^{EL} : \mathbf{AlgSig} \to \mathbf{Class}$ assigns to every algebraic signature Σ the class of all Σ-algebras, and to every algebraic signature morphism $\sigma : \Sigma \to \Sigma'$ – the usual σ-reduct operation.

For any Σ-context X and every Σ-algebra A the set $\mathbf{Val}_{\Sigma,A}^{EL}(X)$ is the set of all (S-indexed) functions from X to $|A|$ – the *carrier* of A. For any Σ-context morphism (substitution) $f : X \to Y$ (i.e., function $f : X \to |T_\Sigma(Y)|$), the function $\mathbf{Val}_{\Sigma,A}^{EL}(f) : \mathbf{Val}_{\Sigma,A}^{EL}(Y) \to \mathbf{Val}_{\Sigma,A}^{EL}(X)$ for every valuation $v : Y \to |A|$ gives the valuation $f; |v^\sharp| : X \to |A|$.

For every morphism $\sigma : \langle \Sigma, A \rangle \to \langle \Sigma', A' \rangle$ in $\mathbf{Elts}(\mathbf{Mod}^{EL})$ its image under $\mathbf{Val}^{EL}$ is a pair $\langle (\mathbf{Ctxt}_\sigma^{EL})^{op}, \mathbf{val}_\sigma^{EL} \rangle$, where the natural transformation $\mathbf{val}_\sigma^{EL} : \mathbf{Val}_{\Sigma,A}^{EL} \Rightarrow (\mathbf{Ctxt}_\sigma^{EL})^{op} ; \mathbf{Val}_{\Sigma',A'}^{EL}$ is given by:

$$\mathbf{val}_\sigma^{EL}(X) \mathrel{\widehat{=}} (\!\!(\, \_ \,)\!\!)_\sigma^{X,|A'|}$$

for every context $X \in |\mathcal{T}_\Sigma|$, where $(\!\!(\, \_ \,)\!\!)_\sigma^{X,|A'|}$ is an appropriate instance of the natural isomorphism described in Sect. 2.

For every signature $\Sigma \in |\mathbf{AlgSig}|$, algebra $A \in |\mathbf{Alg}(\Sigma)|$ and context $X \in |\mathbf{Ctxt}_\Sigma^{EL}|$, the satisfaction relation $A[\,\_\,] \models_{\Sigma,X}^{EL} \_$ is defined by:

$$A[v] \models_{\Sigma,X}^{EL} t_1 \equiv t_2 \quad \text{iff} \quad (t_1)_A^v = (t_2)_A^v,$$

where $v : X \to |A|$, $t_1 \equiv t_2$ is a Σ-equation with variables from X, and $(t_1)_A^v$ denotes the *interpretation* of the term t extending the valuation v.

The Coherence Condition for EL holds by construction. The Satisfaction Condition is a simple consequence of the fact that for any algebraic signature morphism $\sigma : \Sigma \to \Sigma'$, every S-sorted set of variables X, every Σ'-algebra A' and every valuation $v : X \to |\mathbf{Alg}_\sigma(A')|$:

$$v^\sharp = (\iota_X^\sigma)^\sharp ; \mathbf{Alg}_\sigma((\!\!(\, v \,)\!\!)_\sigma^\sharp).$$

---

[2] In both $\mathbf{ISet}_\sigma$ and $\mathbf{SSet}_\sigma$, for readability reasons, we have used signature morphism σ instead of its *sort function*. We shall use this notational convention in other places as well, without mentioning it. We hope it will not lead to confusion

The Substitution Condition follows directly from the fact, that for every context morphism $f : X \to Y$ in $\mathbf{Ctxt}_{\Sigma}^{\mathrm{EL}}$, every, Σ-algebra A and every valuation $v : Y \to |A|$:

$$(f; |v^{\sharp}|)^{\sharp} = f^{\sharp}; v^{\sharp}.$$

3.3   Category of Abstract Context Institutions

The notion of a morphism between abstract context institutions is a specialization of the corresponding notion for ordinary institutions (cf. [6]). Informally speaking, in typical cases, the existence of a morphism between given abstract context institutions expresses the fact, that one of them (the source of the morphism) "extends" the other.

Let $\mathfrak{C}$ and $\mathfrak{D}$ be abstract context institutions. A morphism $\mathfrak{f} : \mathfrak{C} \to \mathfrak{D}$ will be given by four mappings binding together *signatures*, *formulae*, *models* and *valuations* of $\mathfrak{C}$ and $\mathfrak{D}$.

Before giving the formal definition, let us informally introduce the components of the morphism $\mathfrak{f}$ one-by-one. We shall use the above mentioned "extension" metaphor for $\mathfrak{f}$, and call its source $\mathfrak{C}$ and target $\mathfrak{D}$ "richer" and "poorer" respectively. We would like to stress that this naming convention is based on "typical examples" and as such is purely informal.

Signatures. The signature part of the morphism $\mathfrak{f}$, as in the case of ordinary institutions, will be given by a functor $\mathfrak{f}^{\mathrm{S}} : \mathbf{Sig}^{\mathfrak{C}} \to \mathbf{Sig}^{\mathfrak{D}}$, "extracting" the *poorer* signatures of $\mathfrak{D}$ from the *richer* ones of $\mathfrak{C}$.

Formulae. The "inclusion" of *poorer* formulae of $\mathfrak{D}$ into *richer* formulae of $\mathfrak{C}$ will be given by a natural transformation $\mathfrak{f}^{\mathrm{F}} : \mathfrak{f}^{\mathrm{S}}; \mathbf{Frm}^{\mathfrak{D}} \Rightarrow \mathbf{Frm}^{\mathfrak{C}}$. Since the formula functors for abstract context institutions "include" context functors, the components of $\mathfrak{f}^{\mathrm{F}}$ have to deal with contexts as well as with formulae.

For an arbitrary signature Σ of the *richer* institution $\mathfrak{C}$, the component $\mathfrak{f}_{\Sigma}^{\mathrm{F}} : \mathbf{Frm}_{\mathfrak{f}^{\mathrm{S}}(\Sigma)}^{\mathfrak{D}} \to \mathbf{Frm}_{\Sigma}^{\mathfrak{C}}$ (as a morphism in $\mathbf{sDgm(Set)}$) will consist of a functor $\mathfrak{f}_{\Sigma}^{\mathrm{C}} : \mathbf{Ctxt}_{\mathfrak{f}^{\mathrm{S}}(\Sigma)}^{\mathfrak{D}} \to \mathbf{Ctxt}_{\Sigma}^{\mathfrak{C}}$ and a natural transformation $\mathbf{frm}(\mathfrak{f}_{\Sigma}^{\mathrm{F}}) : \mathbf{Frm}_{\mathfrak{f}^{\mathrm{S}}(\Sigma)}^{\mathfrak{D}} \Rightarrow \mathfrak{f}_{\Sigma}^{\mathrm{C}}; \mathbf{Frm}_{\Sigma}^{\mathfrak{C}}$.

Models. The model component of the morphism $\mathfrak{f}$, as in the case of ordinary institutions, will be given by a natural transformation $\mathfrak{f}^{\mathrm{M}} : \mathbf{Mod}^{\mathfrak{C}} \Rightarrow \mathfrak{f}^{\mathrm{S}}; \mathbf{Mod}^{\mathfrak{D}}$, "extracting" from each model of $\mathfrak{C}$ its "submodel" belonging to the *poorer* institution $\mathfrak{D}$.

Valuations. Similarly as in the case of formulae, the morphism $\mathfrak{f}$ "embeds" valuations of the *poorer* institution $\mathfrak{D}$ into the valuations of the *richer* institution $\mathfrak{C}$. Thus, we shall require existence of a natural transformation $\mathfrak{f}^{\mathrm{V}} : \mathbf{Elts}(\langle \mathfrak{f}^{\mathrm{S}}, \mathfrak{f}^{\mathrm{M}} \rangle); \mathbf{Val}^{\mathfrak{D}} \Rightarrow \mathbf{Val}^{\mathfrak{C}}$. For every pair $\langle \Sigma, M \rangle$, where $\Sigma \in |\mathbf{Sig}^{\mathfrak{C}}|$ and $M \in \mathbf{Mod}_{\Sigma}^{\mathfrak{C}}$ (i.e., an object of $\mathbf{Elts}(\mathbf{Mod}\mathfrak{C})$), the component $\mathfrak{f}_{\Sigma, M}^{\mathrm{V}}$ of this transformation is a morphism in the category of diagrams $\mathbf{sDgm(Set)}$. We shall assume (see the *Coherence Condition* below), that the first element of the pair $\mathfrak{f}^V_{\Sigma,M}$ is the functor $(\mathfrak{f}^C_\Sigma)^{op} : (\mathbf{Ctxt}^{\mathfrak{D}}_{\mathfrak{f}^S(\Sigma)})^{op} \to (\mathbf{Ctxt}^{\mathfrak{C}}_\Sigma)^{op}$. The second element is a natural transformation $\mathbf{val}(\mathfrak{f}^V_{\Sigma,M}) : \mathbf{Val}^{\mathfrak{D}}_{\mathfrak{f}^S(\Sigma),\mathfrak{f}^M(M)} \Rightarrow (\mathfrak{f}^C_\Sigma)^{op} ; \mathbf{Val}^{\mathfrak{C}}_{\Sigma,M}$, defining the "inclusion" between the corresponding sets of valuations of the abstract context institutions $\mathfrak{D}$ and $\mathfrak{C}$.

Satisfaction Condition. The Satisfaction Condition for $\mathfrak{f} : \mathfrak{C} \to \mathfrak{D}$ relates mappings given by the natural transformations $\mathfrak{f}^F$, $\mathfrak{f}^M$ and $\mathfrak{f}^V$ to the semantic consequence relations of $\mathfrak{C}$ and $\mathfrak{D}$.

Informally, it says that for every Σ-model M in $\mathfrak{C}$, a formula of the *poorer* institution $\mathfrak{D}$, is satisfied by a valuation v in the "submodel" $\mathfrak{f}^M_\Sigma(M) \in \mathbf{Mod}^{\mathfrak{D}}_{\mathfrak{f}^S(\Sigma)}$ if and only if this formula "embedded" into the set of formulae of the *richer* institution $\mathfrak{C}$ (via $\mathfrak{f}^F$) is satisfied by the image of v wrt. $\mathfrak{f}^V$.

Definition 2. *An abstract context institution morphism $\mathfrak{f} : \mathfrak{C} \to \mathfrak{D}$ consists of:*

- *a functor $\mathfrak{f}^S : \mathbf{Sig}^{\mathfrak{C}} \to \mathbf{Sig}^{\mathfrak{D}}$,*
- *a natural transformation $\mathfrak{f}^F : \mathfrak{f}^S ; \mathbf{Frm}^{\mathfrak{D}} \Rightarrow \mathbf{Frm}^{\mathfrak{C}}$,*
- *a natural transformation $\mathfrak{f}^M : \mathbf{Mod}^{\mathfrak{C}} \Rightarrow (\mathfrak{f}^S)^{op}; \mathbf{Mod}^{\mathfrak{D}}$*
- *a natural transformation $\mathfrak{f}^V : \mathbf{Elts}(\langle \mathfrak{f}^S, \mathfrak{f}^M \rangle); \mathbf{Val}^{\mathfrak{D}} \Rightarrow \mathbf{Val}^{\mathfrak{C}}$,*

such that the following two conditions are satisfied:

- Coherence Condition: $\mathbf{bas}(\mathbf{ind}(\mathbf{Mod}^{\mathfrak{C}}) * \mathfrak{f}^F) = \mathbf{bas}(\mathfrak{f}^V) * (\_)^{op}$
  which says, that for every signature Σ and model $M \in \mathbf{Mod}^{\mathfrak{C}}_\Sigma$, the first component of the pair $\mathfrak{f}^V_{\Sigma,M}$ is the functor $(\mathfrak{f}^C_\Sigma)^{op} : (\mathbf{Ctxt}^{\mathfrak{D}}_{\mathfrak{f}^S(\Sigma)})^{op} \to (\mathbf{Ctxt}^{\mathfrak{C}}_\Sigma)^{op}$,[3] where $\mathfrak{f}^C \cong \mathbf{bas}(\mathfrak{f}^F)$),
- Satisfaction Condition: *for every signature Σ in $\mathfrak{C}$, and every: model $M \in \mathbf{Mod}^{\mathfrak{C}}_\Sigma$, context $\Gamma \in |\mathbf{Ctxt}^{\mathfrak{D}}_{\mathfrak{f}^S(\Sigma)}|$, formula $\psi \in \mathbf{Frm}^{\mathfrak{D}}_{\mathfrak{f}^S(\Sigma)}(\Gamma)$ and valuation v from the set $\mathbf{Val}^{\mathfrak{D}}_{\mathfrak{f}^S(\Sigma),\mathfrak{f}^M(M)}(\Gamma)$:*

$$M[\mathbf{val}(\mathfrak{f}^V_{\Sigma,M})(\Gamma)(v)] \models^{\mathfrak{C}}_{\Sigma,\mathfrak{f}^C_\Sigma(\Gamma)} \mathbf{frm}(\mathfrak{f}^F_\Sigma)(\Gamma)(\psi) \quad \text{iff} \quad \mathfrak{f}^M_\Sigma(M)[v] \models^{\mathfrak{D}}_{\mathfrak{f}^S(\Sigma),\Gamma} \psi.$$

The composition of abstract context institution morphisms $\mathfrak{f} : \mathfrak{A} \to \mathfrak{B}$ and $\mathfrak{g} : \mathfrak{B} \to \mathfrak{C}$ is defined in a straightforward way:

- $(\mathfrak{f};\mathfrak{g})^S \cong \mathfrak{f}^S ; \mathfrak{g}^S$,
- $(\mathfrak{f};\mathfrak{g})^F \cong (\mathfrak{f}^S * \mathfrak{g}^F) ; \mathfrak{f}^F$,
- $(\mathfrak{f};\mathfrak{g})^M \cong \mathfrak{f}^M ; (\mathfrak{f}^S * \mathfrak{g}^M)$,
- $(\mathfrak{f};\mathfrak{g})^V \cong (\mathbf{Elts}(\langle \mathfrak{f}^S, \mathfrak{f}^M \rangle) * \mathfrak{g}^V) ; \mathfrak{f}^V$.

Abstract context institutions and their morphisms form a category which we shall denote by $\mathfrak{AbsConJns}$.

Due to the lack of space, instead of giving a "direct" example, let us only briefly mention, how we can, using abstract context institution morphisms, generalize the construction of the institution of *abstract Hoare logic*, described in [12].

---

[3] Note the similarity of this condition and the Coherence Condition from the definition of abstract context institution.

As the basis for the construction, we take a morphism $\mathfrak{f} : \mathfrak{A} \to \mathfrak{B}$, between abstract context institutions of the *logic of assertions* $\mathfrak{A}$, and the *logic of Boolean expressions* $\mathfrak{B}$. Using context morphisms of $\mathfrak{B}$ as atomic commands, we define an abstract imperative programming language, and give its "institutional" semantics. Then, we can construct an (ordinary) institution $\mathcal{H}(\mathfrak{f})$, of *Hoare logic for* $\mathfrak{f}$, taking appropriately defined *abstract Hoare triples* as sentences.

For the institution $\mathcal{H}(\mathfrak{f})$ we can define Hoare-like inference system, and using only the structural properties of $\mathfrak{A}$, $\mathfrak{B}$, and the morphism $\mathfrak{f}$, we can show, that this system is sound and Cook-complete (cf. [14], Chapt. 6).

4 Presentations

4.1 Metasignatures and Metastructures

The aim of this section is to introduce some auxiliary notions, needed for the definition of *presentation*. *Metasignatures* and *metastructures* will be used for "presenting" the syntactic and semantic aspects of a logical system respectively.[4]

Definition 3. *A* metasignature *is a six-tuple* $L = \langle\, S, \Omega, \Pi, V, C, Q \,\rangle$, *such that* $\langle\, S, \Omega, \Pi \,\rangle$ *is a relational signature,* $V \subseteq S$, *C is a family of sets indexed by natural numbers, and Q is a set.*

The intended interpretation of the respective components will become clear later. For the moment, we can informally say, that metasignatures are just relational signatures enriched by symbols of *logical connectives*, (the family C), *quantifier symbols* (the set Q), and having a distinguished subset of sort names (the set V), for which we want to talk about *variables*.

Definition 4. *A* metasignature morphism $\ell : L \to L'$ *is a triple* $\ell = \langle\, \ell_r, \ell_c, \ell_q \,\rangle$, *such that:*

- $\ell_r : \langle\, S, \Omega, \Pi \,\rangle \to \langle\, S', \Omega', \Pi' \,\rangle$ *is a relational signature morphism s.t.* $\ell_r[V]$ *is a subset of* V',
- $\ell_c = \langle\, \ell_{c,k} : C_k \to C'_k \mid k \in \mathbb{N} \,\rangle$ *and*
- $\ell_q : Q \to Q'$ *is a function.*

It is easy to see, that metasignatures and their morphisms (with an obvious definition of composition) constitute a category. We shall denote it by **MSig**.

For every metasignature $L = \langle\, S, \Omega, \Pi, V, C, Q \,\rangle$, by **Syn**$(L)$ we shall denote an algebraic signature defined as follows:

- *sorts:* $S \uplus \{\star\}^5$

---

[4] The notions of *metasignature* and *metastructure* defined in this paper are slightly different from the analogous notions presented in [12]. The main difference is a separation of *predicate symbols* from *operation symbols* in metasignatures.

[5] The injection of $s \in S$ and $\star$ into the disjoint union $S \uplus \{\star\}$ will be denoted by $T(s)$ and $\mathbb{B}$ respectively. By $T(S)$ we shall denote the set $\{\, T(s) \mid s \in S \,\}$.

– *operation symbols:* $\Omega^{\mathrm{Syn}} \cup \Pi^{\mathrm{Syn}} \cup C^{\mathrm{Syn}}$, where:

$$\Omega^{\mathrm{Syn}}_{t_1 \ldots t_n \to t_0} \;\hat{=}\; \begin{cases} \Omega_{s_1 \ldots s_n \to s_0} & \text{if } t_i = T(s_i) \text{ for } i = 0 \ldots n \\ \emptyset & \text{otherwise} \end{cases}$$

$$\Pi^{\mathrm{Syn}}_{t_1 \ldots t_n \to t_0} \;\hat{=}\; \begin{cases} \Pi_{s_1 \ldots s_n} & \text{if } t_i = T(s_i) \text{ for } i = 1 \ldots n \text{ and } t_0 = \mathbb{B} \\ \emptyset & \text{otherwise} \end{cases}$$

$$C^{\mathrm{Syn}}_{t_1 \ldots t_n \to t_0} \;\hat{=}\; \begin{cases} C_n & \text{if } t_i = \mathbb{B} \text{ for } i = 0 \ldots n \\ \emptyset & \text{otherwise} \end{cases} .$$

As it is easy to check, the above construction can be extended to a functor **Syn** : **MSig** $\to$ **AlgSig**. In what follows, we shall also use two "sub-functors" of **Syn** – **Atm** : **MSig** $\to$ **AlgSig** and **Trm** : **MSig** $\to$ **AlgSig**, such that:

– $\mathbf{Atm}(\langle S, \Omega, \Pi, V, C, Q \rangle) \;\hat{=}\; \langle T(S) \cup \{\mathbb{B}\}, \Omega^{\mathrm{Syn}} \cup \Pi^{\mathrm{Syn}} \rangle$,
– $\mathbf{Trm}(\langle S, \Omega, \Pi, V, C, Q \rangle) \;\hat{=}\; \langle T(S), \langle \Omega^{\mathrm{Syn}}_{w \to t} \mid w \in T(S)^* \wedge t \in T(S) \rangle \rangle$,

The morphisms $\mathbf{Atm}_\ell$ and $\mathbf{Trm}_\ell$ are given by the respective components of the morphism $\mathbf{Syn}_\ell$. For every metasignature L, the signatures $\mathbf{Atm}(L)$ and $\mathbf{Trm}(L)$ are *sub-signatures* of $\mathbf{Syn}(L)$ – hence the informal term "sub-functor".

Definition 5. *An L-metastructure $\mathcal{A}$ consists of:*

– *a $\mathbf{Syn}(L)$-algebra A,*
– *a $T(V)$-indexed set $V_{\mathcal{A}}$, such that for every $s \in V$ $(V_{\mathcal{A}})_{T(s)} \subseteq |A|_{T(s)}$,*
– *a set $D_{\mathcal{A}}$, such that $D_{\mathcal{A}} \subseteq |A|_{\mathbb{B}}$,*
– *for every symbol $q \in Q$, a partial function $q_{\mathcal{A}} : \mathcal{P}(|A|_{\mathbb{B}}) \rightharpoonup |A|_{\mathbb{B}}$.*

The set $|\mathcal{A}| \;\hat{=}\; |A|$ will be called the carrier *of the metastructure $\mathcal{A}$, the subset $V_{\mathcal{A}}$ – its* set of values, *and the set $D_{\mathcal{A}}$ – its* set of designated elements. *The set $|\mathcal{A}|_{\mathbb{B}}$, corresponding to the distinguished sort $\mathbb{B}$ of the "syntactic" signature $\mathbf{Syn}(L)$, will play the rôle of the set of* logical values.

Essentially, L-metastructures are just many-sorted algebras enriched by *generalized operations* – the partial functions corresponding to the symbols from the set Q. In what follows, these operations will be used for giving the semantics of quantifiers.[6]

Definition 6. *Let $\mathcal{A}$ and $\mathcal{B}$ be L-metastructures. A metastructure morphism $h : \mathcal{A} \to \mathcal{A}'$ is a $\mathbf{Syn}(L)$-homomorphism $h : A \to A'$, satisfying the following conditions:*

– *for every symbol $q \in Q$, whenever B belongs to the domain of the generalized operation $q_{\mathcal{A}}$, then $h[B]$ (the image of B wrt. h) belongs to the domain of $q_{\mathcal{A}'}$ and $h(q_{\mathcal{A}}(B)) = q_{\mathcal{A}'}(h[B])$,*
– *$h[V_{\mathcal{A}}] \subseteq V_{\mathcal{A}'}$,*
– *$h[D_{\mathcal{A}}] \subseteq D_{\mathcal{A}'}$.*

---

[6] A similar idea, although in a slightly different context, appeared in the work of H. Rasiowa and R. Sikorski on *algebraization of logic* [16,15].

For every metasignature L, the class of all L-metastructures together with their morphisms form a category, which we shall denote by $\mathbf{MStr}(L)$. The composition of morphisms in $\mathbf{MStr}(L)$ is defined as the composition of $\mathbf{Syn}(L)$-homomorphisms in $\mathbf{Alg}(\mathbf{Syn}(L))$ – the category of $\mathbf{Syn}(L)$-algebras.

Using the functor $\mathbf{Syn}$, for any metasignature morphism $\ell : L \to L'$, we can define a ℓ-reduct functor $\mathbf{MStr}_\ell : \mathbf{MStr}(L') \to \mathbf{MStr}(L)$. This construction extends to an indexed category $\mathbf{MStr} : \mathbf{MSig}^{op} \to \mathbf{Cat}$.

Below, using notions of metasignature and metastructure, we shall define *interpretation structures*, which will play a fundamental rôle in the definition of *presentation*.

4.2 Interpretation Structures

Let us assume, that there is a class of (abstract) objects, which we shall denote by $\mathcal{M}$, and whose elements will be called *models*. An *interpretation function for* $\mathcal{M}$ is a function, which for every element of $\mathcal{M}$ returns a metastructure over a fixed "*metalanguage*" metasignature (the same for the whole class $\mathcal{M}$). More formally:

Definition 7. *An* interpretation structure *is a triple* $\langle L, \mathcal{M}, Int \rangle$, *consisting of:*

- *a metalanguage signature* $L \in |\mathbf{MSig}|$
- *a class of models* $\mathcal{M} \in |\mathbf{Class}|$,
- *an interpretation function (functor)* $Int : \mathcal{M} \to \mathbf{MStr}(L)$.

Definition 8. *A triple* $\langle \ell, m, int \rangle$ *is an* interpretation structure morphism *from* $\langle L, \mathcal{M}, Int \rangle$ *to* $\langle L', \mathcal{M}', Int' \rangle$ *iff:*

- $\ell : L \to L'$ *is a metasignature morphism,*
- $m : \mathcal{M}' \to \mathcal{M}$ *is a function,*
- $int : m \,; Int \Rightarrow Int' \,; \mathbf{MStr}_\ell$ *is a natural transformation.*

Since $\mathcal{M}'$ is a discrete category (a class) the natural transformation int is simply an $\mathcal{M}'$-indexed family of metastructure morphisms, such that for every $M' \in \mathcal{M}'$:

$$int_{M'} : Int(m(M')) \to \mathbf{MStr}_\ell(Int'(M')).$$

The composition of $\langle \ell, m, int \rangle : IS_1 \to IS_2$ and $\langle \ell', m', int' \rangle : IS_2 \to IS_3$ is defined by:

$$\langle \ell, m, int \rangle \,; \langle \ell', m', int' \rangle \cong \langle \ell \,; \ell',\ m' \,; m,\ (m' * int) \,; (int' * \mathbf{MStr}_\ell) \rangle.$$

As it is easy to check, interpretation structures and their morphisms constitute a category. We shall denote it by $\mathbf{IntStr}$.

4.3   Presentations

A *presentation* is an arbitrary functor into the category of interpretation structures:
$$\mathfrak{P} : \mathbf{Sig}^{\mathfrak{P}} \to \mathbf{IntStr}.$$

We shall call $\mathbf{Sig}^{\mathfrak{P}}$ the *category of signatures* of the presentation $\mathfrak{P}$.

For every morphism $\sigma : \Sigma \to \Sigma'$ in $\mathbf{Sig}^{\mathfrak{P}}$, let $\mathfrak{P}(\Sigma) = \langle L_\Sigma, \mathcal{M}_\Sigma, Int_\Sigma \rangle$ and $\mathfrak{P}(\sigma) = \langle \ell_\sigma, m_\sigma, int_\sigma \rangle$. The presentation $\mathfrak{P}$ induces two functors: a *metalanguage functor* and a *model functor*, such that:

- *metalanguage functor* $\mathbf{Lan}^{\mathfrak{P}} : \mathbf{Sig}^{\mathfrak{P}} \to \mathbf{MSig}$
  - $\mathbf{Lan}^{\mathfrak{P}}(\Sigma) \hat{=} L_\Sigma$,
  - $\mathbf{Lan}^{\mathfrak{P}}(\sigma) \hat{=} \ell_\sigma$;
- *model functor* $\mathbf{Mod}^{\mathfrak{P}} : (\mathbf{Sig}^{\mathfrak{P}})^{op} \to \mathbf{Class}$
  - $\mathbf{Mod}^{\mathfrak{P}}(\Sigma) \hat{=} \mathcal{M}_\Sigma$,
  - $\mathbf{Mod}^{\mathfrak{P}}(\sigma) \hat{=} m_\sigma$.

In what follows, by $\mathbf{Syn}^{\mathfrak{P}}$, $\mathbf{Atm}^{\mathfrak{P}}$ and $\mathbf{Trm}^{\mathfrak{P}}$, we shall denote the compositions: $\mathbf{Lan}^{\mathfrak{P}} ; \mathbf{Syn}$, $\mathbf{Lan}^{\mathfrak{P}} ; \mathbf{Atm}$ and $\mathbf{Lan}^{\mathfrak{P}} ; \mathbf{Trm}$ (see Sect. 4.1). We shall call $\mathbf{Atm}^{\mathfrak{P}}$ and $\mathbf{Trm}^{\mathfrak{P}}$ the *atomic formula functor* and the *term functor* for $\mathfrak{P}$ respectively. The meaning of these names shall be clarified in Sect. 4.4, where we describe a construction of an abstract context institution out of a (logical) presentation. Before going any further let us give some examples of presentations.

Model presentations. Let us start with a very simple one. Since (abstract context) institutions advocate a *model-centric* view of logic, we shall show how to "present" a *model part* of a logical system. We shall do this for the case of (many-sorted) algebras.

Let $\mathcal{ALG} : \mathbf{AlgSig} \to \mathbf{IntStr}$ be a presentation such that:

- For every algebraic signature $\Sigma = \langle S, \Omega \rangle$, the metalanguage signature for $\mathcal{ALG}(\Sigma)$ is defined as $\mathbf{Lan}^{\mathcal{ALG}}(\Sigma) \hat{=} \langle S, \Omega, \varnothing, \emptyset, \varnothing, \emptyset \rangle$.[7] The class of models $\mathbf{Mod}^{\mathcal{ALG}}(\Sigma)$ is the class of all Σ-algebras. The interpretation function maps every Σ-algebra A to a $\mathbf{Lan}^{\mathcal{ALG}}(\Sigma)$-metastructure, extending A by a Boolean sort, defined as: $|Int_\Sigma^{\mathcal{ALG}}(A)|_{\mathbb{B}} \hat{=} \{\, \mathtt{tt}, \mathtt{ff} \,\}$, and taking $\{\, \mathtt{tt} \,\}$ as its set of designated elements.
- For every algebraic signature morphism $\sigma : \Sigma \to \Sigma'$, the morphism $\mathcal{ALG}_\sigma$ consists of: an obvious metasignature morphism $\mathbf{Lan}_\sigma^{\mathcal{ALG}}$ "induced" by σ, the $\mathbf{Syn}_\sigma^{\mathcal{ALG}}$-reduct operation, and a trivial natural transformation, whose all components are identities.

In a very similar way one can define a model presentation for the case of *relational structures* $\mathcal{STR}$: $\mathbf{RelSig} \to \mathbf{IntStr}$. The only interesting difference is that the interpretation functions have to map the predicates occurring in relational structures to Boolean-valued operations in metastructures.

---

[7] The symbol $\varnothing$ denotes an appropriately indexed family of empty sets.

Equational logic. Let us now describe a presentation $\mathcal{EL}$, for the (many-sorted) equational logic (an abstract context institution for it has been introduced in Sect. 3.2).

The category of signatures $\mathbf{Sig}^{\mathcal{EL}}$ is the category of algebraic signatures $\mathbf{AlgSig}$. The metalanguage functor $\mathbf{Lan}^{\mathcal{EL}} : \mathbf{AlgSig} \to \mathbf{MSig}$, and the model functor $\mathbf{Mod}^{\mathcal{EL}} : \mathbf{AlgSig}^{op} \to \mathbf{Class}$ are defined as follows:

- $\mathbf{Lan}^{\mathcal{EL}}(\langle S, \Omega \rangle) \stackrel{\frown}{=} \langle S, \Omega, \Pi, S, C, \emptyset \rangle$, where $\Pi_{ss} \stackrel{\frown}{=} \{\equiv\}$ for $s \in S$, and all other elements of Π, and all C_n for $n \geq 0$ are empty.
- for every $\sigma : \Sigma \to \Sigma'$ in $\mathbf{AlgSig}$, the morphism $\mathbf{Lan}^{\mathcal{EL}}_{\sigma} : \mathbf{Lan}^{\mathcal{EL}}(\Sigma) \to \mathbf{Lan}^{\mathcal{EL}}(\Sigma')$ is defined as σ, for the symbols coming from Σ, and as the identity for the *equations* from Π_{ss}, for $s \in S$.
- as the class $\mathbf{Mod}^{\mathcal{EL}}_{\Sigma}$, we take the class of all Σ-algebras $\mathbf{Alg}[\Sigma]$. For every $\sigma : \Sigma \to \Sigma'$ in $\mathbf{AlgSig}$, a function $\mathbf{Mod}^{\mathcal{EL}}_{\sigma} : \mathbf{Mod}^{\mathcal{EL}}_{\Sigma'} \to \mathbf{Mod}^{\mathcal{EL}}_{\Sigma}$ is the algebraic σ-reduct operation.

Let $\Sigma = \langle S, \Omega \rangle$ be an algebraic signature. For every Σ-model (algebra) M, we shall now define the metamodel $Int_{\Sigma}(M)$. Let A_M be a $\mathbf{Syn}^{\mathcal{EL}}(\Sigma)$-algebra such that:

- for every sort $s \in S$, let $|A_M|_{T(s)} \stackrel{\frown}{=} |M|_s$,
- $|A_M|_{\mathbb{B}} \stackrel{\frown}{=} \{\mathtt{tt}, \mathtt{ff}\}$,
- for every $\omega : s_1 \ldots s_n \to s_0$, $\omega_{A_M} \stackrel{\frown}{=} \omega_M$,
- for every $s \in S$ and $a_1, a_2 \in |A_M|_{T(s)}$,

$$\equiv_{A_M}(a_1, a_2) \stackrel{\frown}{=} \begin{cases} \mathtt{tt} \text{ if } a_1 = a_2 \\ \mathtt{ff} \text{ otherwise} \end{cases}$$

For every $s \in S$, as the *set of values* $|V^{Int_{\Sigma}(M)}|_{T(s)}$ for the sort $T(s)$, let us take the whole set $|A_M|_{T(s)}$, and as the set of *designated elements* $D^{Int_{\Sigma}(M)}$ – the singleton set $\{\mathtt{tt}\}$.

For every algebraic signature morphism $\sigma : \Sigma \to \Sigma'$ and every algebra $M' \in \mathbf{Mod}^{\mathcal{EL}}_{\Sigma'}$, let $int^{\sigma}_{M'}$ be the identity morphism.

Partial First-Order Logic. As an even more interesting example, we shall describe a presentation $\mathcal{PFOL}$ for the partial first-order logic (cf. [4]), adding two interesting features: *quantifiers* and *partiality* to the picture.

The category of signatures for $\mathcal{PFOL}$ is the category of *partial relational signatures* $\mathbf{PRelSig}$, whose objects are quadruples $\langle S, \Omega, \mathrm{p}\Omega, \Pi \rangle$, such that $\langle S, \Omega, \Pi \rangle$ is a relational signature, and $\mathrm{p}\Omega$ is an $S^* \times S$-indexed set of *partial operation names*. Morphisms in $\mathbf{PRelSig}$ are defined in the expected way (see [4]).

The metalanguage functor $\mathbf{Lan}^{\mathcal{PFOL}} : \mathbf{PRelSig} \to \mathbf{MSig}$ is defined as follows:

- $\mathbf{Lan}^{\mathcal{PFOL}}(\langle S, \Omega, \mathrm{p}\Omega, \Pi \rangle) \stackrel{\frown}{=} \langle S, \Omega \cup \mathrm{p}\Omega, \Pi(\stackrel{e}{\equiv}), S, C, \{\forall\} \rangle$, where the family $\Pi(\stackrel{e}{\equiv})$ extends Π by adding a symbol $\stackrel{e}{\equiv}$ to every Π_{ss}, for $s \in S$,[8] $C_1 \stackrel{\frown}{=} \{\neg\}$, $C_2 \stackrel{\frown}{=} \{\wedge\}$, and the sets: C_0 and C_n for $n \geq 3$ are empty,

---

[8] We assume here, that the symbol $\stackrel{e}{\equiv}$ did not belong to any of the sets Π_{ss}.

– for every partial relational signature morphism $\sigma : \Sigma \to \Sigma'$, the morphism $\mathbf{Lan}^{\mathcal{PFOL}}_\sigma : \mathbf{Lan}^{\mathcal{PFOL}}(\Sigma) \to \mathbf{Lan}^{\mathcal{PFOL}}(\Sigma')$ is given by σ, for symbols coming from Σ, and is defined as the identity for the symbols of *existential equality*, *connectives*, and *quantifiers* – i.e., for the symbols: $\overset{e}{=}, \neg, \wedge$ and $\forall$.

The functor $\mathbf{Mod}^{\mathcal{PFOL}} : \mathbf{PRelSig}^{op} \to \mathbf{Class}$, for every partial relational signature Σ, returns the class of all *partial relational Σ-structures* (defined in an obvious way). For every $\sigma : \Sigma \to \Sigma'$ in $\mathbf{PRelSig}$, the function $\mathbf{Mod}^{\mathcal{PFOL}}_\sigma$ is the corresponding σ-reduct operation.

Let us now define the interpretation function for $\Sigma = \langle S, \Omega, \mathrm{p}\Omega, \Pi \rangle$. Let M be an arbitrary partial relational Σ-structure. We have to define the metastructure $Int_\Sigma(M)$. Let A_M be a $\mathbf{Syn}^{\mathcal{PFOL}}(\Sigma)$-algebra, such that:

– for every $s \in S$, the set $|A_M|_{T(s)}$, is obtained from $|M|_s$, by adding an extra "undefined" element (different from all the elements of $|M|_s$), which shall always denote by $\bot$,

– $|A_M|_{\mathbb{B}} \mathrel{\widehat{=}} \{\mathtt{tt}, \mathtt{ff}\}$,

– for every $\omega : s_1 \ldots s_n \to s_0$ and $a_i \in |A_M|_{T(s_i)}, i = 0 \ldots n$

$$\omega_{A_M}(a_1, \ldots, a_n) \mathrel{\widehat{=}} \begin{cases} \omega_M(a_1, \ldots, a_n) & \text{if } \langle a_1, \ldots, a_n \rangle \in \mathbf{dom}(\omega_M) \\ \bot & \text{otherwise} \end{cases}$$

– for every $\pi : s_1 \ldots s_n$ and $a_i \in |A_M|_{T(s_i)}, i = 1 \ldots n$

$$\pi_{A_M}(a_1, \ldots, a_n) \mathrel{\widehat{=}} \begin{cases} \mathtt{tt} & \text{if } \langle a_1, \ldots, a_n \rangle \in \pi_M \\ \mathtt{ff} & \text{otherwise} \end{cases}$$

– for every sort $s \in S$, and every $a_1, a_2 \in |A_M|_{T(s)}$

$$\overset{e}{=}_{A_M}(a_1, a_2) \mathrel{\widehat{=}} \begin{cases} \mathtt{tt} & \text{if } a_1 = a_2 \text{ and } a_i \neq \bot \text{ for } i = 1,2 \\ \mathtt{ff} & \text{otherwise} \end{cases}$$

– for every $b, b' \in |A_M|_{\mathbb{B}}$

$$\neg_{A_M}(b) \mathrel{\widehat{=}} \begin{cases} \mathtt{tt} & \text{if } b = \mathtt{ff} \\ \mathtt{ff} & \text{otherwise} \end{cases} \qquad \wedge_{A_M}(b, b') \mathrel{\widehat{=}} \begin{cases} \mathtt{tt} & \text{if } b = \mathtt{tt} = b' \\ \mathtt{ff} & \text{otherwise} \end{cases}$$

For every $s \in S$, as the *set of values* $|V^{Int_\Sigma(M)}|_{T(s)}$, let us take the set $|M|_s$. In other words, the values are all elements of the carrier of M. The "undefined elements" $\bot$ – are not values. As the *set of designated elements* $D^{Int_\Sigma(M)}$, as in the algebraic case, let us take the set $\{\mathtt{tt}\}$.

We still have to define the generalized operation $\forall_{Int_\Sigma(M)}$, interpreting the $\forall$ quantifier. Let us take, for every $B \subseteq |A_M|_{\mathbb{B}}$:

$$\forall_{Int_\Sigma(M)}(B) \mathrel{\widehat{=}} \begin{cases} \mathtt{ff} & \text{if } \mathtt{ff} \in B \\ \mathtt{tt} & \text{otherwise.} \end{cases}$$

For every signature morphism $\sigma : \Sigma \to \Sigma'$ in $\mathbf{PRelSig}$, and every model $M' \in \mathbf{Mod}^{\mathcal{PFOL}}_{\Sigma'}$, let us take the identity as the morphism:

$$int^\sigma_{M'} : Int_\Sigma(\mathbf{Mod}^{\mathcal{PFOL}}_\sigma(M')) \to \mathbf{MStr}_{\mathbf{Lan}^{\mathcal{PFOL}}(\sigma)}(Int_{\Sigma'}(M')).$$

Boolean Presentations. In the case of *model functor presentations*, their main nontrivial part was a model functor, For *Boolean presentations* it will be an *algebra of truth values.*

By a *signature of truth values* we shall mean an arbitrary metasignature of the form $\langle \emptyset, \emptyset, \emptyset, \emptyset, C, Q \rangle$ (in short, we shall denote it by $\langle C, Q \rangle$). A metastructure B over $\langle C, Q \rangle$ will be called an *algebra of truth values*, if all the generalized operations $q_B : \mathcal{P}(|B|) \rightharpoonup |B|$, for $q \in Q$, are total.

For every $\langle C, Q \rangle$-algebra of truth values B, let us define the corresponding presentation $\mathcal{B}(C, Q)$. Its category of signatures is an arbitrary discrete, singleton category, whose only object shall be denoted by „$\heartsuit$". The other components of $\mathcal{B}(C, Q)$ are defined as follows:

- $\mathbf{Lan}_{\heartsuit}^{\mathcal{B}(C,Q)} \cong \langle \emptyset, \emptyset, \emptyset, \emptyset, C, Q \rangle$,
- $\mathbf{Mod}_{\heartsuit}^{\mathcal{B}(C,Q)} \cong \{\spadesuit\},$[9]
- $Int_{\heartsuit}^{\mathcal{B}(C,Q)} \cong B.$

Presentations of the above form shall be called *Boolean.*

The "canonical" example of a Boolean presentation is a presentation for the *classical truth values.* It corresponds to the $\langle \emptyset, \emptyset \rangle$-algebra of truth values $\mathtt{Bool}$, whose carrier is a set $\{\mathtt{tt}, \mathtt{ff}\}$, and $\{\mathtt{tt}\}$ is the set of designated elements. We shall denote this presentation by $\mathcal{B}ool$. Presentations corresponding to "enrichments" of the algebra $\mathtt{Bool}$ by connectives C and quantifiers Q (with a fixed semantics) shall be denoted by $\mathcal{B}ool(C, Q)$. For example, a Boolean presentation for the classical truth values with negation, conjunction and the universal quantifier (the corresponding generalized operation was defined in the previous section – example of $\mathcal{PFOL}$) shall be denoted by $\mathcal{B}ool(\{\neg, \wedge\}, \{\forall\})$.

In what follows a presentation $\mathfrak{P}$ will be called *logical* iff:

- for every signature Σ in $\mathbf{Sig}^{\mathfrak{P}}$, and every model $M \in \mathbf{Mod}_{\Sigma}^{\mathfrak{P}}$, all the generalized operations in the metastructure $Int_{\Sigma}(M)$ are *total,*
- for every morphism $\sigma : \langle \Sigma, M \rangle \rightarrow \langle \Sigma', M' \rangle$ in $\mathbf{Elts}(\mathbf{Mod}^{\mathfrak{P}})$, the morphism $int_{M'}^{\sigma} : Int_{\Sigma}(M) \rightarrow \mathbf{MStr}_{\mathbf{Lan}_{\sigma}^{\mathfrak{P}}}(Int_{\Sigma'}(M'))$ satisfies the following conditions:
  - for every element $b \in |Int_{\Sigma}(M)|_{\mathbb{B}}$, whenever $int_{M'}^{\sigma}(b)$ is a designated element in $\mathbf{MStr}_{\mathbf{Lan}_{\sigma}^{\mathfrak{P}}}(Int_{\Sigma'}(M'))$, then also b is a designated element in $Int_{\Sigma}(M)$,[10]
  - the morphism $int_{M'}^{\sigma}$ *surjectively* maps the set of values of the metastructure $Int_{\Sigma}(M)$ onto the set of values of $\mathbf{MStr}_{\mathbf{Lan}_{\sigma}^{\mathfrak{P}}}(Int_{\Sigma'}(M'))$.

In an obvious way all the examples of presentations given above are logical. A very important feature of logical presentations is that they *generate* abstract context institutions.

---

[9] The only element of the $\heartsuit$-model class of $\mathcal{B}(C, Q)$ shall be denoted by „$\spadesuit$" (similarly, as the only signature, which was denoted by „$\heartsuit$").

[10] In other words, $int_{M'}^{\sigma}$ has to both *preserve* (as every metastructure morphism), and *reflect* the designated elements.

4.4   From Presentations to Abstract Context Institutions

Below, we shall sketch a construction, which for an arbitrary presentation $\mathfrak{P}$ yields an abstract context institution $\mathcal{I}(\mathfrak{P})$.

Signatures, Contexts and Formulae. As the category of signatures for $\mathcal{I}(\mathfrak{P})$, we take the category $\mathbf{Sig}^{\mathfrak{P}}$. For an arbitrary signature $\Sigma \in |\mathbf{Sig}^{\mathfrak{P}}|$, let $\mathbf{Lan}^{\mathfrak{P}}(\Sigma) = \langle\, S, \Omega, \Pi, V, C, Q \,\rangle$.

For every signature Σ in $\mathbf{Sig}^{\mathfrak{P}}$, as $\mathbf{Ctxt}^{\mathcal{I}(\mathfrak{P})}_{\Sigma}$, we take the full subcategory of the category of substitutions $\mathcal{T}_{\mathbf{Trm}^{\mathfrak{P}}(\Sigma)}$ generated by (S-sorted) sets of *variables* X, s.t. $X_{T(s)} = \emptyset$, for $s \notin V$. This restriction corresponds to the intuitive rôle of the subset $V \subseteq S$, as the set of sorts for which we want to talk about variables. The morphism part of the context functor $\mathbf{Ctxt}^{\mathcal{I}(\mathfrak{P})}$ is given by the functor $\mathcal{T}_{\mathbf{Trm}^{\mathfrak{P}}_{-}}$.

To define, for any Σ-context X, the set of formulae $\mathbf{Frm}^{\mathcal{I}(\mathfrak{P})}_{\Sigma}(X)$, we use (almost) the usual first-order syntax approach. First, we take the set $|T_{\mathbf{Atm}^{\mathfrak{P}}(\Sigma)}(X)|_{\mathbb{B}}$ as the set of *atomic formulae* $A_{\Sigma}(X)$. Then, we close $A_{\Sigma}(X)$ wrt. the connectives and quantifier symbols, obtaining the set $F_{\Sigma}(X)$ of Σ-*pre-formulae with variables from X*. To avoid working with equivalence classes (wrt. α-conversion) of pre-formulae, we use the approach of [17]. Eventually, we define the set $\mathbf{Frm}^{\mathcal{I}(\mathfrak{P})}_{\Sigma}(X)$ as the set of pre-formulae, "normalized" wrt. suitably defined notion of syntactic substitution.

The syntax translation along signature morphisms is defined in the standard way, yielding the formula functor $\mathbf{Frm}^{\mathcal{I}(\mathfrak{P})} : \mathbf{Sig}^{\mathfrak{P}} \to \mathbf{sDgm}(\mathbf{Set})$.

Models and Valuations. As the model functor $\mathbf{Mod}^{\mathcal{I}(\mathfrak{P})}$, we take the model functor for the presentation $\mathfrak{P}$ (i.e., $\mathbf{Mod}^{\mathfrak{P}}$ – see Sect. 4.3).

For an arbitrary $\Sigma \in |\mathbf{Sig}^{\mathfrak{P}}|$, let $\mathfrak{P}(\Sigma) = \langle\, \mathbf{Lan}^{\mathfrak{P}}(\Sigma), \mathbf{Mod}^{\mathfrak{P}}(\Sigma), Int_{\Sigma} \,\rangle$. Let $X \in |\mathbf{Ctxt}^{\mathcal{I}(\mathfrak{P})}_{\Sigma}|$ be an arbitrary Σ-context. As the set $\mathbf{Val}^{\mathcal{I}(C)}_{\Sigma, M}(X)$ we shall take the set of all $T(S)$-indexed functions of the form $v : X \to \langle\, |Int_{\Sigma}(M)|_{\tau} \mid \tau \in T(S) \,\rangle$, where $|Int_{\Sigma}(M)|$ is the carrier of the metastructure $Int_{\Sigma}(M)$.

For every morphism $\xi : X \to Y$ in $\mathbf{Ctxt}^{\mathcal{I}(\mathfrak{P})}_{\Sigma}$, the function $\mathbf{Val}^{\mathcal{I}(\mathfrak{P})}_{\Sigma, M}(\xi) : \mathbf{Val}^{\mathcal{I}(\mathfrak{P})}_{\Sigma, M}(Y) \to \mathbf{Val}^{\mathcal{I}(\mathfrak{P})}_{\Sigma, M}(X)$ is defined in the standard way – as in the "algebraic" case.[11] Also the translation of valuations "along" signature morphisms is defined in a similar "algebraic" way.

Satisfaction Relation. The definition of the satisfaction relation for $\mathcal{I}(\mathfrak{P})$ is based upon a suitable notion of a *semantic interpretation of formulae*. For every: signature $\Sigma \in |\mathbf{Sig}^{\mathcal{I}(\mathfrak{P})}|$, model $M \in \mathbf{Mod}^{\mathcal{I}(C)}(\Sigma)$, and context $X \in |\mathbf{Ctxt}^{\mathcal{I}(C)}_{\Sigma}|$ this interpretation is a function:

$$[\![\_]\!]^{\Sigma, M, X}_{-} : \mathbf{Frm}^{\mathcal{I}(C)}_{\Sigma}(X) \times \mathbf{Val}^{\mathcal{I}(C)}_{\Sigma, M}(X) \to |Int_{\Sigma}(M)|_{\mathbb{B}}$$

---

[11] Because, at the level of valuations, metastructures "are" many-sorted algebras.

Let $I(M)$ and $V(M)$ denote the metastructure $Int_\Sigma(M)$ and its *set of values* $V^{Int_\Sigma(M)}$ respectively. We know, that $V(M)$ is a $T(V)$-indexed set, s.t. for every $s \in V$ there is an inclusion $V(M)_{T(s)} \subseteq |I(M)|_{T(s)}$. Let us define:

- for every atomic formula $\phi \in A_\Sigma(X)$, $[\![\phi]\!]_v^{\Sigma, M, X} \cong (\phi)_{I(M)}^{\bar{v}}$, where $(\_)_{I(M)}^{\bar{v}}$ denotes the unique extension of the valuation v to the set of terms for the *atomic formulae* signature $\mathbf{Atm}^\mathfrak{P}(\Sigma)$,
- $[\![c(\phi_1, \ldots, \phi_n)]\!]_v^{\Sigma, M, X} \cong c_{I(M)}([\![\phi_1]\!]_v^{\Sigma, M, X}, \ldots, [\![\phi_n]\!]_v^{\Sigma, M, X})$,
- $[\![qu^\tau.\phi]\!]_v^{\Sigma, M, X} \cong q_{I(M)}(\{[\![\phi]\!]_{v[m/u]}^{\Sigma, M, X \cup \{u\}_\tau} \mid m \in V(M)_\tau\})$.

The third clause, defining the semantic interpretation for quantified formulae, perhaps needs some explanation. To obtain the meaning of a quantified formula $qu^\tau.\phi$ for the valuation $v \in \mathbf{Val}_{\Sigma, M}^{\mathcal{I}(\mathfrak{P})}(X)$, we first interpret the formula ϕ for all the extensions $v[m/u]$ of v, where m ranges over the appropriate set of values. Then, we apply the generalized operation $q_{I(M)}$ to the subset of $|I(M)|_\mathbb{B}$ obtained in this way. Since $\mathfrak{P}$ is logical, $q_{I(M)}$ is total, and thus yields an element of the set $|I(M)|_\mathbb{B}$ as a result.

Using the above semantic interpretation for formulae, we can define the *satisfaction relation* for $\mathcal{I}(\mathfrak{P})$. For every signature $\Sigma \in |\mathbf{Sig}^{\mathcal{I}(\mathfrak{P})}|$, model $M \in \mathbf{Mod}^{\mathcal{I}(\mathfrak{P})}(\Sigma)$ and context $X \in |\mathbf{Ctxt}_\Sigma^{\mathcal{I}(\mathfrak{P})}|$ let us define:

$$M[v] \models_{\Sigma, X}^{\mathcal{I}(\mathcal{C})} \phi \quad \text{iff} \quad [\![\phi]\!]_v^{\Sigma, M, X} \in D_{I(M)}.$$

In other words, a formula ϕ is *satisfied* by a valuation v of the context X in the model M, if and only if, the value of its semantic interpretation for v – $[\![\phi]\!]_v^{\Sigma, M, X}$, belongs to the set of *designated elements* of the metastructure $I(M)$.

It can be shown (cf. [14], Th. 23), that the construction sketched above, for every *logical* presentation $\mathfrak{P}$, yields an abstract context institution $\mathcal{I}(\mathfrak{P})$ as a result.

5   Categories of Presentations

In this section we shall introduce a notion of a *presentation morphism* and its refinement, called a *logical presentation morphism*. We shall also describe some structural properties of the corresponding categories of presentations, and discuss how these categories can be used for modular construction of logical systems.

5.1   Presentation Morphisms

An informal idea behind the notion of *presentation morphisms* will be the same as for *abstract context institution morphisms* – we shall think of the source presentation as being an "extension" of the target one.

The reason why morphisms defining such "extensions" are important for us is very simple. The main application of presentation morphisms we have in mind, is a *modular construction* of logical systems (abstract context institutions). Typically, in this process we "enrich" some "atomic" presentations (i.e., presentations having sufficiently simple internal structure), by adding new features and mechanisms, and then "put together" such enrichments.

Definition 9. *Let* $\mathfrak{P} : \mathbf{Sig}^{\mathfrak{P}} \to \mathbf{IntStr}$ *and* $\mathfrak{Q} : \mathbf{Sig}^{\mathfrak{Q}} \to \mathbf{IntStr}$ *be presentations. A morphism from* $\mathfrak{P}$ *to* $\mathfrak{Q}$ *is a pair* $\langle \Phi, \mu \rangle : \mathfrak{P} \to \mathfrak{Q}$, *such that:*

- $\Phi : \mathbf{Sig}^{\mathfrak{P}} \to \mathbf{Sig}^{\mathfrak{Q}}$ *is a functor,*
- $\mu : \Phi ; \mathfrak{Q} \Rightarrow \mathfrak{P}$ *is a natural transformation.*

For every signature $\Sigma \in |\mathbf{Sig}^{\mathfrak{P}}|$, the Σ-component of the natural transformation μ, as a morphism in $\mathbf{IntStr}$, consists of:

- a metalanguage signature morphism $\mu_{\Sigma}^{\mathrm{Lan}} : \mathbf{Lan}^{\mathfrak{Q}}(\Phi(\Sigma)) \to \mathbf{Lan}^{\mathfrak{P}}(\Sigma)$,
- a functor $\mu_{\Sigma}^{\mathrm{Mod}} : \mathbf{Mod}^{\mathfrak{P}}(\Sigma) \to \mathbf{Mod}^{\mathfrak{Q}}(\Phi(\Sigma))$,
- a natural transformation $\mu_{\Sigma}^{\mathrm{Int}} : \mu_{\Sigma}^{\mathrm{Mod}} ; Int_{\Phi(\Sigma)}^{\mathfrak{Q}} \Rightarrow Int_{\Sigma}^{\mathfrak{P}} ; \mathbf{MStr}(\mu_{\Sigma}^{\mathrm{Lan}})$.

We shall call a morphism $\langle \Phi, \mu \rangle : \mathfrak{P} \to \mathfrak{Q}$ *logical*, iff for every signature $\Sigma \in |\mathbf{Sig}^{\mathfrak{P}}|$ and every model $M \in \mathbf{Mod}_{\Sigma}^{\mathfrak{P}}$ the morphism:

$$\mu_{\Sigma}^{\mathrm{Int}}(M) : Int_{\Phi(\Sigma)}^{\mathfrak{Q}}(\mu_{\Sigma}^{\mathrm{Mod}}(M)) \to \mathbf{MStr}_{\mu_{\Sigma}^{\mathrm{Lan}}}(Int_{\Sigma}^{\mathfrak{P}}(M))$$

satisfies the following conditions:

- the morphism $\mu_{\Sigma}^{\mathrm{Int}}(M)$ has to both *preserve* (as every metastructure morphism), and *reflect* designated elements.
- $\mu_{\Sigma}^{\mathrm{Int}}(M)$ has to be *surjective* on values.[12]

Presentations and their morphisms constitute a category, which we shall call the *category of presentations*, and denote by $\mathfrak{Pres}$. The composition of morphisms $\langle \Phi, \mu \rangle : \mathfrak{P} \to \mathfrak{Q}$ and $\langle \Phi', \mu' \rangle : \mathfrak{Q} \to \mathfrak{R}$ in $\mathfrak{Pres}$, is defined as $\langle \Phi ; \Phi', (\Phi * \mu') ; \mu \rangle$.

It is easy to see, that both the identity morphisms and the composition of *logical morphisms* are logical. The subcategory of $\mathfrak{Pres}$, consisting of logical presentations and logical morphisms between them, will be called the *category of logical presentations* and denoted by $\mathfrak{LogPres}$.

Th notion of *logicality* for presentation morphisms, plays the same rôle as it did for presentations. One can show (cf. [14], Th. 24), that the construction of an abstract context institution $\mathcal{I}(\mathfrak{P})$ out of a logical presentation $\mathfrak{P}$, can be extended to logical presentation morphisms, giving for any logical presentation morphism $\langle \Phi, \mu \rangle : \mathfrak{P} \to \mathfrak{Q}$, an abstract context institution morphism $\mathcal{I}(\langle \Phi, \mu \rangle) : \mathcal{I}(\mathfrak{P}) \to \mathcal{I}(\mathfrak{Q})$. It can be shown, that $\mathcal{I}(\_) : \mathfrak{LogPres} \to \mathfrak{AbsConIns}$ is actually a functor.

---

[12] It has to map the set of values of the metastructure $Int_{\Phi(\Sigma)}^{\mathfrak{Q}}(\mu_{\Sigma}^{\mathrm{Mod}}(M))$ onto the set of values of the metastructure $\mathbf{MStr}_{\mu_{\Sigma}^{\mathrm{Lan}}}(Int_{\Sigma}^{\mathfrak{P}}(M))$.

5.2   Limits in Categories of Presentations

As we have mentioned in the Introduction, presentation-like structures called *parchments*, originally invented as a tool for proving the satisfaction condition for ordinary institutions [5], have been later suitably redefined, and proposed as a tool for modular construction of logics [9,10,11,13]. In all cases, universal categorical constructions in the appropriate categories of parchments, have been used as a tool. In this section, we would like to show, how this approach can be extended to the case of presentations.

Since *logical presentations* and *logical presentation morphisms* generate abstract context institutions and their morphisms, it would be quite natural to expect, that limits in the category $\mathfrak{Log}\mathfrak{Pres}$ can be used for modular construction of (logical) presentations, and hence – abstract context institutions.

Unfortunately, the category of logical presentations is not complete, so not every diagram in it has a limit. Incompleteness of $\mathfrak{Log}\mathfrak{Pres}$ (cf. [14], Sect. 10.4.3), can be shown along the same lines as incompleteness of the category of *model-theoretic parchments* [11], and *context parchments* [13].

As it turns out, the category $\mathfrak{Pres}$ of presentations is complete (cf. [14], Cor. 14). Since only logical presentations generate abstract context institutions, the question is whether limits in $\mathfrak{Pres}$ can be used as a tool for "putting logics together". Before we answer this question, let us give an example. Let us consider a diagram in $\mathfrak{Pres}$ consisting of four morphisms: f_1, f_2, f_3, f_4:

The morphisms $f_1,...,f_4$ are the obviously defined "extensions", and $\mathcal{ALG}$ denotes a presentation for the (many-sorted) *equational logic*.

To get a limit in $\mathfrak{Pres}$ of the above diagram, we can first take a limit of the diagram consisting of the morphisms f_1 and f_2. This gives us the presentation $\mathcal{EL}(\neg, \wedge, \forall)$, for an "almost first-order logic with equality", having algebras as models (i.e., without predicates). Taking a limit of the diagram consisting of f_3 and f_4, we obtain "equational logic for relational structures". Finally, taking the limit of f_1' and f_4' gives us a presentation $\mathcal{FOL}$, for the first-order logic.

All four morphisms from the above diagram (i.e., $f_1,...,f_4$), are *logical*. Also the result, the presentation $\mathcal{FOL}$, is logical. The example shows, that taking a limit of a logical diagram (i.e., a diagram consisting of logical morphisms between logical presentations), in the category $\mathfrak{Pres}$ *may* lead to a logical presentation as a result.

Unfortunately, it is not always like this. As a very simple example, let us take a pullback of two (logical) extensions of the Boolean presentation $\mathcal{B}ool$: one adding the usual universal quantifier, and the other extending the set of truth values to three elements. In the pullback presentation, the universal quantifier symbol is interpreted by a partial generalized operation – undefined for sets of truth values containing the "third value". Hence, the pullback presentation is not logical.

Four out of five presentations occurring in the above diagram can be seen as "atomic". It is certainly so for the model presentations $\mathcal{ALG}$ and $\mathcal{STR}$, and the Boolean presentation $\mathcal{B}ool$. The presentation $\mathcal{EL}$ for the equational logic may also be seen as "atomic" since there does not seem to exist a meaningful way of constructing it from simpler presentations (equality needs terms, and terms come from algebraic structures). The only "non-atomic" presentation in this diagram is $\mathcal{B}ool(\{\neg, \wedge\}, \{\forall\})$ (it can be constructed from $\mathcal{B}ool$ by "adding" connectives and the quantifier one by one). The pullback of f_1 and f_2 can be seen however as a special case of a more general construction, which we shall call *completion*.

A presentation morphism $\langle \Phi, \mu \rangle : \mathfrak{P} \to \mathfrak{B}$ will be called *Boolean* iff $\mathfrak{B}$ is Boolean, and for every signature $\Sigma \in |\mathbf{Sig}^{\mathfrak{B}}|$, and every Σ-model M, the metastructure morphism:

$$\mu_\Sigma^{\mathrm{Int}}(M) : Int_{\Phi(\Sigma)}^{\mathfrak{B}}(\mu_\Sigma^{\mathrm{Mod}}(M)) \to \mathbf{MStr}_{\mu_\Sigma^{\mathrm{Lan}}}(Int_\Sigma^{\mathfrak{P}}(M))$$

is an isomorphism. In other words, a Boolean morphism $\langle \Phi, \mu \rangle : \mathfrak{P} \to \mathfrak{B}$ shows, that the "algebra of truth values" of the presentation $\mathfrak{P}$ is "built over" the algebra of truth values of the Boolean presentation $\mathfrak{B}$. A Boolean morphism $\langle \Phi, \mu \rangle : \mathfrak{P} \to \mathfrak{B}$ will be called a *Boolean extension* if the presentation $\mathfrak{P}$ is Boolean as well.

Let $f : \mathfrak{P} \to \mathfrak{B}$ and $g : \mathfrak{B}' \to \mathfrak{B}$ be a Boolean morphism, and a Boolean extension respectively. We shall call the pullback of f along g a g-*completion of* f. For example, if $g : \mathcal{B}ool(\neg, \wedge) \to \mathcal{B}ool$ is the obvious extension, and $f : \mathfrak{P} \to \mathcal{B}ool$ is a Boolean morphism, then the g-completion of f enriches $\mathfrak{P}$ by adding negation and conjunction. In general, it can be shown that pullbacks of Boolean morphisms along Boolean extensions in $\mathfrak{P}res$ always exist, and that they are logical.

As it has been already pointed out in [9], and later in [10,11,13], one should not expect the "categorical nonsense" to always produce the "expected" result. The main advantage of presentations over other "parchment-like" notions is a clear separation of "logical" and "non-logical" parts. This enables us to define "modularization" techniques, which can always be applied – such as the *completion* construction described above. In general however, the process of modular construction of logics in not (and perhaps will never be) "fully automatic".

6   Concluding Remarks

In the paper, we have described a notion of an *abstract context institution*, generalizing the corresponding notion from [12]. Building upon the work on *parch-*

ments [5,10,11] and *context parchments* [13], we have also introduced a notion of *presentation*.

Abstract context institutions enrich institutions [6] by notions of *context*, *substitution* and *valuation*. The valuations are "abstract" objects[13], satisfying some mild technical assumptions. Thanks to their high level of generality, abstract context institutions have numerous examples (see [14]), including systems of *partial*, *modal* and *higher-order* logics. At the same time however, abstract context institutions and their morphisms are *rich enough* to permit some nontrivial constructions. One such construction, of an institution of abstract Hoare logic, has been mentioned in Sect. 3.3 (see also [12,14]).

Presentations, defined in Sect. 4.3, are structures, which simplify the task of defining abstract context institutions. The notion of presentation is based on the idea of *model-theoretic parchment* [11] and *context parchment* [13]. Many examples of presentations can be found in [14], among them for *total, partial, modal, many-valued* and *higher-order* logics. In Sect. 4.4 we have shown, that for any *logical* presentation $\mathfrak{P}$ we can construct an abstract context institution $\mathcal{I}(\mathfrak{P})$ "generated" by $\mathfrak{P}$. This construction extends to *logical* presentation morphisms.

As it turns out, the category of presentations $\mathfrak{Pres}$, can be used as a framework for modular construction of logics. Presentations share advantages of both model-theoretic parchments and context parchments in this respect.

Thanks to their similarity to model-theoretic parchments, presentations cleanly separate model-theoretic part of a logic from its "syntactic" part. This feature allows for more refined handling of logical syntax in the process of putting presentations together. Furthermore, because the interpretation function takes individual models into account, presentations are capable of describing logics which hardly fit into the standard (context) parchment framework. One example could be a general modal logic, where the space of truth values varies from model to model and depends on a *frame* being a part of the model (cf. [11,14]).

Similarly to context parchments, presentations allow to "separate" the *algebra of truth values* from the rest of the structure, making it possible to define operations which cannot be defined and performed within the standard (model-theoretic) parchment framework. An interesting example of such a construction – the *closure* operation has been described in Sect. 5.2. The same "separation" has another important consequence wrt. (non-context) parchments. Namely, the structure of a presentation for a given logic is usually much simpler than the structure of a parchment for the same logic (even though, presentations contain an extra information for contexts and substitutions).

We believe, that presentations can be used not only for modular construction of (abstract context) institutions, but also for "putting together" inference systems.

---

[13] In the case of context institutions [12], valuations were required to be *total* functions from *variables* to *model carriers*.

References

1. M. Bidoit and A. Tarlecki. Behavioural satisfaction and equivalence in concrete model categories. In H. Kirchner, editor, *Proceedings, 20th Colloquium on Trees in Algebra and Computing CAAP'96, Linköping, April 1996*, volume 1059 of *LNCS*, pages 241–256. Springer-Verlag, 1996.
2. F. Borceux. *Handbook of Categorical Algebra 1, Basic Category Theory*, volume 50 of *Encyclopedia of Mathematics and its Applications*. Cambridge University Press, 1994.
3. R. M. Burstall and J. A. Goguen. Algebras, theories and freenes: an introduction for computer scientist. In *Proc. 1981 Marktoberdorf NATO Summer School*. Reidel, 1982.
4. M. Cerioli, T. Mossakowski, and H. Reichel. From total equational to partial first order logic. In E. Astesiano, H.-J. Kreowski, and B. Krieg-Brückner, editors, *Algebraic Foundations of System Specification*, IFIP State-of-the-Art Reports, pages 31–104. Springer-Verlag, 1999.
5. J. A. Goguen and R. M. Burstall. A study in the foundations of programming methodology: Specifications, institutions, charters and parchments. In D. Pitt, S. Abramsky, A. Poigné, and D. Rydeheard, editors, *Proc. Conference on Category Theory and Computer Programming*, volume 240 of *LNCS*, pages 313–333. Springer-Verlag, 1986.
6. J. A. Goguen and R. M. Burstall. Institutions: Abstract model theory for specification and programming. *Journal of the Association for Computing Machinery*, 39:95–146, 1992.
7. S. MacLane. *Categories for the Working Mathematician*, volume 5 of *Springer Graduate Texts in Mathematics*. Springer-Verlag, 1971.
8. K. Meinke and J. V. Tucker. Universal algebra. In S. Abramsky, D. M. Gabbay, and T. S. E. Maibaum, editors, *Handbook of Logic in Computer Science, Volume 1, Background: Mathematical Structures*, pages 189–411. Clarendon Press, Oxford, 1992.
9. T. Mossakowski. Using limits of parchments to systematically construct institutions of partial algebras. In M. Haveraaen and O.-J. Dahl, editors, *Recent Trends in Data Type Specifications. 11th Workshop on Specification of Abstract Data Types joint with the 8th General COMPASS Workshop. Oslo, Norway, September 1995. Selected Papers*, volume 1130 of *LNCS*, pages 379–393. Springer-Verlag, 1996.
10. T. Mossakowski, A. Tarlecki, and W. Pawłowski. Combining and representing logical systems. In E. Moggi, editor, *Category Theory and Computer Science, CTCS'97, Santa Margherita Ligure, Italy, 1997, Proceedings*, volume 1290 of *LNCS*, pages 177–198. Springer-Verlag, 1997.
11. T. Mossakowski, A. Tarlecki, and W. Pawłowski. Combining and representing logical systems using model-theoretic parchments. In F. Parisi Presicce, editor, *Recent Trends in Algebraic Development Techniques. 12th International Workshop WADT'97. Tarquinia, Italy, June 1997. Selected papers*, volume 1376 of *LNCS*, pages 349–364. Springer-Verlag, 1998.
12. W. Pawłowski. Context institutions. In M. Haveraaen, O. Owe, and O.-J. Dahl, editors, *Recent Trends in Data Type Specifications. 11th Workshop on Specification of Abstract Data Types joint with the 8th general COMPASS workshop. Oslo, Norway, September 1995. Selected papers*, volume 1130 of *LNCS*, pages 436–457. Springer-Verlag, 1996.

13. W. Pawłowski. Context parchments. In F. Parisi Presicce, editor, *Recent Trends in Algebraic Development Techniques. 12th International Workshop WADT'97. Tarquinia, Italy, June 1997. Selected papers*, volume 1376 of *LNCS*, pages 381–401. Springer-Verlag, 1998.

14. W. Pawłowski. *Contextual Logical Systems for the Foundations of Software Specification and Development (in Polish)*. PhD thesis, Inst. of Comp. Sci. PAS, Warsaw, 2000. ftp://ftp.ipipan.gda.pl/wiesiek/papers/thesis.ps.gz.

15. H. Rasiowa and R. Sikorski. *The Mathematics of Metamathematics*. Państwowe Wydawnictwo Naukowe, 1963.

16. R. Sikorski. Algebra of formalized languages. *Colloquium Mathematicum*, 9:2–32, 1962.

17. A. Stoughton. Substitution revisited. *Theoretical Computer Science*, 59:317–325, 1988.

18. A. Tarlecki. Bits and pieces of the theory of institutions. In *Proc. Workshop on Category Theory and Computer Programming*, volume 240 of *LNCS*, pages 334–363. Springer-Verlag, 1986.

19. A. Tarlecki. Institutions: An abstract framework for formal specifications. In E. Astesiano, H.-J. Kreowski, and B. Krieg-Brückner, editors, *Algebraic Foundations of System Specification*, IFIP State-of-the-Art Reports, pages 105–130. Springer-Verlag, 1999.

20. A. Tarlecki, R. M. Burstall, and J. A. Goguen. Some fundamental algebraic tools for the semantics of computation: Part 3. Indexed categories. *Theoretical Computer Science*, 91(2):239–264, 1991.

Programs, Proofs and Parametrized Specifications

Iman Poernomo[1,*], John N. Crossley[1], and Martin Wirsing[2]

[1] School of Computer Science and Software Engineering, Monash University,
Clayton, Victoria, Australia 3800
{ihp,jnc}@csse.monash.edu.au
[2] Institut für Informatik, Ludwig-Maximilians-Universität,
Oettingenstraße 67, 80538 München, Germany.
wirsing@informatik.uni-muenchen.de

Abstract. In a series of papers we have been using a modification of the ideas of Curry and Howard to obtain reliable programs from formal proofs. In this paper we extend our earlier work by presenting a new approach for constructing correct SML structures and SML functors from CASL structured and parametrized specifications by extracting the SML programs from constructive proofs of the axioms of the specifications. We provide a novel formal calculus with rules corresponding to the construction and instantiation of parametrized specifications and then a program extraction procedure which produces *SML* programs that meet their specifications.

Keywords: Parametrized specifications, *CASL*, *SML*, Curry-Howard isomorphism.

1 Introduction

Producing correct programs, that is to say, programs that satisfy their specifications, is a primary goal of software engineering. In a series of papers (beginning with Peterreins's thesis [16] and continuing with [21] and [6]) we have been developing a system that does this by using a combination of standard constructive mathematical proof methods together with structural modifications deriving from Wirsing's original calculi for structured specifications (see e.g. [19]). In this paper we extend our earlier work by presenting a new approach to constructing correct *SML* structures and *SML* functors from *CASL* structured and parametrized specifications by extracting the *SML* programs from constructive proofs of the axioms of the specifications. Others have provided constructive systems for producing correct programs: for example, Schwichtenberg [1], Hayashi [9], Constable [4] and Coquand and Huet [10], but these systems do not involve structured specifications. We are not the first to propose developing correct *SML* programs. Sannella and Tarlecki proposed a stepwise

---

* Research partly supported by ARC grant A 49230989.

M. Cerioli and G. Reggio (Eds.): WADT/CoFI 2001, LNCS 2267, pp. 280–304, 2002.

development process and have designed *Extended ML*, [11], as a language for expressing specifications and *SML* programs by one single syntax; but they do not use proof extraction techniques. Closest to our approach is the one of Doug Smith [18] who also uses program synthesis techniques but not with the emphasis on logical calculi.

Briefly, our approach is as follows. Given a parametrized structured algebraic specification SN with universal axioms and given any function symbol f of the body of SN, we collect all the axioms for f into a formula, say $\forall x \bullet A(x, f(x))$. Then we build a formal (constructive) proof of the unSkolemization $\forall x \bullet \exists y \bullet A(x, y)$ using rules in a natural deduction calculus augmented to include rules for structuring specifications. This proof is encoded in a *Curry-Howard term*, or *proof term*. From this term, we automatically extract a function declaration in Standard *ML* (see e.g. [13] or [15]) which, given x, yields a y such that $A(x, y)$. By collecting the extracted function declarations for all functions in the body of SN we obtain an *SML* functor and this *SML* functor is our program.

In order to minimize the technical problems involved in bringing together two full specification and programming languages such as *CASL* and *SML* we confine ourselves to first order specifications with total functions.[1]

Structured specifications in *CASL* are built from basic (or flat) specifications by means of translation (or renaming), written **with**, taking unions, written **and**, and hiding signatures, written **hide**. (Extension, written **then**, is also used.) In the present paper we extend our earlier work [6], which accommodated such specification construction mechanisms, to parametrized specifications. In *CASL*, the notion of parametrization is a generalization of the notion of enrichment: a parametrization is an enrichment where the enriched part is not fixed. Parameter passing is performed using the push-out approach (see [20] or [14]).

We augment the formal logical calculus of [6]. First we extend the set of (individual) terms to allow *SML* programs to be expressed within *CASL* and secondly we add three new rules because of parametrized specifications. In *CASL* a parametrized specification is given through the notion of a *named* specification, that is, a specification name with a definition for it. Correspondingly we give a rule, (Defn), for defining named, and consequently parametrized, specifications. To accommodate the new name we have to expand the global environment of names. The second rule, (Fit), corresponds to the instantiation of a parametrized specification by fitting arguments. In addition we have an extension of the axiom rule of [6] that allows us to use axioms from the new, named specifications. (We also add new rules of lesser significance that cover extensions in *CASL*. These rules may be regarded as variations of the rules for unions.)

From the proofs in our formal calculus we extract programs using a development of the Curry-Howard isomorphism (see e.g. [8] or [7]). These programs are automatically converted (cf. [5]) to programs in *SML* (and they could be converted into programs in other familiar languages such as C++).

---

[1] So functions are made total by giving a default value if they would otherwise be undefined.

Throughout the paper we illustrate our techniques by a simple example (ultimately deriving from one of Sannella and Tarlecki in [17]) of a specification for a warehouse parametrized by a specification of a catalogue of parts stored in the warehouse. We finally produce an executable refinement of the original parametrized specification and a *SML* functor which realizes the parametrized specification.

Section 2 introduces the *CASL* concepts involving parametrized specifications. Section 3 presents the extensions to the formal calculus of [6]. Section 4 introduces new proof reductions and establishes the strong normalization theorem for the whole calculus. Section 5 extends our extraction process of [6]. In section 6 we continue with our example and obtain an executable refinement of our original specification. Finally, in section 7 we extract an *SML* functor from an executable parametrized specification. In the future we intend to implement the extensions of the present paper in our system **Fred**, see [5].

2   Parametrized Specifications

In addition to the basic constructions **then, and, with** and **hide**, *CASL* also allows the construction of parametrized, *generic* specifications (see [3], section 6.2). These are *named specifications* in which certain of the specifications used are labelled as parameters.

A named specification is a specification with an associated name, by which it may be referred to in other specifications. In this way, named specifications permit the reuse of specifications. A named specification is written

> **spec** $\text{Sn}[\text{Sp\_1}] \dots [\text{Sp\_n}]$ **given** $\text{Sp}''\_1, \dots, \text{Sp}''\_m =$
> Sp
> **end**

where the Sp_i $(i = 1, \dots, n)$ are specifications, called the *parameters* for the specification Sn, and the Sp''_j $(j = 1, \dots, m)$ are specifications, called the *imports* for Sn. The specification Sp is called the *body* of Sn. If $n = 0$, there are no parameter specifications Sp_i, and Sn is said to be a *non-generic* definition, otherwise it is called *generic*.

Example. We consider a very simple example of a specification of a warehouse which houses parts. The warehouse keeps track of the parts by means of a catalogue that contains a list of each part's possible replacements, indexed by the part's name. We specify the warehouse as a parametrized specification, which is generic over a specification of a catalogue. The parameter specification CATALOGUE itself makes use of a generic specification of lists. Different warehouses may use different catalogues, and involve different types of part. So, in our extracted program, we are able to abstract from the features common to a range of warehouses.

Because a catalogue is a list of parts, we need a specification LISTS. We provide this as a generic named specification. First we assume NAT. We shall build LISTS, which is a generic specification of lists of elements of sort *Elem* and contains an operation *hd* that returns the head of a list, on top of NAT. (For convenience we often omit axioms not explicitly used in this paper. This is denoted by ... .)

> **spec** LISTS[**sorts** *Elem*] **given** NAT =
> **sorts** *List(Elem)*
> **ops** *nil* : *List(Elem)*; *cons* : *Elem* × *List(Elem)* → *List(Elem)*;
> *hd* : *List(Elem)* → *Elem*; *size* : *List(Elem)* → *Nat*
> **preds** ∈: *Elem* × *List(Elem)*
> **axioms** ∀k : *List(Elem)* • $size(k) > 0 \Rightarrow hd(k) \in k$
> ...

The *global environment* for *CASL* declarations is a mutable set of named specification declarations. When a named specification for SN is defined, the global environment is extended to include the definition, provided a definition for SN has not been given previously. In the example we add CATALOGUE and LISTS.

A non-generic specification may be used as a structured specification in place of its definition body. However, a generic specification must be instantiated in order to be used as a structured specification. This instantiation is done through *fitting arguments*. Fitting is a process of providing arguments that "fit" a specification, FA_i, in place of the original SP_i by means of an appropriate fitting morphism, FM_i, where the signatures match (i.e. fit). The union, FM, of the fitting morphisms and the identity morphisms on the imports and on the body of the specification name have to be compatible. (See [3], section 6.2.2.. In our example we shall use a map that sends *Elem* to *Parts*, see below.) Instantiation of a generic specification with name SN is written

$$\text{SN}[FA_1]\ldots[FA_n]$$

where $FA_1, \ldots, FA_n$ are fitting arguments from the parameters of SN to the *instantiating* specifications SP'_1 ... SP'_n. The instantiation is valid provided that SN has been defined in the global environment. The semantics of this instantiation can be described as

$$\text{SP}^* \text{ with } FM \text{ and } \{\text{SP}'\_1 \text{ and } \ldots \text{ and } \text{SP}'\_n\}$$

where, if SN is of the form

> **spec** SN[SP_1] ... [SP_n] **given** SP''_1, ..., SP''_m =
>   SP
> **end**

then SP* stands for

$$\{\text{SP}''\_1 \text{ and } \ldots \text{ and } \text{SP}''\_m\} \text{ then } \{\text{SP}\_1 \text{ and } \ldots \text{ and } \text{SP}\_n\} \text{ then } \text{SP} \quad (1)$$

and FM is the push-out of the morphisms to the fitting arguments $FA_1, \ldots, FA_n$ together with the identity morphisms over SP''_1, ..., SP''_m and SP (and therefore is a morphism that "fits" the instantiating specifications to the parameter specifications as described above).

$$\frac{}{\{x : A\} \vdash_{\langle \text{sig}(A),\ \emptyset \rangle} \mathbf{ass}(A, x) : A} \text{ (Ass I)} \qquad \frac{}{\emptyset \vdash_{\langle \Sigma, Ax \rangle} \mathbf{ax}(\langle \Sigma, Ax \rangle, a) : A} \text{ (Ax I)}$$

$$\text{for } \{a : A\} \in Ax$$

$$\frac{\Gamma \cup \{x : A\} \vdash_{\text{SP}} d : B}{\Gamma \vdash_{\text{SP}} \lambda x : A.d : (A \Rightarrow B)} \ (\Rightarrow \text{I}) \qquad \frac{\Gamma_1 \vdash_{\text{SP}} d : A \quad \Gamma_2 \vdash_{\text{SP}} e : B}{\Gamma_1 \cup \Gamma_2 \vdash_{\text{SP}} \langle d, e \rangle : (A \wedge B)} \ (\wedge \text{I})$$

$$\frac{\Gamma \vdash_{\text{SP}} d : A}{\Gamma \vdash_{\text{SP}} \mathbf{inl}(d) : (A \vee B)} \ (\vee_1 \text{I}) \qquad \frac{\Gamma \vdash_{\text{SP}} e : B}{\Gamma \vdash_{\text{SP}} \mathbf{inr}(e) : A \vee B} \ (\vee_2 \text{I})$$

$$\frac{\Gamma \vdash_{\text{SP}} d : A}{\Gamma \vdash_{\text{SP}} \lambda x : s.d : \forall x : s \bullet A} \ (\forall \text{I}) \qquad \frac{\Gamma \vdash_{\text{SP}} d : A[t/x]}{\Gamma \vdash_{\text{SP}} (t, d) : \exists x : s \bullet A} \ (\exists \text{I})$$

$$\frac{\Gamma \vdash_{\text{SP}} d : (A_1 \wedge A_2)}{\Gamma \vdash_{\text{SP}} \mathbf{fst}(d) : A_1} \ (\wedge_1 \text{E}) \qquad \frac{\Gamma \vdash_{\text{SP}} d : (A_1 \wedge A_2)}{\Gamma \vdash_{\text{SP}} \mathbf{snd}(d) : A_2} \ (\wedge_2 \text{E})$$

$$\frac{\Gamma \vdash_{\text{SP}} d : \forall x : s \bullet A}{\Gamma \vdash_{\text{SP}} dt : A[t/x]} \ (\forall \text{E}) \qquad \frac{\Gamma \vdash_{\text{SP}} d : \bot}{\Gamma \vdash_{\text{SP}} dA : A} \ (\bot \text{E})$$

$$\text{provided } A \text{ is Harrop}$$

$$\frac{\Gamma_1 \vdash_{\text{SP}} d : (A \Rightarrow B) \quad \Gamma_2 \vdash_{\text{SP}} r : A}{\Gamma_1 \cup \Gamma_2 \vdash_{\text{SP}} (dr) : B} \ (\Rightarrow \text{E})$$

$$\frac{\Gamma_1 \cup \{x : A\} \vdash_{\text{SP}} d : C \quad \Gamma_2 \cup \{y : B\} \vdash_{\text{SP}} e : C \quad \Gamma \vdash_{\text{SP}} f : (A \vee B)}{\Gamma_1 \cup \Gamma_2 \cup \Gamma \vdash_{\text{SP}} \mathbf{case}(x : A.d : C, y : B.e : C, f : (A \vee B)) : C} \ (\vee \text{E})$$

$$\frac{\Gamma_1 \vdash_{\text{SP}} d : \exists x : s \bullet A \quad \Gamma_2 \cup \{y : A[z/x]\} \vdash_{\text{SP}} e : C}{\Gamma_1 \cup \Gamma_2 \vdash_{\text{SP}} \mathbf{select}(z : s.y : A[z/x].e : C, d : \exists x : s \bullet A) : C} \ (\exists \text{E})$$

Fig. 1. Logical Rules. See notes in the text, section 3.

Example (*continued*). We instantiate and extend LISTS to give the specification of a catalogue:[2]

> **spec** CATALOGUE = LISTS[*Elem* ↦ *Part*] **then** BODYCAT

where BODYCAT stands for

> **sorts** *Catalogue*
> **ops** *myCat* : *Catalogue*; *lOR* : *Part* → *List*(*Part*)
> **preds** *Rep* : *Part* × *Part* × *Catalogue*; *In* : *Part* × *Catalogue*
> **axioms** $\forall i, e : Part \bullet e \in lOR(i) \wedge In(e, myCat) \Rightarrow Rep(e, i, myCat)$;
> $\forall i : Part \bullet size(lOR(i)) > 0$

and the generic specification of the warehouse is then[3]

spec WAREHOUSE[CATALOGUE] =
ops *rep* : *Part* → *Part*
axioms $\forall i : Part \bullet size(lOR(i)) > 0 \Rightarrow (In(rep(i), myCat) \Rightarrow Rep(rep(i), i, myCat))$

---

[2] We shall use *lOR* for the operation *listOfReplacements* and *Rep* for the predicate *Replaces* with their obvious semantics.

[3] The body of this parametrized specification will be denoted by BODYWARE.

$$\frac{\Gamma \vdash_{\text{SP}} d : A}{\rho^{\prime}\Gamma \vdash_{\text{SP with }\rho} \rho \bullet d : \rho \bullet (A)} \text{ (trans)}$$

$$\frac{\Gamma \vdash_{\text{SP}} d : A}{\Gamma \vdash_{\text{SP hide }\Sigma} d \text{ hide } \Sigma : A} \text{ (hide)}$$
$$\text{if } \Gamma \text{ is a } (\text{sig}(\text{SP}) - \Sigma)\text{-context}$$
$$\text{and } A \text{ is a } (\text{sig}(\text{SP}) - \Sigma)\text{-formula}$$

$$\frac{\Gamma \vdash_{\text{SP\_1}} d : A}{\Gamma \vdash_{\text{SP\_1 and SP\_2}} union_1(d, \text{ SP\_2}) : A} \text{ (union}_1)$$

$$\frac{\Gamma \vdash_{\text{SP\_2}} d : A}{\Gamma \vdash_{\text{SP\_1 and SP\_2}} union_2(d, \text{ SP\_1}) : A} \text{ (union}_2)$$

$$\frac{\Gamma \vdash_{\text{SP\_1}} d : A}{\Gamma \vdash_{\text{SP\_1 then SP\_2}} ext_1(d, \text{ SP\_2}) : A} \text{ (ext}_1)$$

$$\frac{\Gamma \vdash_{\text{SP\_2}} d : A}{\Gamma \vdash_{\text{SP\_1 then SP\_2}} ext_2(d, \text{ SP\_1}) : A} \text{ (ext}_2)$$

Fig. 2. Structural Rules. (The translation of the context Γ by ρ is written $\rho^{\prime}\Gamma$.)

3   The Formal Calculus and Proof Terms

The formal calculus of [6] deals with judgments of the form $\Gamma \vdash_{\text{SP}} d : A$. The context, Γ, is a set of labelled formulae including all the (labelled) assumptions. (We assume that all the axioms and assumptions are *uniquely* labelled.) The formulae are those of an ordinary many-sorted first order logic which includes function symbols and predicates. SP is the specification in which we are working and A is the formula being proved. d is the *proof term* which "remembers" the construction of the proof of the formula A. We shall also require certain other functions, first noted by Peterreins in his thesis [16], the most important of these being $sp(d)$ which, from a proof term d, yields the specification SP.

The following notes apply to the rules. 1. The usual *eigenvariable* restrictions apply, e.g. in ($\forall$ I) and ($\exists$ E), the variable x (of sort s), must not occur free in any formula in the context. 2. We assume that all undischarged assumptions are collected and included in the Γs to the left of the $\vdash$ sign. Observe that we are not concerned if there are also other elements in the Γs that are not actually used. (This is a form of *Weakening*.) 3. We always assume the specification contains the appropriate equality axioms. 4. A Harrop formula is a formula which has no computational information – see [6] for an exact definition. 5. The eliminations of $\vee$ ("or") and of $\exists$ imitate proof by cases and use the operators **case** and **select**, respectively.

Besides logical rules our calculus also contains structural rules which are for operating in structured specifications using *CASL* constructions. In this paper we slightly modify the rules of [6] and add rules for **then** which is used to build extensions but the modified rules are logically equivalent to the original ones. The rules are given in Figs 1 and 2. As in our earlier papers, [21] and [6], we use push-outs for constructing unions. The rule (ext$_1$) simply records that the specification SP_1 has been extended to a new (complete) specification SP_2 and the formula A is still provable. The rule (ext$_2$) is the same except that the theorem has now been derived in the specification SP_2 rather than in SP_1.

$$\frac{\Gamma \vdash_{\text{SP}^*} p : A \quad \text{SN} \notin \text{Global}}{\Gamma \vdash_{\text{SN}} \text{name}(p, \text{SN}; \text{SP\_1}, \ldots, \text{SP\_}n; \text{SP}''\_1, \ldots, \text{SP}''\_m) : A} \quad (\text{Defn})$$

Global := Global ∪ { **spec** SN[SP_1] ... [SP_n] **given** SP''_1, ... SP''_m =
SP
end }

$$\frac{\Gamma \vdash_{\text{SN}} p : B \quad \text{SN} \in \text{Global}}{FM\,{}^{\backprime}\Gamma \vdash_{\text{SN}[FA_1]\ldots[FA_n]} \text{instantiate}(p, \text{SN}; FM) : FM \bullet B} \quad (\text{Fit})$$

where FM is a fitting morphism for the FA_i.

$$\frac{\text{SN} \in \text{Global} \quad \{a : A\} \in Axioms(\text{SN})}{\emptyset \vdash_{\text{SN}} \mathbf{ax}(\text{SN}, a) : A} \quad (\text{Ax}^* \text{ I})$$

Fig. 3. The new rules with their proof terms. Recall that SP* is defined in formula (1).

In the case of ordinary intuitionistic logic, we can view the calculus as a type theory with dependent product and sum types, and the rules of the calculus as a type inference system (see e.g. [12]). The same applies to our calculus. Formulae are types: implications correspond to functional types, conjunctions to product types, disjunctions to disjoint unions, universal quantification corresponds to dependent products and existential quantification to dependent sums. Terms of the type theory represent proofs.

Here (just as in our earlier papers, see e.g. [6] and [21]), we use the standard simplification rules such as associativity, commutativity, idempotence and absorption of a sub-signature (equivalences [A], [C], [I], [Sig] of [21]) and we only consider structured specifications up to equivalence under these laws.

The new rules. We now extend the calculus to cover parametrized specifications. We work in the *global environment*, Global. For *CASL* declarations this is a mutable set of named specification declarations. We add three new rules for parametrized specifications. These are shown in Fig. 3. We also introduce two new proof term formation operators name for (Defn) and instantiate for (Fit). At the same time we also extend the function sp(p) in the obvious way.

Notation and terminology. We write $Axioms(\text{SP})$ to denote the set of axioms in a structured specification SP. If SN names a generic specification, we write $Axioms(\text{SN})$ to denote the set of axioms in its body and imports. We call the instantiation SN$[FA_1] \ldots [FA_n]$ an *unevaluated instantiation*. We call its expansion, given by the semantics above, an *evaluated instantiation*. We write Global to denote the global environment, and then Global := Global ∪ {SN} means that the global environment is extended by the specification definition for the name SN. We write SN ∈ Global if, and only if, SN is defined in Global.

Definition rule. The definition rule (Defn), see Fig. 3, corresponds to the introduction of a new named specification. The formula is not affected but the global environment must now include the new name. Note that the final specification obtained in proving a given formula will depend on the order in which the applications of (Defn) are introduced. Once the proof has been constructed, Strong Normalization (see Theorem 2) and program extraction are carried out in a purely declarative way as usual.

Fitting rule.[4] The fitting rule (Fit) in Fig. 3 corresponds to instantiating the specification parameters $SP\_1, \ldots, SP\_n$ by the specifications $SP'\_1, \ldots, SP'\_n$. The new specification is then computed according to the *CASL* rules (see [3] section 6.2.2) using a fitting morphism FM (see the previous section). (i) $FM \bullet B$ denotes the translation of the formula B under FM and (ii) $FM'\Gamma$ denotes the translation of the set of formulae Γ under the fitting morphism FM.

Axiom Introduction rule. The third rule of Fig. 3 allows us to use an axiom from the named specification, even if the specification is generic. It extends the axiom rule of [6] but we now have to ensure that SN is in Global.

It is easy to see that the proof term uniquely determines the form of the proof. In particular, the specification of the conclusion of a proof can be determined from the proof term by means of a function sp. We now extend this function as defined in [21] and extended in [6] by

$$sp(\mathbf{ax}(SN, x)) = SN$$
$$sp(\mathsf{instantiate}(p, SN; FM)) = SN[FA_1] \ldots [FA_n]$$
$$sp(\mathsf{name}(p, SN; SP\_1, \ldots, SP\_n; SP''\_1, \ldots, SP''\_m)) = SN$$

where $FA_1, \ldots FA_n$ are the fitting argument specifications (which can be recovered from the fitting morphism FM).

N.B. In *CASL* a generic specification name can only be used in a structured specification when instantiated. Consequently the original structural rules in [6] cannot be used with generic specifications.

Theorem 1. *The system of logical and structural rules is sound and complete.*

The proof is like those in [6] and [2].   □

4   Proof Reductions and Strong Normalization

Constructive proofs can be reduced by discarding unnecessary applications of rules such as where one rule immediately negates the effect of the preceding rule. The basic process is described in many places (e.g. [8] or [7]).

---

[4] At WADT2001, Sannella pointed out that one can use the earlier rules to simulate the effects of (Fit). While this is true our aim has always been to accommodate our system to actual practice. Including the (Fit) rule explicitly allows the user to use the standard apparatus of *CASL* directly.

The Curry-Howard isomorphism shows that such proof reductions correspond to lambda calculus reductions (whose transitive closure we denote by $\succ$) over the proof terms. Thus an implication introduction followed by an implication elimination:

$$\cfrac{\cfrac{\{x:A\} \vdash_{\text{SP}} b : B}{\vdash_{\text{SP}} (\lambda x : A.b : B) : (A \Rightarrow B)} \; (\Rightarrow \text{I}) \qquad \{x:A\} \vdash_{\text{SP}} x : A}{\{x:A\} \vdash_{\text{SP}} \{(\lambda x : A.b : B) : (A \Rightarrow B)\}(x:A) : B} \; (\Rightarrow \text{E})$$

can be reduced to the triviality $\{x:A\} \vdash_{\text{SP}} b : B$ because $(\lambda x : A.b : B) : (A \Rightarrow B)(x:A) \succ b : B$.

In [21] and [6] we extended our reduction procedures to include structural rules for reasoning about structured specifications. We now do the same with respect to the new rules for generic specifications.

Reducing instantiations by evaluation. We wish to reduce proofs in which a generic specification is defined (by an application of rule (Defn)) and then immediately instantiated by an application of the rule (Fit). The form of such a proof is

$$\cfrac{\cfrac{\begin{array}{c}\vdots\end{array}}{\Gamma \vdash_{\text{SP*}} d : A}}{\cfrac{\Gamma \vdash_{\text{SN}} \text{name}(d, \text{SN}; \text{SP\_1}, \ldots, \text{SP\_}n; \text{SP''\_1}, \ldots, \text{SP''\_}n) : A}{FM'\Gamma \vdash_{\text{SN}[FA_1]\ldots[FA_n]} \text{instantiate}(d', \text{SN}; FM) : FM \bullet A} \; \text{(Fit)}} \; \text{(Defn)} \qquad (2)$$

where d' is $\text{name}(d, \text{SN}; \text{SP\_1}, \ldots, \text{SP\_}n; \text{SP}''_1, \ldots, \text{SP''\_}n)$ and
$$\begin{aligned}\text{Global} := \text{Global} \cup \{ \; & \textbf{spec} \; \text{SN}[\text{SP\_1}] \ldots [\text{SP\_}n] \quad \textbf{given} \quad \text{SP''\_1}, \ldots \text{SP''\_}m = \\ & \text{SP} \\ & \textbf{end} \; \}\end{aligned}$$

Such a proof can be reduced to a proof that does not use (Defn) nor (Fit):[5]

$$\cfrac{\cfrac{\cfrac{\begin{array}{c}\vdots\end{array}}{\Gamma \vdash_{\text{SP*}} d : A}}{FM'\Gamma \vdash_{\text{SP* with } FM} d_2' : FM \bullet A} \; \text{(trans)}}{FM'\Gamma \vdash_{\{\text{SP* with } FM\} \text{ and } \{\text{SP'\_1 and } \ldots \text{ and SP'\_N}\}} d_3' : FM \bullet A} \; \text{(union}_1) \qquad (3)$$

where $d_2' = FM \bullet d$ and $d_3' = union_1(d_2', \{\text{SP'\_1 and } \ldots \text{ and SP'\_}n\})$.

We justify this reduction by the semantics for instantiation: The conclusion of the unreduced proof (2) is equivalent to the conclusion of the reduced proof (3). To see this, first observe that the formulae are identical. Also, the unreduced proof (2) concludes with the specification $\text{SN}[FA_1]\ldots[FA_n]$. By the semantics for instantiation, this specification denotes the concluding specification of the reduced proof (3). That is, the reduction transforms the concluding specification from an *unevaluated* instantiation to an equivalent, *evaluated* instantiation.

We also note that the premise of the unreduced proof (2) is identical to the premise of the reduced proof (3). Thus the reduction yields a proof that does not use (Defn) or (Fit), yet proves the same conclusion from the same premise.

---

[5] Recall that SP* was defined in formula (1).

This reduction may be formalized by extending the reduction relation $\succ$ over proof terms by:

$$\text{instantiate}(\text{name}(p, \text{SN}; \text{SP}\_1, \ldots, \text{SP}\_n; \text{SP}''\_1, \ldots, \text{SP}''\_m), \text{SN}; FM) \succ INST_1$$

where $INST_1$ is the proof term

$$union_1((FM \bullet d), \{\text{SP}' \_1 \text{ and } \ldots \text{ and } \text{SP}'\_n\})$$

N.B. We assume that all values in Global in the unreduced proof remain defined in the reduced proof, including any definitions made in discarded occurrences of (Defn). We are required to make this assumption, because it is possible that these definitions are used in other proofs. For example, if we have sub-proofs:

$$\frac{\dfrac{\vdots \quad \Gamma_3 \vdash_{\text{SP}^*} c_3 : C}{\Gamma_3 \vdash_{\text{SN}} \text{name}(c_3, \text{SN}; \text{SP}\_1, \ldots, \text{SP}\_n; \text{SP}''\_1, \ldots, \text{SP}''\_m) : C} \text{(Defn)}}{FM'\Gamma_3 \vdash_{\text{SN}[FA]} FM \bullet \text{name}(c_3, \text{SN}; \text{SP}\_1, \ldots, \text{SP}\_n; \text{SP}''\_1, \ldots, \text{SP}''\_m) : FM \bullet C} \text{(Fit)}$$

$$\vdots$$

$$\Gamma \vdash_{\text{SP}\_2} b : B$$

and

$$\frac{\text{SN} \in \text{Global} \quad \{a : A\} \in Axioms(\text{SN})}{\Gamma_1 \vdash_{\text{SN}} \mathbf{ax}(\text{SN}, a) : C} \text{(Ax* I)}$$

$$\vdots$$

$$\Gamma_2 \vdash_{\text{SN}} d : A$$

the upper proof defines SN to be the name of a generic specification over the specification body SP. That proof may be reduced, because the definition is immediately followed by an instantiation. However, the definition of SN must still remain in Global, because it is required by the lower proof.

We also wish to reduce proofs in which a generic specification is defined in Global, an axiom of the generic specification is introduced, and the generic specification is then instantiated (by an application of the rule (Fit)). The form of such a proof is

$$\frac{\dfrac{\text{SN} \in \text{Global} \quad \{a : A\} \in Axioms(\text{SN})}{\Gamma \vdash_{\text{SN}} \mathbf{ax}(\text{SN}, a) : A} \text{(Ax* I)}}{FM'\Gamma \vdash_{\text{SN}[FA_1]\ldots[FA_n]} \text{instantiate}(\mathbf{ax}(\text{SN}, a), \text{SN}; FM) : FM \bullet A} \text{(Fit)}$$

Here $\{a : A\}$ must be an axiom of the body of SN, or of one of its parameters or imports. We only present the first case here as they are very similar. Such a proof can be reduced to a proof with the body SP of SN instead of SN and which does not use (Fit):

$$\frac{\dfrac{\dfrac{\{a : A\} \in Axioms(\text{SP})}{\Gamma \vdash_{\text{SP}} \mathbf{ax}(\text{SP}, a) : A} \text{(Ax I)}}{\dfrac{\Gamma \vdash_{\{\text{SP}''\_1 \text{ and } \ldots \text{ and } \text{SP}''\_m\} \text{ then } \{\text{SP}\_1 \text{ and } \ldots \text{ and } \text{SP}\_n\} \text{ then } \text{SP}} d_1 : A}{FM'\Gamma \vdash_{\text{SP}^* \text{ with } FM} FM \bullet d_1 : FM \bullet A} \text{(trans)}}{FM'\Gamma \vdash_{\{\text{SP}^* \text{ with } FM\} \text{ and } \{\text{SP}'\_1 \text{ and } \ldots \text{ and } \text{SP}'\_N\}} d_3 : FM \bullet A} \text{(union}_1\text{)}} \text{(ext}_2\text{)}$$

where

$$d_1 = ext_2(\mathbf{ax}(\text{SP}, a), \{\text{SP}''\_1 \text{ and } \ldots \text{ and } \text{SP}''\_m\}, \{\text{SP}\_1 \text{ and } \ldots \text{ and } \text{SP}\_n\})$$
$$d_3 = union_1(FM \bullet d_1, \{\text{SP}'\_1 \text{ and } \ldots \text{ and } \text{SP}'\_n\})$$

The justification for this reduction is similar to that given for (Defn)/(Fit) pairs. The reduction may be formalized by extending the reduction relation $\succ$ on proof terms by:

$$\text{instantiate}(Axiom(\text{SN}, a), \text{SN}; FM) \succ INST_2$$

where $INST_2$ is the proof term

$$union_1(FM \bullet ext_2(\mathbf{ax}(\text{SP}, a),$$
$$\{\text{SP}''\_1 \text{ and } \ldots \text{ and } \text{SP}''\_m\} \text{ then } \{\text{SP}\_1 \text{ and } \ldots \text{SP}\_n\}),$$
$$\text{SP}'\_1 \text{ and } \ldots \text{ and } \text{SP}'\_n)$$

There are changes to all specification labels that follow either of the reductions above. This is due to the fact that, unlike the proof reductions for structural rules given in [21], the proposed reductions result in conclusions with different (but equivalent) specification labels. (Both reductions result in a change from unevaluated to evaluated instantiations.)

Theorem 2 (Strong Normalization). *The calculus is strongly normalizing.*

Proof. This follows easily from the Strong Normalization theorem of [21] since the above reductions replace occurrences of (Defn) and (Fit) by rules of our earlier calculus and the other occurrences of the new rules do not change the character of the proof terms.   □

5   Program Extraction

In [6], we adapted techniques for extracting programs from constructive proofs to produce executable refinements of specifications. This gave a method that used the (Sk) rule and unSkolemization to derive an executable refinement for a given specification. We now extend this to the extraction of *SML* programs and to our calculus extended by the new rules.

First we need to define the *SML* types that are needed in the extraction from proof terms. This we do in a straightforward way by a map which we call ϕ. (Note that the sorts of our *CASL* specifications are represented in *SML* by types of the same name: we differentiate the latter by use of the typewriter font.) Then we define an extraction map extract from proof terms to programs. This map discards computationally irrelevant logical and specification information and retains the computational information from a constructive proof. The cases are shown in Figs 5, 6 and 7 for the rules of Figs 1 and 2. For the new rules we now modify the extract function (of [6]) and extend it to proofs involving generic specifications. For the new proof terms we have:

$$\text{extract}(\text{instantiate}(p, \text{SN}; FM)) = FM \bullet \text{extract}(p)$$
$$\text{extract}(\text{name}(p, \text{SN}; \text{SP}\_1, \ldots, \text{SP}\_n; \text{SP}''\_1, \ldots, \text{SP}''\_m)) = \text{extract}(p)$$

and for (Ax* I) the obvious extension of (Ax I).

$$\phi(A) = \text{void} \qquad\qquad \text{if } A \text{ is Harrop}$$

$$\phi(A \wedge B) = \begin{cases} \phi(A) & \text{if } B \text{ is Harrop} \\ \phi(B) & \text{if } A \text{ is Harrop} \\ \phi(A)*\phi(\text{B}) & \text{otherwise} \end{cases}$$

$$\phi(A \vee B) = (\phi(A), \phi(B))\text{DisjointUnion}$$

$$\phi(A \ \text{->} \ B) = \begin{cases} \phi(B) & \text{if } A \text{ is Harrop} \\ \phi(A) \ \text{->} \ \phi(B) & \text{otherwise} \end{cases}$$

$$\phi(\forall x : s \bullet A) = s \ \text{->} \ \phi(A)$$

$$\phi(\exists x : s \bullet A) = \begin{cases} s & \text{if } A \text{ is Harrop} \\ s*\phi(\text{A}) & \text{otherwise} \end{cases}$$

where we define datatype $('\text{a},'\text{b})$ DisjointUnion $=$ inl of $'\text{a}$ | inr of $'\text{b}$;

Fig. 4. The definition of the map ϕ used in defining extract in Figs 5, 6 and 7. (The formula $\bot$ is Harrop so comes under the first clause.)

Thus the extraction map "ignores" the occurrence of (Defn) rules. So, extracting a program from a proof that ends in an application of (Defn) is the same as extracting a program from the proof without this application. In the case of (Fit) we simply apply the components of the fitting morphism FM recursively.

Our reasons for the form of the definition over name terms are like those for the definition over structural unions. The new rules do not affect the computational nature of the extracted program in the sense that the extracted program is executable (*modulo* references to other operations of the specification).

Our ultimate aim is to obtain *SML* programs. However the specifications from *CASL* often do not uniquely define a function and other axioms applying to the same function may be added later. In order to define unique (executable) functions we apply unSkolemization. (One has to be careful about the use of **hide** but this is dealt with in the same way as in [6].)

Definition 1. *A proof term d is said to be* critical *with respect to a symbol list Σ if 1. d is of the form e **hide** Σ and 2. if the term e depends on symbols that are in Σ: that is, if $\text{sig}(\text{sp}(e)) \cap \Sigma \neq \emptyset$. A proof-term is said to be* modular *if it contains no critical (sub-)proof-terms.*

Theorem 3 (Program Extraction). *Given a proof $\vdash_{SP} p : \forall x : s_1 \bullet \exists y : s_2 \bullet A(x, y)$ such that p is modular, the program $e = \text{extract}(p)$ is a modified realizer (see [1]) of $\forall x : s_1 \bullet \exists y : s_2 \bullet A(x, y)$. That is to say, we can compute a proof term q and prove*

$$\vdash_{SP} q : \forall x : s_1 \bullet A(x, f_A(x))[\text{extract}(p)/f_A]$$

$$\textbf{extract}(\textbf{ass}(A,x)) \quad = \begin{cases} () & \text{if } \mathfrak{H}(A) \\ \texttt{x} & \text{otherwise} \end{cases}$$

$$\textbf{extract}(\textbf{ax}(\langle \Sigma, Ax \rangle, x) : A) \quad = () \quad \text{if } \mathfrak{H}(A)$$

$$\textbf{extract}((\lambda x : A.d) : (A \Rightarrow B)) = \begin{cases} \textbf{extract}(d) & \text{if } \mathfrak{H}(A) \\ () & \text{if } \mathfrak{H}(B) \\ \texttt{fn x :}\phi(A) \quad \texttt{->} & \text{otherwise} \\ \textbf{extract}(d) & \end{cases}$$

$$\textbf{extract}(\langle a,b \rangle : (A \wedge B)) \quad = \begin{cases} \textbf{extract}(b) & \text{if } \mathfrak{H}(A) \text{ and not } \mathfrak{H}(B) \\ \textbf{extract}(a) & \text{if } \mathfrak{H}(B) \text{ and not } \mathfrak{H}(A) \\ () & \text{if } \mathfrak{H}((A \wedge B)) \\ (\textbf{extract}(a), \textbf{extract}(b)) & \text{otherwise} \end{cases}$$

$$\textbf{extract}(\textbf{inl}(d) : (A_1 \vee A_2)) \quad = \texttt{inl}(\textbf{extract}(d))$$

$$\textbf{extract}(\textbf{inr}(d) : (A_1 \vee A_2)) \quad = \texttt{inr}(\textbf{extract}(d))$$

$$\textbf{extract}((\lambda x : s.d) : \forall x : s \bullet A) = \begin{cases} () & \text{if } \mathfrak{H}(\forall x : s \bullet A) \\ \texttt{fn x : s} \quad \texttt{->} \textbf{extract}(d) & \text{otherwise} \end{cases}$$

$$\textbf{extract}((t,d) : \exists x : s \bullet A) \quad = \begin{cases} \textsf{program}(t) & \text{if } \mathfrak{H}(A) \\ (\textsf{program}(t), \textbf{extract}(d)) & \text{otherwise} \end{cases}$$

Fig. 5. The definition of **extract** for the logical introduction rules except we have written $\mathfrak{H}(A)$ for "A is Harrop". The definition of ϕ is given in Fig. 4 and $\textsf{program}(t)$ is the *SML* code for the individual term t.

where f_A is the Skolem function for the formula. (The Skolemized version of the formula A is $\forall x : s_1 \bullet A(x, f_A(x))$ and is denoted by $Sk(A)$.) If SP *is non-generic (resp. generic), then* SP *can be extended to a consistent extension of* SP *(resp. the body of* SP*) by* {**axioms** $Sk(A); f_A = \textsf{extract}(p)$}.

Proof. Similar to that in [6]. □

Now although the application of (Fit) does not affect the nature of the extracted program, this rule is important in program synthesis. It provides a specification that can be *consistently* extended by the extracted program together with a new axiom about the program. This situation is identical to that of [6], where such a result was shown to hold for structured specifications.

Note that there is a slight difference in the statement of the theorem for generic specifications. This is because it is not possible to extend a generic specification by a structured specification. However, it is possible to apply a similar

$$\textbf{extract}((d : (A \Rightarrow B)) \atop (r : A) : B) \quad = \begin{cases} \textbf{extract}(d) & \text{if } \mathfrak{H}(A) \\ () & \text{if } \mathfrak{H}(B) \\ \textbf{extract}(d)\textbf{extract}(r) & \text{otherwise} \end{cases}$$

$$\textbf{extract}(\textbf{fst}(d : (A_1 \wedge A_2)) : A_1) \quad = \begin{cases} () & \text{if } \mathfrak{H}(A_i), i = 1, 2 \\ \textbf{extract}(d) & \text{if } \mathfrak{H}(A_2) \text{ and not } \mathfrak{H}(A_1) \\ \texttt{fst}(\textbf{extract}(d)) & \text{otherwise} \end{cases}$$

$$\textbf{extract}(\textbf{snd}(d : (A_1 \wedge A_2)) : A_2) \quad = \begin{cases} () & \text{if } \mathfrak{H}(A_i), i = 1, 2 \\ \textbf{extract}(d) & \text{if } \mathfrak{H}(A_1) \text{ and not } \mathfrak{H}(A_2) \\ \texttt{snd}(\textbf{extract}(d)) & \text{otherwise} \end{cases}$$

$$\textbf{extract}(\textbf{case}(x : A.d : C, \atop y : B.e : C, f : (A \vee B)) : C) \quad = \begin{cases} () & \text{if } \mathfrak{H}(C) \\ (\texttt{function} & \text{otherwise} \\ \quad \texttt{inl(x)} \ \texttt{->} \ \textbf{extract}(d) \\ \quad | \ \texttt{inr(y)} \ \texttt{->} \ \textbf{extract}(e)) \\ \qquad \textbf{extract}(f) \end{cases}$$

$$\textbf{extract}((d : \forall x : s \bullet A(x))(r : s) : \atop A[r/x]) \quad = \begin{cases} () & \text{if } \mathfrak{H}(A[d/x]) \\ \textbf{extract}(d)r & \text{otherwise} \end{cases}$$

$$\textbf{extract}(dA : A) \quad = \quad () \quad \text{for any } A$$

$$\textbf{extract}(\textbf{select}(z : s.y : A[z/x]. \atop e : C, d : \exists x : s \bullet A) : C) \quad = \begin{cases} () & \text{if } \mathfrak{H}(C) \\ (\texttt{fun x -> }\textbf{extract}(e))\textbf{extract}(d) & \text{if } \mathfrak{H}(A) \\ (\texttt{function z y -> }\textbf{extract}(e)) & \text{otherwise} \\ \quad \texttt{fst}(\textbf{extract}(d))\texttt{snd}(\textbf{extract}(d)) \end{cases}$$

Fig. 6. The definition of extract for the logical elimination rules except we have written $\mathfrak{H}(A)$ for "A is Harrop". The definition of ϕ is given in Fig. 4.

extension to the *body* of the generic specification. For instance, from a proof that ends in (Defn)

$$\frac{\vdash_{\text{SP}^*} p : A}{\vdash_{\text{SN}} p : A} \ (\text{Defn})$$

we can extract a program extract(p) such that $Sk(A)[\textsf{extract}(p)/f_A]$, a Skolemized form of A, is provable from the axioms of SN. The (Defn) rule tells us that we can conservatively extend the body of SN by the Skolemized form of A.

$$\textbf{extract}(union_1(d, Sp)) = \textbf{extract}(d)$$
$$\textbf{extract}(union_2(d, Sp)) = \textbf{extract}(d)$$
$$\textbf{extract}(ext_1(d, Sp)) = \textbf{extract}(d)$$
$$\textbf{extract}(ext_2(d, Sp)) = \textbf{extract}(d)$$
$$\textbf{extract}(\rho \bullet d) = \rho \bullet (\textbf{extract}(d))$$
$$\textbf{extract}(d \textbf{ hide } l) = \textbf{extract}(d)$$

Fig. 7. The definition of extract for the structural rules.

6   Executable Refinements

By an executable specification we simply mean a specification in which every function or sort in the specification has an executable definition (in *SML*).

Definition 2 (Executable specification). *Let* Sp_A *be a well-formed, consistent, structured specification in which every symbol is declared at most once.*

1. *If* Sp_A *is non-generic then* Sp_A *is called* executable *if, and only if,*
   (a) *for each visible or hidden sort symbol s used in* Sp_A,
       – *either* Sp_A *contains a basic or partial specification in which s has a free data type declaration. (This corresponds to a SML datatype declaration* sdec(s).*)*[6]
       – *or s corresponds to a basic SML datatype, which we denote by* s *(using the* typewriter *font).*[7]
   (b) *for each visible or hidden function symbol f of* Sp_A, *we can prove that* $\vdash_{Sp\_B} f = \textsf{dec}(f)$ *for some SML term* dec(f), *where* Sp_B *is a subspecification of* Sp_A.
   (c) *For each fitting morphism FA in the specification* Sp_A, *each visible or hidden function symbol occurring in the image of FA is executable.*
2. *If* Sp_A *is generic of the form* **spec** Sp_B[Sp_1] ... [Sp_n] **given** Sp″_1 ... Sp″_m = Sp, *then* Sp_A *is called* executable *if, and only if,* Sp_A *is such that* Sp *is executable.*

Item (1b) of the definition says that we can find a program definition for every function of the given specification. This permits us to have axioms which define functions quantifying over applications (for example, $\forall x : Int \bullet f(x) = x + x$), because we can use such axioms to determine a (non-applicative) definition (for example, dec(f)= (fn f x : int => x + x)).

We now expand the process of producing executable refinements of [6]. Refinements of non-generic specifications are defined as usual (see, for example, the introduction to [6]).

---

[6] This translation can always be computed. For instance, the free datatype declaration free type Nat ::= 0 | suc(Nat) can be translated as datatype Nat = 0 | suc of Nat

[7] For example, the basic *SML* datatype int corresponds to a sort *int* of Sp_A if, and only if, the specification contains a generated datatype definition of *int* that generates *int* using the basic operations of int, with the operations of int satisfying all the constraint axioms of *int*.

Definition 3 (Executable refinement of a generic specification). *Let* SN *be the name of a generic specification defined by* **spec** SN[SP_1]...[SP_n] **given** SP″_1...SP″_m = SP. *Now define a new generic specification* SN′ *by* **spec** SN′[SP_1]...[SP_n] **given** SP″_1...SP″_m = SP′. *Then* SN′ *is said to be a* refinement *of* SN *if, and only if, for all fitting arguments* $FA_1, \ldots, FA_n$, SN′[FA_1]...[FA_n] *is a refinement of* SN[FA_1]...[FA_n]. *We write* SN′ ⊑ SN *in this case.*

SN′ *is an* executable refinement *of* SN *when* SN′ *is an executable specification.*

We now extract programs for every function declared in the imports SP″_1, ..., SP″_m or in the body SP.

Theorem 4 (Executable refinements from a generic specification). *Suppose* SN *names a consistent, generic specification of the form*

$$\textbf{spec } \text{SN}[\text{SP\_1}]\ldots[\text{SP\_}n] \textbf{ given } \text{SP}''\_1\ldots\text{SP}''\_m = \text{SP},$$

where SP″_1...SP″_m *are executable, all the sorts declared in the body of* SP *are basic or have free datatype declarations and the unSkolemized versions of all the axioms of* SP *can be proved using only executable specifications and the parameter specifications. Then we can obtain a consistent, executable refinement,* SN_EXEC, *which is a conservative extension of* SN.

Proof. We sketch the process for producing a chain of refinements SN = SN_0 ⊑ SN_1 ⊑ ...SN_n = SN_EXEC. We shall assume that each SN_i is of the form

$$\textbf{spec } \text{SN\_}i[\text{SP\_1}]\ldots[\text{SP\_}n] \textbf{ given } \text{SP}''\_1\ldots\text{SP}''\_m = \text{SPBODY\_}i.$$

1. Given the specification SN_i, take any function f in SPBODY _i for which there is no executable definition.
2. Prove the unSkolemization, D_f, of the conjunction, A_f, of all the axioms for f to give a proof

$$\vdash_{\{\text{SP}''\_1 \text{ and } \ldots \text{ and } \ldots\text{SP}''\_n\} \text{ then } \{\text{SP\_1 and } \ldots \text{ and } \ldots\text{SP\_}m\} \text{ then } \text{SP}'} \; p : D_f$$

where SP′ is the specification corresponding to the shortest sub-expression of SPBODY_i containing all of A_f. Note that SPBODY_i is of the form $\mathbf{C}[\text{SP}'/X]$ where $\mathbf{C}$ is a specification expression formed using the specification meta-variable X and the structuring operators **and**, etc.
3. Apply the extraction map **extract** to the proof and add the axioms $Sk(D_f); f_D(x) = \textbf{extract}(p)(x)$ and finally rename f_D to f. This will result in a conclusion of the form

$$\vdash \{\text{SP}''\_1 \textbf{ and } \ldots \textbf{ and } \ldots\text{SP}''\_n\} \qquad Sk(D_f)[\textbf{extract}(p)/f_D]$$
$$\textbf{then } \{\text{SP\_1 and } \ldots \textbf{ and } \ldots\text{SP\_}m\}$$
$$\textbf{then } \text{SP}' \textbf{ then}$$
$$\textbf{axioms } Sk(D_f); f_D(x) = \textbf{extract}(p)(x)$$
$$\textbf{with } f_D \mapsto f$$

4. Now set

$$\text{SN\_}(i+1) = \textbf{spec } \text{SN}'\_i[\text{SP\_1}]\dots[\text{SP\_}n] \textbf{ given } \text{SP}''\_1\dots\text{SP}''\_m$$
$$= \textbf{C}[\text{SP}'/X] \textbf{ then axioms } Sk(D_f); f_D(x) = \textbf{extract}(p)(x) \textbf{ with } f_D \mapsto f$$

Then $\text{SN\_}(i+1)$ is a refinement of $\text{SN\_}i$ which, by construction is a conservative extension. Each $\text{SN\_}(i+1)$ contains one less non-executable function than the previous $\text{SN\_}i$, so the chain of refinements must terminate at some finite stage r, say, so that $\text{SN\_}r = \text{SN\_EXEC}$, as required.   □

Example (*continued*). To derive an executable refinement of WAREHOUSE, we need to extract a program definition for *rep*. We shall show that the definition that we shall get is $rep = \texttt{fn i => hd(lOR(i))}$ which returns the first element in the list of replacements for i. Here the function `listOfReplacements` is a parameter depending on the particular catalogue.

For convenience we shall omit the contexts and Global.

We first unSkolemize the axiom for *rep* as

$$\forall i : Part \bullet \exists y : Part \bullet size(lOR(i)) \Rightarrow (In(y, myCat) \Rightarrow Rep(y, i, myCat))$$

This is our goal formula and we shall use extract on the proof, with the (Sk) rule, to obtain the definition for *rep*. In constructing the proof we shall build a specification which refines WAREHOUSE.

We begin with the axiom for $\in$ from the generic specification LISTS:

$$\vdash_{\text{LISTS}} \textbf{ax}(\text{LISTS}, l) : \forall k : List(Elem) \bullet size(k) > 0 \Rightarrow hd(k) \in k$$

We instantiate the specification to obtain a theorem about $\in$ for elements of $List(Parts)$ by applying the rule (Fit), with the fitting argument $[Elem \mapsto Parts]$, which yields the fitting morphism FM:

$$\vdash_{\text{LISTS}[Elem \mapsto Parts]} \textbf{instantiate}(\textbf{ax}(\text{LISTS}, l), \text{LISTS}; FM) :$$
$$\forall k : List(Parts) \bullet size(k) > 0 \Rightarrow hd(k) \in k$$

We next apply (ext_1) to this obtaining the same theorem over CATALOGUE and (trivially) the (Defn) rule to get

$$\vdash_{\text{CATALOGUE}} \textbf{name}(ext_1(\textbf{instantiate}\,(\textbf{ax}(\text{LISTS}, l), \text{LISTS}; FM), \text{BODYCAT}))(lOr(i))$$
$$(\textbf{ax}(\text{CATALOGUE}, c), \text{CATALOGUE}; ;) : \forall k : List(Parts) \bullet size(k) > 0 \Rightarrow hd(k) \in k$$

and call this proof term q_4. We apply ($\forall$-E) to this theorem with $lOR(i)$ for k and the formula obtained from the CATALOGUE axiom by ($\forall$-E) to get:

$$\vdash_{\text{CATALOGUE}} \textbf{ax}(\text{CATALOGUE}, c)lOR(i) : size(lOR(i)) > 0$$

So to these two theorems we apply ($\Rightarrow$-E) obtaining the formula

$$\vdash_{\text{CATALOGUE}} q_5 : hd(lOR(i)) \in lOR(i) \tag{4}$$

where $q_5 = (q_4(lOr(i)))((\textbf{ax}(\text{CATALOGUE}, c)lOR(i)))$.

Now we use the axiom for lOR given by WAREHOUSE, labelling it with w. Applying $\forall$-E twice, with $hd(lOR(i))$ for e, we obtain:

$$\vdash_{\text{CATALOGUE}} q_6 : hd(lOR(i)) \in lOR(i)$$
$$\Rightarrow (In(hd(lOR(i)), myCat) \Rightarrow Rep(hd(lOR(i)), i, myCat))$$

where $q_6 = (\mathbf{ax}(\text{WAREHOUSE}, w)\, i)(hd(lOR(i))))$.

Applying $\Rightarrow$-E to this formula and (4), we get

$$\vdash_{\text{CATALOGUE}} q_6 q_5 : In(hd(lOR(i)), myCat) \Rightarrow Rep(hd(lOR(i)), i, myCat)$$

We apply (ext_1) to the axiom for lOR, extending CATALOGUE by the body of the specification WAREHOUSE, that is: BODYWARE (see footnote 3). We then apply ($\exists$-I):

$$\vdash_{\text{CATALOGUE then BODYWARE}} p_2 : \exists y : Part \bullet In(y, myCat) \Rightarrow Rep(y, i, myCat)$$

where $p_2 = (hd(lOR(i)), ext_1(q_6 q_5, \text{BODYWARE}))$.

Applying ($\forall$-I) over the sequent gives us the goal formula with proof term p_3 but with the specification CATALOGUE then BODYWARE where $p_3 = \lambda i : Part.(hd(lOR(i)), ext_2(q_6 q_5, \text{CATALOGUE}))$.

We apply the (Defn) rule to the sequent, abstracting over CATALOGUE to obtain the specification WAREHOUSE, and our goal theorem

$$\vdash_{\text{WAREHOUSE}} p_4 : \forall i : Part \bullet \exists y : Part \bullet In(y, myCat) \Rightarrow Rep(y, i, myCat)$$

where $p_4 = \mathbf{name}(\lambda i : Part.(hd(lOR(i)),$
$$ext_2(q_6 q_5, \text{BODYWARE})), \text{WAREHOUSE}; \text{CATALOGUE};)$$

Now applying extract to the proof term p_4, gives an *SML* function

```
val rep = fn i => hd(listOfReplacements(i))
```

which returns the first element in the list of replacements for i. Here the function `listOfReplacements` is a parameter depending on the particular catalogue.

Theorem 3 ensures this program satisfies the goal formula, in the sense that $\forall i : Part \bullet In(\text{extract}(p_4)(i), myCat) \Rightarrow Rep(\text{extract}(p_4)(i), i, myCat)$ is true for WAREHOUSE. We obtain the executable refinement, WAREHOUSE′, of WAREHOUSE, by the algorithm of Theorem 4:

> **spec** WAREHOUSE[CATALOGUE] =
> BODYCAT **then**
>   **ops** $f_A : Part \to Part$
>   **axioms** $\forall i : Part \bullet In(f_A(i), myCat) \Rightarrow Rep(f_A(i), i, myCat)$;
> $f_A = $ **fn** i => hd(listOfReplacements(i))

where A stands for the goal formula and f_A for the Skolem function for A. Now, by applying the morphism $[f_A \mapsto rep]$ to the body of WAREHOUSE′, we obtain the required executable refinement WAREHOUSE_EXEC of WAREHOUSE. xxx

7   Module Extraction

In this section we show how to extract *SML* structures from specifications and *SML* functors from parametrized specifications. To do this we consider first non-parametrized structured specifications:

For any structured specification SP_A we construct its *SML* signature by the operation getSig (see Fig. 8) and, if SP_A is executable, its realizing *SML* structure by the operation getBody (see Fig. 9) and its realizing *SML* structure declaration by the operation getStruct (see Fig. 10). The former two operations are defined using auxiliary operations innerSig(Sp_A) and innerBody(Sp_A), where innerSig(Sp_A) collects all the sorts and all the functions symbols of Sp_A with their types in *SML* notation and innerBody(Sp_A) collects all the type declarations of the *executable* sorts of Sp_A and all the function declarations of the *executable* functions of Sp_A. Then getSig(Sp_A) = sig innerSig(Sp_A) end and getBody(Sp_A) = struct innerBody(Sp_A) end. The operation getStruct uses these to give: getStruct(Sp_A) = structure sname(Sp_A) : getSig(Sp_A) = getBody(Sp_A), where sname assigns an *SML* structure name to a *CASL* specification.

The operations innerSig(Sp_A) and innerBody(Sp_A) are defined by structural recursion on the form of Sp_A. If Sp_A is basic then innerSig(Sp_A) collects all sorts and functions symbols of Sp_A with their types and innerBody(Sp_A) collects all type declarations of *executable* sorts of Sp_A and all function declarations of *executable* functions of Sp_A. If Sp_A has the form Sp_B **and** Sp_C then common symbols of Sp_B and Sp_C are shared in *CASL*. In *SML* we define them only once by conjoining innerSig(Sp_B) with those symbols of innerSig(Sp_C) which do not occur in Sp_B. Analogously we define innerBody(Sp_B **then** Sp_C). This is all possible because we assume that Sp_A is consistent. Here two executable definitions of the same symbol must be equivalent. Renaming and hiding are defined recursively. For instantiation we use recursion on the "macro-expansion" definition of *CASL*, (formula (1)), except for innerBody in the case where the instantiated parametrized specification, the given specifications and the fitting morphisms are executable and, moreover, the renaming maps of the fitting morphisms are empty. Then we can use the instantiation mechanism of *SML* , but we have to adapt the resulting signature to the *CASL* conventions.

After all these preparations we can define the operation **getStruct** which constructs the realizing functor of an executable parametrized specification Sn. Let Sn be of the form **spec** Sn[Sp$_1$]...[Sp$_n$] **given** Sp''_1...Sp''_m = Sp and assume that Sn and all imported specifications Sp''_1...Sp''_m are executable. By the induction hypothesis getSig is defined for all formal parameters and all imported specifications; getStruct is defined for all imported specifications, and the innerBody is defined for the body of Sn. Then the functor for Sn takes structures with the signature of the Sp_i as formal parameters. The body of the functor is formed by the formal parameter structures, the realizing structures of the imported specifications and innerBody of the body specification. Because of the different naming conventions of *CASL* and *SML* we have to rename the symbols of the parameters and the givens appropriately in order to use them correctly in the realizing structure of Sp.

Notation and terminology. Given a specification Sp_B, we write $\overline{sym}_{\text{Sp\_B}}$ for the list of all visible sort and function symbols of Sp_B. Moreover, we write $\overline{f}_{ex,\text{Sp\_B}}$ for the list of all executable visible function symbols of Sp_B, and $\overline{s}_{ex,\text{Sp\_B}}$ for the list of all visible sort symbols of Sp_B.

Specification, SP_A	innerSig(SP_A)
A basic, possibly partial, specification, SP_A, where $\overline{f}$ are all the function symbols of $\overline{f}_{\text{SP\_A}}$, of sorts $\overline{t}$, and where $\overline{s}$ are the sorts.	sdec($\overline{s}$) val $\overline{f}$: $\overline{t}$
SP_B **and** SP_C	innerSig(SP_B) innerSig(SP_C) \ innerSig(SP_B)
SP_B **then** SP_C	innerSig(SP_B) innerSig(SP_C) \ innerSig(SP_B)
SP_B **hide** Σ	sig(innerSig(SP_B) \ Σ)
SP_B **with** FM	FM'(innerSig(SP_B))
An instantiated specification $\text{SN}[FA_1]\ldots[FA_n]$, where SN names a specification of the form **spec** $\text{SP\_N}[\text{SP}_1]\ldots[\text{SP}_n]$ **given** $\text{SP}''\_1\ldots\text{SP}''\_m = \text{SP}$	innerSig(SP* **with** FM **then** $\{\text{SP}'\_1$ **and** ... **and** $\text{SP}'\_n\})$ (Recall that SP* was defined in formula (1).)

$$\text{getSig(SP\_A)} \quad = \quad \begin{array}{l} \texttt{sig} \\ \quad \text{innerSig(SP\_A)} \\ \texttt{end} \end{array}$$

Fig. 8. Definition of the maps getSig and innerSig. (FM indicates the translation induced in *SML* by FM.)

Given two lists of *SML* types $\mathbf{s} = \{\mathbf{s}_1,\ldots,\mathbf{s}_n\}$ and $\mathbf{t} = \{\mathbf{t}_1,\ldots,\mathbf{t}_n\}$, we abbreviate **type** $\mathbf{s}_1 = \mathbf{t}_1$ ... **type** $\mathbf{s}_n = \mathbf{t}_n$ by **type** $\overline{\mathbf{s}} = \overline{\mathbf{t}}$. Similarly, given two lists of *SML* functions $\mathbf{f} = \{\mathbf{f}_1 : \mathbf{t}_1,\ldots,\mathbf{f}_n : \mathbf{t}_n\}$ and $\mathbf{g} = \{\mathbf{g}_1 : \mathbf{t}_1,\ldots,\mathbf{g}_n : \mathbf{t}_n\}$ which agree on their types, we abbreviate **val** $\mathbf{f}_1 = \mathbf{g}_1$ ... **val** $\mathbf{f}_n = \mathbf{g}_n$ by **val** $\overline{\mathbf{f}} = \overline{\mathbf{g}}$.

Given two lists L and K of *SML* type and function declarations, we denote by L\K the sublist of L where all declarations of types *and functions* are deleted which possess declarations in K.

For a *SML* structure B with *SML* signature $\texttt{sig}_\text{B}$

1. given two lists of symbols $\overline{\text{sym}_1}$ and $\overline{\text{sym}_2}$ consisting of type symbols and function symbols which agree on their types, we write $\text{B}[\overline{\text{sym}_2}/\overline{\text{sym}_1}]$ for the simultaneous substitution in B of each sym_2 by the corresponding sym_1.

2. Given a list of names $\text{N}_1,\ldots,\text{N}_k$ of *SML* structures with symbol lists $\overline{\text{sym}_1},\ldots,$ $\overline{\text{sym}_k}$, we write $\overline{\text{N}.\text{sym}}$ for the list (of lists) $\text{N}_1.\overline{\text{sym}_1},\ldots,\text{N}_k.\overline{\text{sym}_k}$, and analogously $\text{B}[\overline{\text{N}.\text{sym}}/\overline{\text{N}'.\text{sym}'}]$ for the simultaneous substitution.

3. $\text{sig}(\text{B}\backslash\Sigma)$ is the largest *SML* sub-signature of $\texttt{sig}_\text{B}$ *not* containing any elements of Σ. This is computed from $\texttt{sig}_\text{B}$ by deleting any type and any function symbol of Σ, and any function symbol which has types of Σ in its type.

Obviously innerSig(SP_A), as defined in Fig. 8, yields the signature of SP_A in *SML* notation.

Theorem 5 (Signature Extraction). *For any structured specification* SP_A getSig(SP_A) *is equal to the signature of* SP_A *up to simple syntactic modification.*

Obviously, any *SML* structure B where every function is a total first-order function can be considered as a total sig_B algebra. It suffices to define total carrier sets and total functions from the standard operational semantics of *SML* by deriving a carrier set without bottom element for any type of sig_B and a total function for any function symbol of sig_B.

We can now define when an *SML* structure or functor realizes a structured or parametrized specification.

Definition 4 (*SML* Realizer of an Executable Specification).

1. *Let* SP_A *be a structured specification which admits only total functions. An SML structure* S_A *is said to be a SML realizer of* SP_A *if its signature is equivalent to* getSig(SP_A) *and its total algebra semantics is a model of* SP_A.

2. *Let* SP_A *be a parametrized specification* **spec** SP_A[SP_1]...[SP_n] **given** SP''_1...SP''_m = SP *which admits only total functions and assume that* SP* *is a persistent extension of the imported specifications* SP''_i. *A SML functor of the form* **functor** F(**structure** S_1 : sig_1, ... **structure** S_n : sig_n) = Body *realizes* SP_A *if*

   (a) sig_i *is equivalent to* getSig(SP_i), *for* $i = 1, \ldots, n$, *and the resulting signature of* F *is equivalent to* getSig(SP*), *and*

   (b) *if, for any actual parameter structures of* F *which satisfy the parameter requirements of* SP_A, *and for any realizers of the imported specifications, the result of the functor application is a realizer of the corresponding instantiation of* SP_A *(if this is consistent), i.e. if the following holds: Let* $S_1, \ldots, S_n$ *be* $\text{sig}_1, \ldots \text{sig}_n$ *structures which are realizers of* SP_1, ... SP_n. *For* $i = 1, \ldots, n$, *let* FA_i *be the fitting morphisms from* SP_i *to the theory of* S_i, *and let* SP_A[FA_1]...[FA_n] *be consistent. Then* F(**structure** $S_1 = S_1$ ... **structure** $S_n = S_n$) *is a realizer of* SP_A[FA_1]...[FA_n].

The extraction procedures yield *SML* programs as realizers of executable specifications. We first prove a general theorem which asserts that getBody(SP) and getStruct(SP), as defined in Figs 9 and 10, are realizers of some executable sub-specification of SP.

Theorem 6. *1. Let* SP_A *be a consistent specification which admits only total functions and assume, moreover, that any parameter instantiation with empty renaming mapping occurring in the specification expression* SP_A *is built from an executable parametrized specification and executable fitting morphisms. Then* getBody(SP_A) *realizes an executable sub-specification of* SP_A

Specification, Sp_A	innerBody(Sp_A)
A basic specification, Sp_A, where $\bar{s}$ are the executable sorts and where $\bar{f}$ are the executable function symbols of $\bar{f}_{\text{Sp\_A}}$ with types in $\bar{s}$	`sdec`$(\bar{s})$ `val < rec >` $\bar{f} = \text{dec}(\bar{f})$
Sp_B **and** Sp_C	innerBody(Sp_B) innerBody(Sp_C) \ innerBody(Sp_B)
Sp_B **then** Sp_C	innerBody(Sp_B) innerBody(Sp_C) \ innerBody(Sp_B)

Sp_B **hide** Σ	`local getStruct(`Sp_B`)`
	`in`
	   `type` $\bar{s}_{ex,\text{Sp\_B}}$ **hide** Σ = `sname(`Sp_B`).`$\bar{s}$
	   `val` $\bar{f}_{ex,\text{Sp\_B}}$ **hide** Σ = `sname(`Sp_B`).`$\bar{f}$
	`end`

Sp_B **with** FM	`local getStruct(`Sp_B`)`
	`in`
	   `type` FM$(\bar{s}_{ex,\text{Sp\_B}})$ = `sname(`Sp_B`).`$\bar{s}$
	   `val` FM$(\bar{f}_{ex,\text{Sp\_B}})$ = `sname(`Sp_B`).`$\bar{f}$
	`end`

An instantiated specification $\text{Sn}[FA_1]\dots[FA_n]$, where Sn names a specification of the form
 	
    spec $\text{Sn}[\text{Sp\_1}]\dots[\text{Sp\_}n]$
  given $\text{Sp}''\_1\dots\text{Sp}''\_m = \text{Sp}$
	If Sn, $\overline{\text{Sp}''}$, $\overline{FA}$ are executable, and, for each i, FA_i is an empty list of renaming maps from Sp_i to Sp'_i, and getStruct(Sp'_i) is of the same signature as getStruct(Sp_i), then innerBody(Sp_A) is:
	
	   `local structure H =`
	      `Sn (structure` $\overline{\text{Sp}}$ = `sname(`$\overline{\text{SP}}$`)`
	  `in`
	     `type` $\overline{s_{\text{Sp}}}$ = `H.`$\overline{\text{Sp}}$`.`$\bar{s}$
	     `type` $\overline{s_{\text{Sp}''}}$ = `H.`$\overline{\text{Sp}''}$`.`$\bar{s}$
	     `val` $\overline{f_{\text{Sp}}}$ = `H.`$\overline{\text{Sp}}$`.`$\bar{f}$
	     `val` $\overline{f_{\text{Sp}''}}$ = `H.`$\overline{\text{Sp}''}$`.`$\bar{f}$
	  `end`
	
	Otherwise, innerBody(Sp_A) is:
	innerBody(Sp* **with** FM **then**
	{Sp'_1 **and** ... **and** Sp'_n})
	where Sp* is defined in formula (1).

$$\text{getBody}(\text{Sp\_A}) = \begin{array}{l} \texttt{struct} \\ \text{innerBody}(\text{Sp\_A}) \\ \texttt{end} \end{array}$$

Fig. 9. Mutually recursive definition of the maps innerBody and getBody which are used to define getStruct. (<> indicates `rec` is omitted for constants and non-recursive functions. See Fig. 8 for the definition of getSig. Note that FM indicates the translation induced in *SML* by FM.)

Specification, SP_A	getStruct(SP_A)

spec
 SP_A[SP_1]...[SP_n]
 given
 SP''_1...SP''_m
 = SP
	functor sname(SP_A) (**structure** $\overline{\overline{\mathrm{Sp}}}$: getSig($\overline{\mathrm{Sp}}$))
	 sig
	      structure $\overline{\overline{\mathrm{Sp}}}$: getSig($\overline{\mathrm{Sp}}$)
	      structure sname($\overline{\mathrm{SP''}}$) : getSig($\overline{\mathrm{SP''}}$)
	      innerSig($\overline{\mathrm{SP}}$)
	      [sname($\overline{\mathrm{SP''}}$).sym_sname($\mathrm{SP''}$)]/sym_sname($\mathrm{SP''}$)]
	      [$\overline{\mathrm{Sp}}$.sym_Sp/sym_Sp]
	 end =
	 struct
	      structure $\overline{\overline{\mathrm{Sp}}}$ = $\overline{\overline{\mathrm{Sp}}}$
	      getStruct($\mathrm{SP''}$)
	      innerBody(SP)
	      [sname($\overline{\mathrm{SP''}}$).sym_sname($\mathrm{SP''}$)/sym_sname($\mathrm{SP''}$)]
	      [$\overline{\mathrm{Sp}}$.sym_Sp/sym_Sp]
	 end

Otherwise	**structure** sname(SP_A) : getSig(SP_A) = getBody(SP_A)

Fig. 10. Definition of the map getStruct. (See Fig. 9 for the definition of getBody and Fig. 8 for that of getSig.)

and getStruct(SP_A) *realizes an executable sub-specification of the specification named* sname(SP_A). *If* SP_A *is executable, then* getBody(SP_A) *realizes* SP_A *and* getStruct(SP_A) *realizes the specification named* sname(SP_A).

2. *Let* SP_A *be a parametrized specification* **spec** SP_A[SP_1]...[SP_n] **given** SP''_1...SP''_m = SP *which is executable, consistent and which admits only total functions and let* SP''_1...SP''_m *be executable consistent specifications which admit only total functions. Then* getStruct(SP_A) *is an SML functor that realizes* SP_A.

Proof. Since getBody is only a syntactic modification of innerBody the theorem is proved for innerBody and getStruct by simultaneous structural induction. □

Corollary 1 (Structure Extraction). *Let* SP_A *be a consistent, executable specification which admits only total functions and assume, moreover, that any parameter instantiation with empty renaming maps occurring in the specification expression* SP_A *is built from an executable parametrized specification. Then* getBody(SP_A) *realizes* SP_A *and* getStruct(SP_A) *realizes the specification named* sname(SP_A).

Proof. This follows directly from the above theorem. In the case of a parameter instantiation with empty renaming maps the executability of the fitting morphisms ensures the precondition of the theorem. □

Corollary 2 (Functor Extraction). *Let* SP_A *be a parametrized specification of the form* **spec** SP_A[SP_1] ... [SP_n] **given** SP''_1 ... SP''_m = SP *which is executable, consistent and which admits only total functions. Then the SML functor* getStruct(SP_A) *realizes* SP_A.

Example (*concluded.*) Finally we use getStruct(WAREHOUSE_EXEC) to obtain the functor declaration

```
functor
    Warehouse(structure Catalogue: CatalogueSig): sig
        structure Catalogue : CatalogueSig
            val rep: Part  -> Part  end
= struct
        structure Catalogue = Catalogue
            val rep = fn i : Part => Catalogue.hd (Catalogue.lOR(i));
end;
```

where `CatalogueSig` is

getSig(CATALOGUE) =

```
sig
 type Part;
 datatype List_Part = empty | add of Part * List_Part;
  <<standard operations of List_Part>>
 type Catalogue;
 val myCat: Catalogue;
 val lOR : Part -> List_Part;
 val Rep : Part * Part * Catalogue -> Bool;
end
```

References

1. U. Berger and H. Schwichtenberg, Program development by Proof Transformation, pp. 1–45 in *Proceedings of the NATO Advanced Study Institute on Proof and Computation*, Marktoberdorf, Germany, 1993.
2. M. V. Cengarle, *Formal specifications with higher-order parametrization.* PhD thesis, Ludwig-Maximilians-Universität, Munich, 1995.
3. CoFI Language Design Task Group on Language Design, *CASL, The Common Algebraic Specification Language, Summary, 25 March 2001*, available at http://www.brics.dk/Projects/CoFI/Documents/CASL/Summary/ (accessed 28.05.2001).
4. R. L. Constable, S. F. Allen, H. M. Bromley, W. R. Cleaveland, J. F. Cremer, R. Harper, D. J. Howe, T. B. Knoblock, N. P. Panangaden, J. T. Sasaki, and S. F. Smith, *Implementing Mathematics with the Nuprl Development System.* Prentice-Hall, Englewood Cliffs, New Jersey, 1986.
5. J. N. Crossley and I. Poernomo, **Fred**: An approach to generating real, correct, reusable programs from proofs, FMTOOLS 2000, special issue of *J.U.C.S.*, **7**, no.1, 71–88, available at http://www.jucs.org/jucs_7_1/fred_an_approach_to (accessed 28.05.2001).

6. J. N. Crossley, I. Poernomo and M. Wirsing, Extraction of Structured Programs from Specification Proofs, pp. 419-437 in D. Bert, C. Choppy and P. Mosses (eds), *Recent Trends in Algebraic Development Techniques* (WADT'99), Lecture Notes in Computer Science **1827**, Berlin: Springer, 2000.

7. J. N. Crossley and J. C. Shepherdson, Extracting programs from proofs by an extension of the Curry-Howard process, pp. 222–288 in J. N. Crossley, J. B. Remmel, R. A. Shore, and M. E. Sweedler (eds),*Logical Methods*, Birkhäuser, Boston, 1993.

8. J.-Y. Girard, Y. Lafont and P. Taylor, *Proofs and types*, Cambridge University Press, 1989.

9. S. Hayashi and H. Nakano, *PX, a computational logic*. MIT Press, Cambridge, Mass., 1988.

10. G. Huet, G. Kahn, and C. Paulin-Mohring. *The Coq Proof assistant Reference Manual: Version 6.1*. Coq project research report RT-0203, Inria, 1997.

11. S. Kahrs, D. Sannella and A. Tarlecki, The definition of Extended ML: A gentle introduction. *Theoretical Computer Science*, **173** (1997) 445-484.

12. Per Martin-Löf, *Intuitionistic Type Theory*, Bibliopolis, Naples, Italy, 1984.

13. R. Milner, M. Tofte, and R. Harper, *The definition of Standard ML*. Cambridge, Mass., MIT Press, 1990.

14. F. Orejas. Structuring and modularity, pp. 159–200 in E. Astesiano, H.-J. Kreowski, B. Krieg-Bruckner (eds), *Algebraic Foundations of Systems Specification*, IFIP State-of-the-Art Reports. Berlin, Springer, 1999.

15. L. C. Paulson, ML *for the Working Programmer*, second edition. Cambridge University Press, 1996.

16. H. Peterreins, *A natural-deduction-like calculus for structured specifications*. PhD thesis, Ludwig-Maximilians-Universität, Munich, 1996.

17. D. Sannella and A. Tarlecki. Toward formal development of ML programs: foundations and methodology, pp 375–389 in J. Diaz and F. Orejas (eds), *TAPSOFT '89*, vol. 2, Lecture Notes in Computer Science **352**, Berlin, Springer 1989.

18. D.R. Smith, Constructing Specification Morphisms, *Journal of Symbolic Computation*, **15** (1993) 571–606

19. M. Wirsing, Structured algebraic specifications: a kernel language, *Theoretical Computer Science*, **43** (1986) 123–250.

20. M. Wirsing, Algebraic specification, pp.675–788 in J. van Leeuwen, (ed.), *Handbook of Theoretical Computer Science*, volume B, Amsterdam; New York: Elsevier; Cambridge, Mass.: MIT Press, 1990.

21. M. Wirsing, J. N. Crossley and H. Peterreins, Proof normalization of structured algebraic specifications is convergent, pp. 322-337 in J. Fiadeiro (ed.), *Proceedings of the Twelfth International Workshop on Recent Trends in Algebraic Development Techniques*, Lecture Notes in Computer Science **1589**, Berlin, Springer, 1999.

Towards Trustworthy Specifications I: Consistency Checks

Markus Roggenbach and Lutz Schröder

BISS, Department of Computer Science, Bremen University

Abstract. As the first of two methodological devices aimed at increasing the trust in the 'correctness' of a specification, we develop a calculus for proving consistency of CASL specifications. It turns out to be possible to delegate large parts of the proof load to syntactical criteria by structuring consistency proofs along the given specification structure, so that only in rather few remaining focus points, actual theorem proving is required. The practical usability of the resulting calculus is demonstrated by extensive examples taken from the CASL library of basic data types.

Introduction

The verification of *programs* is a well-established topic in computer science; here, correctness of programs is usually defined w.r.t. requirements given in a more or less formal language, possibly an algebraic specification language such as the language CASL (*Common Algebraic Specification Language*) [8,2] designed by CoFI, the international *Common Framework Initiative for Algebraic Specification and Development* [7]. However, these *requirement specifications* themselves may very well contain errors in the sense that they fail to have exactly the intended class of models; early detection of such errors will substantially reduce the cost of software development.

Since the notion of an 'intended' class of models is not equipped with a formal meaning, the correctness of a requirement specification in this sense can only be 'verified' in an approximative process. This process splits into two parts, corresponding to the two required inclusions between the actual and the intended model class.

In order to increase the trust in the claim that the given specification does not admit unwanted models, one can attempt to prove certain intended consequences from the given axioms. Thus, it is shown that at least those models which fail to have these intended properties are excluded from the actual model class. This program is systematically pursued in the forthcoming [21]. (The opposite strategy — constructing counterexamples for intended properties in order to detect faulty specifications — is developed, e.g., in [1] and in [19].) Conversely, of course, one has to make sure that none of the intended models fails to satisfy the specification. The first step in this direction is to show that the specification has at least one model, i.e. that it is (semantically) consistent. (Note that for CASL specifications, semantical consistency and syntactical consistency, i.e. absence

M. Cerioli and G. Reggio (Eds.): WADT/CoFI 2001, LNCS 2267, pp. 305–327, 2002.

of contradictions, do not coincide, since CASL sort generation constraints and CASL free specifications are higher order concepts). It is important to note that the proof of intended consequences as discussed above is meaningless without a consistency proof — ex falso quodlibet.

Here, we develop a set of rules for checking consistency of CASL specifications in a systematic way. As these rules work along the actual specification text, the need to construct (and prove) actual models of specifications is avoided as far as possible. The given set of rules is certainly far from complete. However, we illustrate its applicability by demonstrating how it can be used to establish the consistency of a large part of the CASL Basic Datatypes [20]. In order to be able to deal with this example and other realistic specifications, we discuss the calculus for full CASL here rather than restrict to a possibly more digestible sublanguage.

The consistency rules are designed in such a way that they can be automated to a large extent: most of them can be statically checked, i.e. they make use of the static semantics only. Just in a few cases, e.g. in rules that assume the correctness of views, actual theorem proving is required to discharge proof obligations that arise from the model semantics.

Besides using certain syntactical criteria, the consistency rules rely heavily on the CASL structuring mechanisms and their semantic annotations. Consequently, consistency proofs follow the structure of the given specification. A simple example for this type of argument is the exploitation of specification morphisms that arise e.g. from instantiations or extensions for transporting consistency. In this way, our rules highlight the (usually few) 'hot spots' of a specification, while the (lengthy) 'trivial' parts of the consistency argument are discharged automatically.

The structuring of proofs along the structure of specifications is a well-established concept. Proof systems of this kind for statements expressed within the given logic are elaborated e.g. in [15]. The *development graph* [3] provides automatic proof support for this type of reasoning, as well as for structured 'meta-reasoning' (although such terminology is deemed unsuitable for minors [12]) about specification refinement; the latter point is also addressed in [5,11]. The most central metapredicate used below is conservativity of extensions; reasoning about this predicate within the development graph is discussed in [18]. To the best of our knowledge, however, the question how to structure consistency proofs along the structure of specifications has not been addressed before.

1   CASL

The specification language CASL has been designed by CoFI, the international *Common Framework Initiative for Algebraic Specification and Development* [7]. Following [2], we present a short overview of those features of CASL which are relevant here. For the full definition of the language, we refer to [8].

Roughly speaking, a CASL *basic specification* consists of a *signature* made up of sorts, operations, and predicates (declared by means of the keywords **sort**, **op**, and **pred**, respectively, optionally equipped with a defining term or formula), and axioms referring to the signature items (keywords **axiom, forall**). Operations can be partial or total. Furthermore, one may declare a *subsort relation* on the sort symbols. Axioms are written in first-order logic. Going one step beyond first order logic, CASL also features sort generation constraints for datatypes (keywords **generated, free type**).

A *model* of such a specification is an algebra which interprets the sorts as (non empty) sets and the operations and predicates as (partial) functions and subsets, respectively, in such a way that the given axioms are satisfied. The subsorting relation is reflected by injective coercion functions between the sets interpreting the involved sorts (*not* by subset inclusion); for a discussion of the difficulties arising from this subtlety and their solution see [22].

Moreover, CASL provides ways of building complex (*structured*) specifications out of simpler ones (the simplest ones being basic specifications) by means of various *specification-building operations*. These include translation, hiding, union, and both free and loose forms of extension.

Translations of declared symbols to new symbols are specified by giving lists of 'maplets' of the form *old* ↦ *new* (keyword **with**).

Reducing a specification means removing symbols from its signature and the corresponding items from its models. CASL provides two ways of specifying a reduction: by listing the symbols to be *hidden* (keyword **hide**), or by listing those to be left visible, i.e., *revealed* (keyword **reveal**). CASL also facilitates the hiding of auxiliary symbols by allowing the local scope of their declarations to be indicated (keyword **local**).

The signature of a *union* of two specifications is the union of their signatures. Given models over the component signatures, the unique model over the union signature that extends each of these models is called their *amalgamation*; a pair of models is called *compatible* if their amalgamation exists. Clearly, not all pairs of models over component signatures amalgamate: an obvious necessary condition is that the models coincide on the common symbols (including subsort embeddings) of the component signatures. The models of a union (keyword **and**) are all amalgamations of the models of the component specifications.

Extensions (keyword **then**) may specify new symbols or merely require further properties of old ones. Extensions can be classified by their effect on the model class specified. For instance, an extension is called *conservative* when no models are lost: every model of the specification being extended is a reduct of some model of the extended specification. CASL provides annotations %**implies**, %**def**, and %**cons** to denote that the model class is not changed, that each model of the specification can be uniquely extended to a model of the extended specification, or that the extension is conservative, resp. It is important to note that these annotations have no effect on the semantics of a specification: a specifier may use them to express his intentions, tools may use them to generate proof obligations. In the consistency proofs of Section 3, we use these annotations as a guideline for selecting appropriate rules.

Structured specifications may be *named*, and a named specification may be *generic* in the sense that it declares *parameters* that need to be *instantiated* when the specification is (re)used. Instantiation is a matter of providing an appropriate *argument specification* together with a *fitting morphism* from the parameter to the argument specification. A generic specification may also declare *imports* (keyword **given**) which behave like immediately instantiated parameters; the formal parameters (as well as the argument specifications in any instantiation) are regarded as extensions of the imports.

To allow reuse of fitting 'views', specification morphisms (from parameters to arguments) may themselves be named (keyword **view**).

Specifications may be declared to be *closed* (keyword **closed**), which means that they behave as though there were no previously declared signature elements (this becomes relevant as soon as there are translations in the scope of a **closed** construct). Instantiations of generic specifications are implicitly closed.

The simplest case of a *free* specification (keyword **free**) is the one where the specification constrained to be interpreted freely is closed. The signature of the specification is unchanged, but its model class is restricted to the initial models. More generally, a free specification may be a free extension, e.g.:

> **sort** *Elem* **then**
> **free**
> **{ type** *Set* ::= {} | {_}(*Elem*) | _ ∪ _(*Set*; *Set*)
>   **op** _ ∪ _ : *Set* × *Set* → *Set*, **assoc**, **comm**, **idem**, **unit** {} **}**

Many structured specifications can be flattened, i.e. transformed into basic specifications (although for general structured specifications, flattening is precluded by the presence of free specifications and hiding operations; cf. [5]). However, the strategy pursued here consists to a large extent in exploiting these structuring operations in order to obtain correspondingly structured proofs; therefore, flattening will not play any rôle.

2   The Calculus

The first point that needs to be stressed is that the calculus developed below is really not about consistency at all, but rather about *conservativity* of extensions. Intuitively speaking, an extension is conservative if it does not 'specify away' any models, i.e. if each model of the original specification can be enlarged to a model of the extended specification [23]. This can be formalized as follows: an extension $\sigma : Sp_1 \hookrightarrow Sp_2$ induces a model reduction

$$\mathbf{Mod}(\sigma) : \mathbf{Mod}(Sp_2) \to \mathbf{Mod}(Sp_1),$$

where $\mathbf{Mod}(Sp)$ may for the purposes of this paper be thought of as denoting the *class* of models of a specification Sp (cf. [6,8,9] for details). σ is *conservative* if $\mathbf{Mod}(\sigma)$ is surjective, and *definitional* if $\mathbf{Mod}(\sigma)$ is bijective. For both these properties, CASL offers *semantic annotations*: conservative and definitional extensions may be indicated by the annotations %**cons** and %**def**, respectively. Note that definitionality of an extension implies that it does not declare any new sorts. There is a third annotation, %**implies**, which applies to definitional extensions that do not affect the signature.

Now it is trivial to observe that a specification is *consistent* (in the sense that its model class is non-empty) iff it conservatively extends the empty specification {}. Thus, it does indeed suffice to provide a calculus for conservativity (which would necessarily have formed a part of a consistency calculus anyway). Conservativity is in itself an important notion in many contexts; for a recent application, see e.g. [18]. Moreover, we sketch the beginnings of a definitionality calculus, which, in this context, serves primarily the purpose of improving the readability of conservativity proofs.

The judgements of the conservativity calculus are of the form $cons(Sp_1)(Sp_2)$, where $Sp_1 \hookrightarrow Sp_2$ is an extension (in a somewhat generalized sense to be made precise below). Similarly, definitionality is expressed by the predicate $def(Sp_1)(Sp_2)$. Moreover, we use a predicate $implies(Sp_1)(Sp_2)$ which corresponds to the annotation %**implies** in CASL, i.e. Sp_1 and Sp_2 have the same signature and class of models. This predicate is not supported by a calculus; instead, its verification will assumed to be discharged by a suitable theorem prover. Finally, we do introduce a consistency predicate $c(Sp)$; however, this is just an abbreviation for $cons(\{\})(Sp)$. Only a few rules refer specifically to the consistency predicate.

The calculus has been minimized as far as possible in order to keep it manageable and understandable. Its actual application, in particular by a tool, will require a catalogue of derived rules in order to avoid overly clumsy proofs; some examples of such rules are discussed at the end of this section.

The rules are organized as follows: Figure 1 contains a simple extension calculus for CASL specifications. The conservativity calculus proper is then subdivided into general conservativity rules (Figure 2), rules for special extensions (Figure 3), rules for structuring constructs (Figure 4), and definitionality rules (Figure 5). Correctness proofs for the rules have been generally omitted.

In the following we use without further explanation the symbol system introduced in the CASL Language Summary [8], where e.g. a specification is denoted by Sp, FM is a fitting morphism, BI is a basic item, etc. *All rules are subject to the silent premise that the specifications they are applied to are well-formed.*

Extension Calculus. In order to keep the number of rules in the conservativity calculus as small as possible, we begin by introducing an *extension calculus* for specifications (Figure 1). This calculus is purely auxiliary and does not have any claims to completeness (nor is it meant as a step towards a calculus for specification refinement); however, it does allow a rather more elegant formulation of many of the conservativity rules. The calculus concerns judgements of the form $Sp_1 \preceq Sp_2$, which are taken to mean that $Sp_1 \hookrightarrow Sp_2$ is an extension. Here, the word *extension* is to be understood as meaning that the signature defined by Sp_1 is a subsignature of that defined by Sp_2, and that the associated signature inclusion is in fact a specification morphism $Sp_1 \to Sp_2$. At the level of CASL, this can be rephrased as correctness of the view

view V: *Sp_1* **to** *Sp_2* **end.**

Most of the extension calculus is really about *equivalence* of CASL specifications. Equivalence of two specifications Sp_1 and Sp_2, denoted $Sp_1 \simeq Sp_2$, is defined as mutual extension; i.e.,

$$Sp_1 \simeq Sp_2 : \iff (Sp_1 \preceq Sp_2 \text{ and } Sp_2 \preceq Sp_1).$$

Of course, $Sp_1 \simeq Sp_2$ is equivalent to $implies(Sp_1)(Sp_2)$. However, the two notations serve different methodological purposes: while $Sp_1 \simeq Sp_2$ is used to transform specifications, $implies(Sp_1)(Sp_2)$ denotes a proof obligation to be discharged by a suitable theorem prover. Again, it should be stressed that the notion $\simeq$ is subsidiary to the conservativity calculus and by no means aimed at normal forms of specifications and the like.

The relation $\preceq$ is reflexive and transitive; moreover, it is compatible with the structuring operations. This implies that the same holds for $\simeq$ (which is, moreover, trivially symmetric).

The union operator **and** is commutative and associative. It is also idempotent in the sense that a union of a specification and an extension of that specification is essentially the same as the extension. There are rules for 'unfolding' instantiations of (parametrized) specifications according to the circumscription in the summary. Since we are really interested in keeping the structure of specifications here, these rules have been included more for the sake of their obviousness than in order to be routinely applied, with the exception of the simplest case, which concerns instantiations of unparametrized named specifications.

The remaining rules of Figure 1 are largely self-explanatory, such as idempotency of the **closed**-construct, equivalence of extension and union under well-formedness of the latter, or introductions of extra **then**s. Of course, certain basic equivalences may be derived; e.g., by rules (inst-rp) and (cl), $SN[FA_1]\dots[FA_n] \simeq \textbf{closed } SN[FA_1]\dots[FA_n]$.

General Rules. Some basic rules of the conservativity calculus are shown in Figure 2. Besides the obvious rules on trivial extensions and composition of extensions and on the relation between the metapredicates *cons, def,* and *implies,* there is a weakening rule (wk), as well as rules (rp1), (rp2) that allow replacing specifications by equivalent ones. Of course, the premises concerning the relations $\preceq$ and $\simeq$ in the latter three rules are meant to be discharged by means of the extension calculus of Figure 1. As discussed above, the main purpose of these rules is the simplification of the calculus; they are intended for 'conservative' use, i.e. mostly for minor syntactical adjustments rather than, say, for wholesale flattening and the like.

Rules for Special Extensions. These rules (Figure 3) provide mechanisms for dealing with features of basic specifications such as axioms or data type definitions; since we are not concerned with actual theorem proving here, their scope is necessarily somewhat limited. Some of the rules listed here are discussed in a similar context in [14,17]. Most of them make use of a predicate *newSorts* which expresses the (easily checkable) fact that a sort s or the sorts declared in signature items $SI_1, \dots, SI_n$ or datatype declarations $DD_1, \dots, DD_n$, respectively, are not contained in the signature of Sp. If a new sort is declared to be (freely) generated, it has to be checked whether the declared signature provides a closed term (i.e. a term without free variables) of this sort. This complication is due to the fact that the CASL semantics requires non-empty carriers.

$$(\text{refl}) \ \frac{}{Sp \preceq Sp} \qquad (\text{trans}) \ \frac{Sp_1 \preceq Sp_2 \quad Sp_2 \preceq Sp_3}{Sp_1 \preceq Sp_3}$$

$$(\text{cong}) \ \frac{Sp_1 \preceq Sp_1'}{\{Sp_1 \text{ then } Sp_2\} \preceq \{Sp_1' \text{ then } Sp_2\} \text{ etc.}} \qquad (\text{ext}) \ \frac{}{Sp_1 \preceq \{Sp_1 \text{ then } Sp_2\}}$$

$$(\text{hdx}) \ \frac{}{Sp \text{ hide } SL \preceq Sp} \qquad (\text{rvx}) \ \frac{}{Sp \text{ reveal } SL \preceq Sp}$$

$$(\text{rveq}) \ \frac{SL \text{ lists the symbols renamed by } SM}{Sp \text{ reveal } SM \simeq Sp \text{ reveal } SL \text{ with } SM}$$

$$(\text{union-c}) \ \frac{}{\{Sp_1 \text{ and } Sp_2\} \simeq \{Sp_2 \text{ and } Sp_1\}}$$

$$(\text{union-a}) \ \frac{}{\{\{Sp_1 \text{ and } Sp_2\} \text{ and } Sp_3\} \simeq \{Sp_1 \text{ and } \{Sp_2 \text{ and } Sp_3\}\}}$$

$$(\text{union-i}) \ \frac{Sp_1 \preceq Sp_2}{Sp_2 \simeq \{Sp_1 \text{ and } Sp_2\}} \qquad (\text{inst-eq1}) \ \frac{SN = Sp}{SN \simeq \text{closed}\{Sp\}}$$

$$(\text{inst-eq2}) \ \frac{\begin{array}{c} SN[Sp_1]\ldots[Sp_n] = Sp \\ Sp' = \{\{Sp_1 \text{ and}\ldots\text{and } Sp_n\} \text{ then } Sp\} \\ FA_i : Sp_i \to Sp_i', \ i = 1,\ldots,n \\ FM \text{ is the extension of the } FA_i \text{ to } Sp' \end{array}}{SN[FA_1]\ldots[FA_n] \simeq \text{closed}\{\{Sp' \text{ with } FM\} \text{ and } Sp_1' \text{ and}\ldots\text{and } Sp_n'\}}$$

$$(\text{inst-eq3}) \ \frac{\begin{array}{c} SN[Sp_1]\ldots[Sp_n] \text{ given } Sp_1'',\ldots,Sp_m'' = Sp \\ Sp' = \{\{Sp_1'' \text{ and}\ldots\text{and } Sp_m''\} \text{ then } \{Sp_1 \text{ and}\ldots\text{and } Sp_n\} \text{ then } Sp\} \\ FA_i{:}\{\{Sp_1'' \text{ and}\ldots\text{and } Sp_m''\} \text{ then } Sp_i\} \to \{\{Sp_1'' \text{ and}\ldots\text{and } Sp_m''\} \text{ then } Sp_i'\}, \\ i = 1,\ldots,n \\ FM \text{ is the extension of the } FA_i \text{ to } Sp' \end{array}}{SN[FA_1]\ldots[FA_n] \simeq \text{closed}\{\{Sp' \text{ with } FM\} \text{ and } Sp_1' \text{ and}\ldots\text{and } Sp_n'\}}$$

$$(\text{cl}) \ \frac{}{\text{closed } \{Sp\} \simeq Sp}$$

$$(\text{double}) \ \frac{Sp \text{ contains } BI}{Sp \simeq \{Sp \text{ then } BI\}}$$

$$(\text{then-and}) \ \frac{Sp_1 \text{ and } Sp_2 \text{ is well-formed}}{\{Sp_1 \text{ then } Sp_2\} \simeq \{Sp_1 \text{ and } Sp_2\}}$$

$$(\text{sep1}) \ \frac{}{\{BI_1\ldots BI_n\} \simeq \{BI_1\ldots BI_k \text{ then } BI_{k+1}\ldots BI_n\}}$$

$$(\text{sep2}) \ \frac{}{Sp \simeq \{\} \text{ then } Sp}$$

$$(\text{perm}) \ \frac{\begin{array}{c} \{BI_1\ldots BI_k \ BI_{k+1}\ldots BI_n\} \text{ is well-formed} \\ \{BI_1\ldots BI_{k+1} \ BI_k\ldots BI_n\} \text{ is well-formed} \end{array}}{\{BI_1\ldots BI_k \ BI_{k+1}\ldots BI_n\} \simeq \{BI_1\ldots BI_{k+1} \ BI_k\ldots BI_n\}}$$

$$(\text{wkfree}) \ \frac{}{\text{generated } \{SI_1\ldots SI_n\} \preceq \text{free } \{SI_1\ldots SI_n\}}$$

Fig. 1. The extension calculus

$$\textbf{(triv)} \ \frac{}{implies(Sp)(Sp)} \qquad \textbf{(comp)} \ \frac{cons/def/implies(Sp_1)(Sp_2)}{cons/def/implies(Sp_2)(Sp_3)}{cons/def/implies(Sp_1)(Sp_3)}$$

$$\textbf{(wk)} \ \frac{Sp_1 \preceq Sp_2 \preceq Sp_3}{cons(Sp_1)(Sp_3)}{cons(Sp_1)(Sp_2)}$$

$$\textbf{(rp1)} \ \frac{Sp_1 \simeq Sp_1'}{cons/def/implies(Sp_1)(Sp_2)}{cons/def/implies(Sp_1')(Sp_2)} \qquad \textbf{(rp2)} \ \frac{Sp_2 \simeq Sp_2'}{cons/def/implies(Sp_1)(Sp_2)}{cons/def/implies(Sp_1)(Sp_2')}$$

$$\textbf{(def)} \ \frac{def(Sp_1)(Sp_2)}{cons(Sp_1)(Sp_2)} \qquad \textbf{(imp)} \ \frac{implies(Sp_1)(Sp_2)}{def(Sp_1)(Sp_2)}$$

Fig. 2. General conservativity rules

Rule (horn) states that extensions in *positive* Horn logic are conservative if they have no effect on the previously declared sorts. The metapredicate $horn(Sp)$ is true for a basic specification iff all its axioms (including the implicit ones) are positive Horn clauses, possibly after performing skolemization, e.g. in the case of the existence axiom for inverses in a group; cf. Section 3.1. (Note that the axiom of choice is explicitly assumed in the CASL semantics.) Subsort declarations and **type** definitions (without sort generation constraints) can be coded by positive Horn clauses and hence are regarded as such. The *horn* predicate is recursively extended to structured specifications in the obvious way, except that hiding must be excluded here; i.e. if Sp is a specification in positive horn logic, then Sp **hide** SM need not be equivalent to a specification in positive Horn logic. (In the case of instantiations of parametrized specifications, the extension of the horn predicate involves additional proof obligations concerning well-formedness of the instantiation, i.e. correctness of the fitting morphism, in the same way as in rule (inst-eq3) of Figure 1.) Note that the phrase 'Conclusions in Sp_2 concern new predicates or equality in new sorts over Sp_1' refers also to implicitly generated axioms arising, e.g., from subsorting and overloading.

Further, rather obvious rules (rules (sub), (free), and (gtd1)) concern the introduction of subsorts, free datatypes, and (unrestricted) generated types, respectively. The somewhat surprising satisfiability constraint expressed by the premise $implies(\ldots)(\ldots)$ in rule (sub) is necessary due to the fact that the CASL semantics requires non-empty carriers; of course, discharging this constraint will in general require some form of theorem proving (even if the formula in question is in positive Horn form!). The case of generated types with a specified equivalence relation is somewhat more complicated. The rules provided here for this purpose (rules (gtd2) and (gtd3)) cover two syntactical patterns (the adherence to which is meant to be mechanically checkable): definition of equality via an observer or recursive definition of equality (where, for the time being, recursion is restricted to primitive recursion over the term structure). An observer is understood to be a function or a predicate which is defined by recursion over the term structure of the relevant data types. Observers may contain additional parameters of types other than the ones introduced in the datatype definition. A typical example of an observer is the elementhood predicate for finite sets, c.f. Section 3.2; using this observer, equality of sets is defined by the usual extensionality axiom. In both cases, the phrase 'defines equality' comes with a proof obligation, namely that the resulting relation on terms is indeed a congruence on the associated term algebra. If the 'equality' is defined by an observer, the equivalence axioms come for free, but the compatibility with the constructors remains as a proof obligation. Note that we have refrained from formulating explicit rules for the **generated types** construct, which, as in [8], is regarded as a special case of the **generated** construct.

Rules for Structuring Constructs. As discussed above, consistency proofs profit from being designed following the specification structure. This requires rules that allow breaking down the structuring constructs of CASL. Rules of this type are listed in Figure 4; some of them are related to rules presented in [18].

The rules on translation, hiding, revealing, and local specifications (rules (tr), (hd1), (hd), (rv1), (rv2), and (local)) hardly require explanation. Rule (free) captures the fact that specifications in positive Horn form have initial models, provided there is a closed term for each sort. Rule (view) uses a 'correctness predicate' for CASL views; discharging this premise will in general require actual theorem proving. The rules on named specifications just state that a named specification is consistent if its 'unnamed version' is consistent.

The last three rules of Figure 4 are somewhat more involved. All of them require certain diagrams of *signatures* to be amalgamable; this is to be understood as follows: a commutative square

of signature morphisms is called *amalgamable* if it is mapped to a pullback (of model classes or categories) under the model functor. Roughly speaking, this means that for each pair (M, N), where M and N are models of Σ_2 and Σ_3, respectively, such that the respective reductions of M and N to Σ_1 agree, there exists a unique model of Σ_4 that extends M and N ('compatible models can be amalgamated'). This requirement is *approximated* by the condition that the above diagram forms a pushout in the signature category. However, there are certain additional difficulties that arise from subsorting; see [22] for a detailed discussion. In fact, whether or not a diagram as above is amalgamable is, in gen-

$$\textbf{(horn)} \quad \frac{\begin{array}{c} horn(Sp_2) \\ \text{Conclusions in } Sp_2 \text{ concern new predicates or equality in new sorts over } Sp_1 \end{array}}{cons(Sp_1)(Sp_1 \textbf{ then } Sp_2)}$$

$$\textbf{(sub)} \quad \frac{\begin{array}{c} newSorts(s)(Sp) \\ implies(Sp)(Sp \textbf{ then axiom } \exists\, v:t \bullet F) \end{array}}{cons(Sp)(Sp \textbf{ then sort } s = \{v:t \bullet F\})}$$

$$\textbf{(gtd1)} \quad \frac{\begin{array}{c} newSorts(SI_1 \ldots SI_n)(Sp) \\ Sp \textbf{ then } SI_1 \ldots SI_n \text{ has a closed term for each new sort} \end{array}}{cons(Sp)(Sp \textbf{ then generated } \{SI_1 \ldots SI_n\})}$$

$$\textbf{(gtd2)} \quad \frac{\begin{array}{c} newSorts(SI_1 \ldots SI_n)(Sp) \\ Sp_1 \textbf{ then } SI_1 \ldots SI_n \text{ has a closed term for each new sort} \\ Sp_2 \text{ recursively defines an observer } f \\ Sp_3 \text{ defines equality on new sorts by } f \end{array}}{cons(Sp_1)(Sp_1 \textbf{ then generated } \{SI_1 \ldots SI_n\} \textbf{ then } Sp_2 \textbf{ then } Sp_3)}$$

$$\textbf{(gtd3)} \quad \frac{\begin{array}{c} newSorts(SI_1 \ldots SI_n)(Sp) \\ Sp_1 \textbf{ then } SI_1 \ldots SI_n \text{ has a closed term for each new sort} \\ Sp_2 \text{ recursively defines equality on new sorts} \end{array}}{cons(Sp_1)(Sp_1 \textbf{ then generated } \{SI_1 \ldots SI_n\} \textbf{ then } Sp_2)}$$

$$\textbf{(free)} \quad \frac{\begin{array}{c} newSorts(DD_1 \ldots DD_n)(Sp) \\ Sp \textbf{ then types } DD_1; \ldots; DD_n \text{ has a closed term for each new sort} \end{array}}{cons(Sp)(Sp \textbf{ then free types } DD_1; \ldots; DD_n)}$$

Fig. 3. Rules for special extensions

eral, undecidable; however, practically relevant cases are covered by a polynomial algorithm which is currently implemented in the CASL tool set [16].

The relevant point here is that conservativity of the morphism $\Sigma_1 \to \Sigma_2$ in the above diagram implies conservativity of the morphism $\Sigma_3 \to \Sigma_4$, provided that the diagram is amalgamable (cf. [10]). A typical example is given by instantiations of parametrized specifications: a parametrized specification $SN[Sp_1] = Sp$ (for the sake of simplicity without imports and with only one parameter) defines, at the level of signatures, a signature extension $\Sigma_F \to \Sigma_B$, where Σ_F is the signature of the formal parameter Sp_1 and Σ_B that of the body. An instantiation of this parametrized specification requires a fitting morphism $FA : Sp_1 \to Sp_1'$; this fitting morphism is, in particular, a signature morphism $\Sigma_F \to \Sigma_A$, where Σ_A is the signature of the actual parameter Sp_1'. The result of the instantiation $SN[FA]$ then has signature Σ_R, where

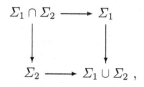

is a pushout in the signature category. In many typical situations, the body turns out to be a conservative extension of Sp_1. Thus, amalgamability of the above diagram allows us to deduce that $SN[FA]$ is conservative over Sp'_1; this is the essence of rules (inst2) and (inst3). (Of course, rule (inst3) may be regarded as subsuming rule (inst2); the latter has been included for sake of its better readability.) Note that this type of reasoning may substantially decrease the proof burden, since the non-trivial task of verifying amalgamability can in practically relevant cases be delegated to the above-mentioned algorithm.

There is an additional twist to the — otherwise similar — rule (union). At the level of signatures, a union Sp_1 **and** Sp_2 corresponds to a diagram

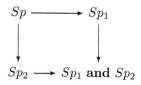

where Σ_i is the signature defined by Sp_i, $i = 1, 2$. By the phrase '$\Sigma_1 \cup \Sigma_2$ is amalgamable' we mean that this diagram is amalgamable. However, there is no standard way of forming intersections of *specifications* in CASL. The model class of such an intersection would consist of all reducts of models of either Sp_1 or Sp_2 (such a structuring operation for specifications has been suggested in [4]); this model class need not be specifiable in CASL. It suffices, however, to lift the above diagram of signatures to a diagram

$$Sp \longrightarrow Sp_1$$
$$\downarrow \qquad\qquad \downarrow$$
$$Sp_2 \longrightarrow Sp_1 \text{ and } Sp_2$$

of specifications where Sp_1 is conservative over Sp; these are precisely the requirements on Sp in the premises of the rule (union).

Example 1. Without amalgamability, instantiations of parametrized specifications may produce nasty surprises. Consider, e.g., the following specifications:

spec SP1 =
   sorts s, t
   op $f : s \to t$
   axiom $\forall x, y : s \bullet f(x) = f(y) \Rightarrow x = y$

spec SP2 =
  SP1
then
  sorts $s < t$
  op $g : s \to t$
  axiom $\forall x : s \bullet x = g(x)$

spec SP3[SP1]=
  sorts $s < t$
  axiom $\forall x : s \bullet x = f(x)$

The instantiation SP3[SP2] fails to be conservative over its argument SP2 (since the interpretations of f and g are forced to be equal), although the parametrized specification SP3[SP1] is conservative over its formal parameter SP1. Indeed, the signature diagram of the instantiation SP3[SP2] is the standard counterexample for amalgamation in CASL (cf. [22]).

A particular difficulty is attached to the **closed** construct, for which there is, as yet, no rule in the conservativity calculus. Dealing with this construct properly would require carrying around the local environment (cf. [8]), which appears, for the moment, to be an undue complication of the calculus. For practical cases, the equivalence rules of Figure 1 have proven to be sufficient.

Derived Rules. One frequently used derived rule is

$$\textbf{(cext)} \quad \frac{c(Sp_1), \; cons(Sp_1)(Sp_1 \textbf{ then } Sp_2)}{c(Sp_1 \textbf{ then } Sp_2)}$$

The derivation of this rule makes use of the rule (comp) of Figure 2. Moreover, using the existing conservativity rule for unions, one may derive

$$\textbf{(union2)} \quad \frac{cons(Sp_1)(Sp_1') \atop cons(Sp_2)(Sp_2') \atop \text{The signatures of } Sp_1' \text{ and } Sp_2' \text{ are disjoint}}{cons(Sp_1 \textbf{ and } Sp_2)(Sp_1' \textbf{ and } Sp_2')}$$

Finally, it is convenient to have the following rule, derived from rules (rp1) and (rp2) of Figure 2:

$$\textbf{(rpl)} \quad \frac{Sp_1 \simeq Sp_1' \atop Sp_2 \simeq Sp_2' \atop cons/def/implies(Sp_1)(Sp_2)}{cons/def/implies(Sp_1')(Sp_2')}$$

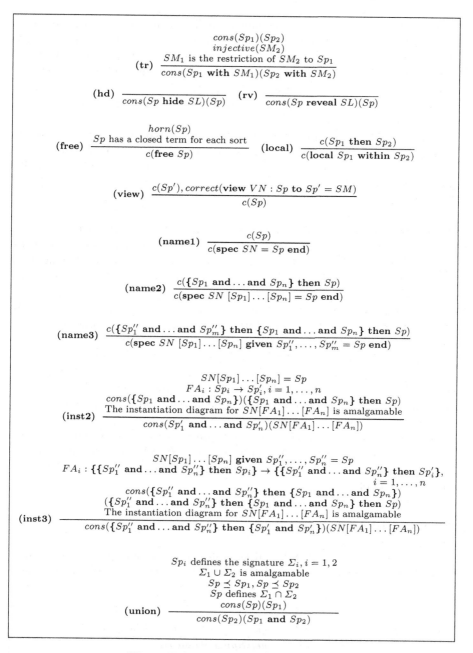

$$\text{(tr)}\ \frac{\begin{array}{c} cons(Sp_1)(Sp_2) \\ injective(SM_2) \\ SM_1 \text{ is the restriction of } SM_2 \text{ to } Sp_1 \end{array}}{cons(Sp_1 \text{ with } SM_1)(Sp_2 \text{ with } SM_2)}$$

$$\text{(hd)}\ \frac{}{cons(Sp \text{ hide } SL)(Sp)} \qquad \text{(rv)}\ \frac{}{cons(Sp \text{ reveal } SL)(Sp)}$$

$$\text{(free)}\ \frac{\begin{array}{c} horn(Sp) \\ Sp \text{ has a closed term for each sort} \end{array}}{c(\text{free } Sp)} \qquad \text{(local)}\ \frac{c(Sp_1 \text{ then } Sp_2)}{c(\text{local } Sp_1 \text{ within } Sp_2)}$$

$$\text{(view)}\ \frac{c(Sp'), correct(\text{view } VN : Sp \text{ to } Sp' = SM)}{c(Sp)}$$

$$\text{(name1)}\ \frac{c(Sp)}{c(\text{spec } SN = Sp \text{ end})}$$

$$\text{(name2)}\ \frac{c(\{Sp_1 \text{ and} \ldots \text{and } Sp_n\} \text{ then } Sp)}{c(\text{spec } SN [Sp_1] \ldots [Sp_n] = Sp \text{ end})}$$

$$\text{(name3)}\ \frac{c(\{Sp_1'' \text{ and} \ldots \text{and } Sp_m''\} \text{ then } \{Sp_1 \text{ and} \ldots \text{and } Sp_n\} \text{ then } Sp)}{c(\text{spec } SN [Sp_1] \ldots [Sp_n] \text{ given } Sp_1'', \ldots, Sp_m'' = Sp \text{ end})}$$

$$\text{(inst2)}\ \frac{\begin{array}{c} SN[Sp_1] \ldots [Sp_n] = Sp \\ FA_i : Sp_i \to Sp_i', i = 1, \ldots, n \\ cons(\{Sp_1 \text{ and} \ldots \text{and } Sp_n\})(\{Sp_1 \text{ and} \ldots \text{and } Sp_n\} \text{ then } Sp) \\ \text{The instantiation diagram for } SN[FA_1] \ldots [FA_n] \text{ is amalgamable} \end{array}}{cons(Sp_1' \text{ and} \ldots \text{and } Sp_n')(SN[FA_1] \ldots [FA_n])}$$

$$\text{(inst3)}\ \frac{\begin{array}{c} SN[Sp_1] \ldots [Sp_n] \text{ given } Sp_1'', \ldots, Sp_n'' = Sp \\ FA_i : \{\{Sp_1'' \text{ and} \ldots \text{and } Sp_n''\} \text{ then } Sp_i\} \to \{\{Sp_1'' \text{ and} \ldots \text{and } Sp_n''\} \text{ then } Sp_i'\}, \\ i = 1, \ldots, n \\ cons(\{Sp_1'' \text{ and} \ldots \text{and } Sp_n''\} \text{ then } \{Sp_1 \text{ and} \ldots \text{and } Sp_n\}) \\ (\{Sp_1'' \text{ and} \ldots \text{and } Sp_n''\} \text{ then } \{Sp_1 \text{ and} \ldots \text{and } Sp_n\} \text{ then } Sp) \\ \text{The instantiation diagram for } SN[FA_1] \ldots [FA_n] \text{ is amalgamable} \end{array}}{cons(\{Sp_1'' \text{ and} \ldots \text{and } Sp_n''\} \text{ then } \{Sp_1' \text{ and } Sp_n'\})(SN[FA_1] \ldots [FA_n])}$$

$$\text{(union)}\ \frac{\begin{array}{c} Sp_i \text{ defines the signature } \Sigma_i, i = 1, 2 \\ \Sigma_1 \cup \Sigma_2 \text{ is amalgamable} \\ Sp \preceq Sp_1, Sp \preceq Sp_2 \\ Sp \text{ defines } \Sigma_1 \cap \Sigma_2 \\ cons(Sp)(Sp_1) \end{array}}{cons(Sp_2)(Sp_1 \text{ and } Sp_2)}$$

Fig. 4. Rules for structuring constructs

Definitionality Rules. A somewhat rudimentary definitionality calculus is shown in Figure 5. Its rules cover two syntactical patterns: operations or predicates that are defined by a term or a formula, respectively, at the time of their declaration (basic items of this type are called operation and predicate definitions, respectively, in [8]) are, of course, definitional. The same goes for operations that come with axioms which amount to a recursive definition. The precise meaning of the phrase 'Sp_2 recursively defines new operations and predicates over Sp_1' is as follows: all signature items newly declared in Sp_2 are either operations or predicates on existing sorts; for all these signature items, Sp_2 contains recursive definitions; Sp_2 does not contain any further axioms; and Sp_2 does not contain hiding or renaming.

Recursion is only possible over constructor-generated sorts. Unless these sorts are explicitly freely generated, it has to be guaranteed that the definitions are independent of the choice of representative (the same goes for the recursive definitions of equality or observers as required in rules (gtd1) and (gtd2) of Figure 3).

Proof obligations. It is quite instructive to briefly summarize what kinds of proof obligations arise from applying the rules of the calculus to a CASL specification. Astonishingly, only a few rules depend on 'real' theorem proving. One such example is the rule (view): here one has to discharge the metapredicate $correct(\textbf{view } VN : Sp \textbf{ to } Sp' = SM)$. Other 'real' proof obligations are generated by the well-definedness condition for recursive definitions on non-freely generated datatypes (rules (gtd1), (gtd2), and (def2)) and the congruence condition for equality on such datatypes (rules (gtd1) and (gtd2)).

The 'next hardest' type of proof obligation arises in the context of instantiations of parametrized specifications. Here, the premises require checking signature diagrams for amalgamability. As discussed above, this is algorithmically hard in theory, but decidable by a polynomial algorithm in the practically relevant cases.

Most of the rules, however, have only premises of a purely syntactical nature: $horn(Sp)$, $newSorts(s)(Sp)$, and $injective(Sp, SM)$ are typical examples. Thus, our calculus indeed reduces the amount of theorem proving required in consistency proofs.

3   Examples

In this section, we present selected consistency proofs for specifications taken from the CASL library of Basic Datatypes [20]. Besides providing a standard

$$(\text{def1}) \quad \frac{BI \text{ is an operation or predicate definition}}{def(Sp)(Sp \textbf{ then } BI)}$$

$$(\text{def2}) \quad \frac{Sp_2 \text{ recursively defines new operations or predicates over } Sp_1}{def(Sp_1)(Sp_1 \textbf{ then } Sp_2)}$$

Fig. 5. Definitionality rules

library, this library also illustrates how to write and structure specifications in CASL. All important features of CASL basic and structured specifications are used. Furthermore, the Basic Datatypes are the largest collection of CASL specifications currently available.

We start with a simple proof within the horn fragment, an example of a consistency proof that does not involve any theorem proving but which is based solely on our calculus and simple syntactical analysis. As a more advanced example, we prove the consistency of a specification for sets. This proof illustrates how to propagate consistency along the CASL structuring constructs, how to handle instantiations, and how to deal with generated types. Finally, we present the consistency proof for a specification of characters — a quite large, if not overly difficult specification which can, using our calculus, be proven to be consistent nearly exclusively by syntactical analysis.

3.1   Consistency Proofs within the Horn Fragment

Figure 6 shows typical specifications from the library ALGEBRA_I in the collection of Basic Datatypes [20]. Nearly all of this library's specifications are in positive Horn form. Their consistency proofs are similar to the one shown in Figure 7 for the specification GROUP: first, we unfold the specification definition using the rule (name1). The next step is to decode the consistency predicate into a proposition on conservativity (decode). Now we add the empty specification to the second argument, justified by rule (sep2) of the extension calculus and the general replace rule (rp2). This finally allows us to apply the rule (horn). As the empty specification has no signature, the second premise of this rule holds trivially, and it remains to prove the metapredicate *horn*. Within the specification GROUP we obtain by skolemization that the axiom *inverse* is in positive Horn form. Since the specification MONOID is also in positive Horn form, we may now conclude that GROUP is consistent. Note that the proof of the metapredicate *horn* just involves syntax checks. Furthermore, note that the conservativity calculus does not flatten the original specification.

spec BINALG = **sort** *Elem*; **op** $\_*\_$: *Elem* * *Elem* → *Elem*

spec SEMIGROUP = BINALG
then op$\_*\_$: *Elem* * *Elem* → *Elem*, **assoc**

spec MONOID = SEMIGROUP
then ops e : *Elem*; $\_*\_$: *Elem* * *Elem* → *Elem*, **unit** e

spec GROUP = MONOID
then forall x : *Elem* • $\exists x'$: *Elem* • $x' * x = e$ %(*inverse*)%

Fig. 6. Specifications within the Horn fragement of CASL

$$
\cfrac{
\cfrac{
\cfrac{
\cfrac{
c(\textbf{spec } \text{GROUP} = \ldots)
}{
c(\text{MONOID } \textbf{then forall } x : Elem \bullet \exists x' : Elem \bullet x' * x = e)
} \text{ (name1)}
}{
\mathrm{cons}(\{\})(\text{MONOID } \textbf{then forall } x : Elem \bullet \exists x' : Elem \bullet x' * x = e)
} \text{ (decode)}
}{
\mathrm{cons}(\{\})(\{\} \textbf{ then } \text{MONOID } \textbf{then forall } x : Elem \bullet \exists x' : Elem \bullet x' * x = e)
} \text{ (sep2, rp2)}
}{
horn(\text{MONOID } \textbf{then forall } x : Elem \bullet \exists x' : Elem \bullet x' * x = e)
} \text{ (horn)}
$$

Fig. 7. Consistency proofs with **(horn)**

3.2   Consistency of the Specification SET

The specification of sets in [20] is split up into two parts: GENERATESET (c.f. Figure 8) is concerned with sort generation, while SET (c.f. Figure 9) provides the typical operations and predicates on sets.

Note that the following consistency proof of SET does not 'import' c(GENERATESET). What we need instead is a statement concerning conservativity, namely that in GENERATESET the specifiation body conservatively extends the parameter. This illustrates again that – although we are primarily interested in showing specifications to be consistent – our proofs are essentially about conservativity.

Again we start by unfolding the specification definition:

$$
\cfrac{
c(\textbf{spec } \text{SET } [\textbf{sort } Elem] \textbf{ given } \text{NAT} = \ldots)
}{
\begin{array}{c}
c(\text{NAT } \textbf{then sort } Elem \textbf{ then } \text{GENERATESET } [\textbf{sort } Elem] \\
\textbf{then } \%\textbf{def} \ldots \textbf{then } \%\textbf{def} \ldots \textbf{then } \%\textbf{def} \ldots \textbf{then } \%\textbf{implies} \ldots \textbf{end})
\end{array}
} \text{ (name3)}
$$

Now we show that the four extensions annotated with %**def** and %**implies**, resp., do not affect the consistency of the specification. To this end, our first step is to produce two goals by applying the derived rule (cext) to the last extension:

$$
\cfrac{
\begin{array}{c}
c(\text{NAT } \textbf{then sort } Elem \textbf{ then } \text{GENERATESET } [\textbf{sort } Elem] \\
\textbf{then } \%\textbf{def} \ldots \textbf{then } \%\textbf{def} \ldots \textbf{then } \%\textbf{def} \ldots \textbf{then } \%\textbf{implies} \ldots \textbf{end})
\end{array}
}{
\begin{array}{c}
c(\text{ NAT} \ldots (M - N) \cup (N - M)), \\
\mathrm{cons}(\text{ NAT} \ldots (M - N) \cup (N - M)) (\text{ NAT} \ldots \textbf{end})
\end{array}
} \text{ (cext)}
$$

The first goal states that the first part of the specification up to the extension operator is consistent; the second claims that the extension preserves models.

For the moment, we consider only the second goal and – guided by the annotation %**implies**– strengthen it using the rules (def) and (imp):

$$\frac{\text{cons}(\ \text{NAT} \dots (M - N) \cup (N - M)\ )\ (\ \text{NAT} \dots \mathbf{end})}{\text{def}(\ \text{NAT} \dots (M - N) \cup (N - M)\ )\ (\ \text{NAT} \dots \mathbf{end})} \quad (\text{def})$$

$$\frac{}{\text{implies}(\ \text{NAT} \dots (M - N) \cup (N - M)\ )\ (\ \text{NAT} \dots \mathbf{end})} \quad (\text{imp})$$

Following our methodology, we assume here that this last predicate is discharged by a suitable theorem prover, which has to show that $\_\cup\_$ and $\_\cap\_$ are associative, commutative, and idempotent.

Thus, for the consistency proof of SET it remains to discharge

$$\text{c}(\ \text{NAT} \dots (M - N) \cup (N - M)\ ).$$

To this end, we apply rule (cext) to the last extension of this subspecification of SET and strengthen the second of the resulting goals using (def):

$$\frac{\text{c}(\ \text{NAT} \dots (M - N) \cup (N - M)\ )}{\substack{\text{c}(\ \text{NAT} \dots \textit{when } x \epsilon M \textit{ else } (\sharp M) + 1\ ), \\ \text{def}(\ \text{NAT} \dots \textit{when } x \epsilon M \textit{ else } (\sharp M) + 1\ )\ (\ \text{NAT} \dots (M - N) \cup (N - M)\ )}} \quad (\text{cext, def})$$

The second goal is discharged by syntactical analysis: as the operator $\_symmDiff\_$ is introduced by an operation definition, we can apply rule (def1).

The remaining proof obligation is

$$\text{c}(\ \text{NAT} \dots \textit{when } x \epsilon M \textit{ else } (\sharp M) + 1\ )$$

The technique for discharging this goal is the same as in the previous two steps:

$$\frac{\text{c}(\ \text{NAT} \dots \textit{when } x \epsilon M \textit{ else } (\sharp M) + 1\ )}{\substack{\text{c}(\ \text{NAT} \dots\ = M + x\ ), \\ \text{def}(\ \text{NAT} \dots\ = M + x\ )\ (\ \text{NAT} \dots \textit{when } x \epsilon M \textit{ else } (\sharp M) + 1\ )}} \quad (\text{cext, def})$$

This time, the resulting proof obligation concerning definitionality is more interesting: using (def2) to show that the operators $\_\cup\_$, $\_\cap\_$, $\_-\_$, $\_-\_$, and $\sharp\_$ are definitional over the previous specification involves — besides syntactical analysis — some theorem proving: e.g., given the axioms for $\_\cup\_$

- $M \cup \{\} = M$
- $M \cup (N + x) = (M \cup N) + x$,

one has to show that

- $\{\} = N + x \Rightarrow M = (M \cup N) + x$
- $N + x = O + y \Rightarrow (M \cup N) + x = (M \cup O) + y$

Next, we have to establish

$$\text{c}(\ \text{NAT} \dots\ = M + x\ ).$$

Here we again apply (cext) to the last extension, strengthen the second resulting goal by (def), and discharge it by syntactical analysis using rule (def1).

Now the consistency problem for the original specification is reduced to c(NAT **then sort** *Elem* **then** GENERATESET [**sort** *Elem*]).

By (cext), this again splits up into a consistency and a conservativity problem:

$$\frac{\text{c(NAT \textbf{then sort} \textit{Elem} \textbf{then} GENERATESET [\textbf{sort} \textit{Elem}])}}{\begin{array}{c}\text{c(NAT \textbf{then sort} \textit{Elem}),}\\\text{cons(NAT \textbf{then sort} \textit{Elem})}\\\text{(NAT \textbf{then sort} \textit{Elem} \textbf{then} GENERATESET [\textbf{sort} \textit{Elem}])}\end{array}} \quad \text{(cext)}$$

Assuming NAT to be consistent, we obtain from the rules (cext) and (horn) the consistency of NAT **then sort** *Elem*.

Thus it remains to deal with the second goal:

cons(NAT **then sort** *Elem*)
(NAT **then sort** *Elem* **then** GENERATESET [**sort** *Elem*])

Using the extension calculus, we obtain the equivalence

NAT **then sort** *Elem* **then** GENERATESET [**sort** *Elem*] $\simeq$
NAT **then** GENERATESET [**sort** *Elem*]

which allows us to modify our goal by rule (rp2) into

cons(NAT **then sort** *Elem*) (NAT **then** GENERATESET [**sort** *Elem*])

The combination of rule (then-and) of the extension calculus and the derived rule (rpl) transforms this into

cons(NAT **and sort** *Elem*) (NAT **and** GENERATESET [**sort** *Elem*])

As the signatures of NAT and GENERATESET [**sort** *Elem*] are disjoint, we can apply the derived rule (union2) in order to obtain:

$$\frac{\text{cons(NAT \textbf{and sort} \textit{Elem}) (NAT \textbf{and} GENERATESET [\textbf{sort} \textit{Elem}])}}{\text{cons(NAT)(NAT), cons(\textbf{sort} \textit{Elem})(GENERATESET [\textbf{sort} \textit{Elem}])}} \quad \text{(union2)}$$

cons(NAT)(NAT) holds if *def* (NAT)(NAT) is true (def), which can be justified by *implies*(NAT)(NAT) (imp), which holds thanks to (triv).

Thus our remaining problem is:

cons(**sort** *Elem*)(GENERATESET[**sort** *Elem*]).

Rule (inst2) deals with the instantiation of GENERATESET:

$$\frac{\text{cons(\textbf{sort} \textit{Elem})(GENERATESET[\textbf{sort} \textit{Elem}])}}{\begin{array}{c}\text{amalg. inst. diagr. for GENERATESET[\textbf{sort} \textit{Elem}],}\\\text{cons(\textbf{sort} \textit{Elem})}\\\text{(\textbf{sort} \textit{Elem} \textbf{then generated type} \textit{FinSet}[\textit{Elem}] ::= \ldots)}\end{array}} \quad \text{(inst2)}$$

Obviously the instantiation diagram is amalgamable. Furthermore, we have to make sure that the specification body of GENERATESET is conservative over its parameter. (From a methodological point of view, in the consistency proof for SET, one would import the following proof as an already established result on GENERATESET.) This is achieved by applying rule (gtd2):

$$\frac{\text{cons}(\textbf{sort } Elem)}{(\textbf{sort } Elem \textbf{ then generated type } FinSet[Elem] ::= \dots)} \quad \text{(gtd2)}$$

$$newSorts(FinSet[Elem])(\textbf{sort } Elem)$$
$$Sp_2 \text{ recursively defines an observer } \_\epsilon\_$$
$$Sp_3 \text{ defines equality on new sort by } \_\epsilon\_$$

where Sp_2 is
pred $\_\epsilon\_$: $Elem * FinSet[Elem]$;
forall $x, y : Elem; M, N : FinSet[Elem]$
- *not* $x\epsilon\{\}$
- $x\epsilon(M + y) \Leftrightarrow x = y \vee x\epsilon M$

and Sp_3 denotes
forall $M, N : FinSet[Elem]$
- $M = N \Leftrightarrow (\forall x : Elem \bullet x\epsilon M \Leftrightarrow x\epsilon N)$

Obviously, $FinSet[Elem]$ is a new sort over $\{\textbf{sort } Elem\}$, and the predicate $\_\epsilon\_$ is a recursively defined observer. Concerning the new equality on $FinSet[Elem]$, the proof obligation
forall $M, N : FinSet[Elem]$
- $(\forall x : Elem \bullet x\epsilon M \Leftrightarrow x\epsilon N) \Rightarrow (\forall x, y : Elem \bullet x\epsilon(M + y) \Leftrightarrow x\epsilon(N + y))$
has to be discharged. As this is a consequence of the definition of $\_\epsilon\_$, the consistency proof for SET is finished.

```
spec GENERATESET [sort Elem] =
    generated type FinSet[Elem] ::= {} | _ + _(FinSet[Elem]; Elem);
then %def
    pred __ε__ : Elem * FinSet[Elem];
    forall  x, y : Elem; M, N : FinSet[Elem]
    •  ¬ xε{}
    •  xε(M + y) ⇔ x = y ∨ xεM
then
    forall  M, N : FinSet[Elem]
    •  M = N ⇔ (∀x : Elem • xεM ⇔ xεN)
end
```

Fig. 8. The specification GENERATESET

spec SET [**sort** *Elem*] **given** NAT = GENERATESET [**sort** *Elem*]
then %def
   preds *isNonEmpty* $(M : FinSet[Elem]) \Leftrightarrow \neg\, M = \{\}$
         $\_\_ \subseteq \_\_ \, (M, N : FinSet[Elem]) \Leftrightarrow \forall x : Elem \bullet x\epsilon M \Rightarrow x\epsilon N$
   ops $\{\_\_\} \, (x : Elem) : FinSet[Elem] = \{\} + x$
         $\_\_ + \_\_ \, (x : Elem; M : FinSet[Elem]) : FinSet[Elem] = M + x$
then %def
   ops $\_\_ \cup \_\_, \_\_ \cap \_\_, \_\_ - \_\_ : \quad FinSet[Elem] * FinSet[Elem] \rightarrow FinSet[Elem];$
         $\_\_ - \_\_ : \qquad\qquad\qquad FinSet[Elem] * Elem \rightarrow FinSet[Elem];$
         $\natural\_\_ : \qquad\qquad\qquad\quad FinSet[Elem] \rightarrow Nat;$
   forall $x, y : Elem; M, N : FinSet[Elem]$

- $M \cup \{\} = M$
- $M \cup (N + x) = (M \cup N) + x$

- $M \cap \{\} = \{\}$
- $M \cap (N + x) = M \cap N \; when \; \neg x\epsilon M \; else \; (M \cap N) + x$

- $M - \{\} = M$
- $M - (N + x) = (M - N) - x$

- $\{\} - x = \{\}$
- $(M + x) - y = M - y \; when \; x = y \; else \; (M - y) + x$

- $\natural\{\} = 0$
- $\natural(M + x) = \natural M \; when \; x\epsilon M \; else \; (\natural M) + 1$

then %def
   op $\_\_ symmDiff \_\_ \, (M, N : FinSet[Elem]) : FinSet[Elem] = (M - N) \cup (N - M)$
then %implies
   ops $\_\_ \cup \_\_, \_\_ \cap \_\_ : FinSet[Elem] * FinSet[Elem] \rightarrow FinSet[Elem],$
                 assoc, comm, idem;

end

Fig. 9. The specification SET

3.3   Consistency of CHAR

The specification CHAR of [20] (c.f. Figure 10) consists of about 1000 lines of
CASL text, most of them operation or predicate definitions. This allows us to
reduce the consistency problem for this specification by systematic use of the
rules (cext), (def), and (def1) to the consistency of
NAT
then sort $Byte = \{n : Nat \bullet n \le 255\}$
then free type $Char ::= chr(ord : Byte)$
Again assuming NAT to be consistent, we can establish this using the rules
(cext), (sub), and (free): obviously, $0 \le 255$, so that there is at least one element
in the carrier of **sort** *Byte*. The premises of (free) are discharged by syntactical
analysis: *Char* is a new sort, and *chr*(0) is a closed term of sort *Char*.
    Thus, theorem proving is necessary in just one step of our consistency proof,
the others require only 'pattern matching', i.e. finding suitable rules to be applied

to the specification text, and syntactical analysis (to discharge the premises of these rules). Dealing with a flat specification with about 800 different axioms just by theorem proving would be nearly impossible. One might argue that an 'intelligent' theorem prover would sort out the same 'core specification' by some kind of syntactical analysis as well — but this just illustrates the necessity of the strategy pursued here: economical proof organization by replacing parts of the necessary theorem proving by the application of 'meta rules'.

spec CHAR = NAT
then sort $Byte = \{n : Nat \bullet n \leq 255\}$
then free type $Char ::= chr(ord : Byte)$
then %def
   ops $'\backslash 000' : Char = chr(0); \ldots; '\backslash 255' : Char = chr(255);$
then %def
   ops $'\,' : Char = '\backslash 032'; \ldots; 'ÿ' : Char = '\backslash 255';$
then %def
   preds $isLetter(c : Char) \Leftrightarrow \quad ((ord('A') \leq ord(c) \wedge ord(c) \leq ord('Z')) \vee$
   $(ord('a') \leq ord(c) \wedge ord(c) \leq ord('z')));$
   $isDigit(c : Char) \Leftrightarrow \quad ord('0') \leq ord(c) \wedge ord(c) \leq ord('9');$
   $isPrintable(c : Char) \Leftrightarrow ((ord('\,') \leq ord(c) \wedge ord(c) \leq ord('\sim')) \vee$
   $(ord('\,') \leq ord(c) \wedge ord(c) \leq ord('ÿ')))$
then %def
   ops $'\backslash o000' : Char = '\backslash 000'; \ldots; '\backslash o377' : Char = '\backslash 255';$
   $'\backslash x00' : Char = '\backslash 000'; \ldots; '\backslash xFF' : Char = '\backslash 255';$
   $NUL : Char = '\backslash 000'; \quad \ldots$
then %def
   ops $NL : Char = LF; \ldots$
then %def
   ops $'\backslash n' : Char = NL; \ldots; '\backslash ?' : Char = '?';$
end

Fig. 10. The specification CHAR

4   Conclusions and Future Work

We have developed a calculus for proving conservativity of specification extensions in CASL, and we have used this calculus to establish the consistency of substantial portions of the CASL Basic Datatypes [20] (which have a good claim to being the largest CASL specification presently in existence). Several examples of such proofs have been presented and discussed. These examples have illustrated how the calculus facilitates the exploitation of the specification structure for the structuring of proofs. In fact, the proofs were 'automatically' directed to the few critical items that required proper theorem proving; by contrast, most of the proof obligations produced along the way were of an entirely syntactical nature.

There is no claim that the calculus as presented here is complete (since the underlying logic of CASL is undecidable, absolute completeness cannot be expected anyway); however, the case study that has been conducted on the library of Basic Datatypes, which makes use of all important features of CASL basic and structured specifications, has shown that the calculus is able to deal with quite substantial specifications. (More precisely, five sublibraries containing more than 80 specifications of an overall length of roughly 2500 lines have been checked for consistency.) Summing up, we believe that our calculus is able to deal with consistency problems that arise in the context of software engineering projects.

Although no explicit mention was made of the concept of institution [13], the parts of the calculus that concern the CASL structuring mechanisms are, just as these mechanisms themselves [9], in fact institution independent. This makes the calculus easily adaptable with respect to, e.g., extensions of the underlying logic.

Future directions of research include the development of the second method of 'verifying' specifications mentioned in the introduction, namely, the testing of intended consequences, as well as the implementation of tool support for the conservativity calculus; this will possibly involve use of the development graph [3,18]. It is expected that a forthcoming tool will allow semiautomatic consistency proofs, with the syntactical premises discharged automatically — via syntactic analysis or the more complex algorithms discussed in [16] — and the 'hard' ones output as formal proof obligations.

References

[1] Wolfgang Ahrendt, *A basis for model computation in free data types*, Proceedings of the CADE-17 Workshop on Model Computation, 2000.

[2] Egidio Astesiano, Michel Bidoit, Hélène Kirchner, Bernd Krieg-Brückner, Peter D. Mosses, Donald Sannella, and Andrzej Tarlecki, CASL: *The common algebraic specification language*, Theoretical Computer Science (to appear).

[3] S. Autexier, D. Hutter, H. Mantel, and A. Schairer, *Towards an evolutionary formal software development using* CASL, Recent Trends in Algebraic Development Techniques, LNCS, vol. 1827, Springer, 1999, pp. 73–88.

[4] H. Baumeister, *Relations between abstract datatypes modeled as abstract datatypes*, Ph.D. thesis, Universität des Saarlandes, 1998.

[5] M. Bidoit, M. V. Cengarle, and R. Hennicker, *Proof systems for structured specifications and their refinements*, Algebraic Foundations of Systems Specification (E. Astesiano et al., eds.), Springer, 1999, pp. 385–433.

[6] M. Cerioli, A. Haxthausen, B. Krieg-Brückner, and T. Mossakowski, *Permissive subsorted partial logic in* CASL, Algebraic Methodology and Software Technology, LNCS, vol. 1349, Springer, 1997, pp. 91–107.

[7] CoFI, *The Common Framework Initiative for algebraic specification and development, electronic archives*, notes and documents accessible from http://www.brics.dk/Projects/CoFI.

[8] CoFI Language Design Task Group, CASL – *The* CoFI *Algebraic Specification Language – Summary, version 1.0.1*, Documents/CASLSummary, in [7], March 2001.

[9] CoFI Semantics Task Group, CASL – *The* CoFI *Algebraic Specification Language – Semantics*, Note S-9 (version 0.96), in [7], July 1999.

[10] R. Diaconescu, J. Goguen, and P. Stefaneas, *Logical support for modularisation*, Logical Environments, Cambridge, 1993, pp. 83–130.

[11] J. Farrés-Casals, *Proving correctness of constructor implementations*, Mathematical Foundations of Computer Science, LNCS, vol. 379, Springer, 1989, pp. 225–236.

[12] J.-Y. Girard, *Locus solum*, Math. Struct. Comput. Sci., To appear.

[13] J. Goguen and R. Burstall, *Institutions: Abstract model theory for specification and programming*, J. ACM **39** (1992), 95–146.

[14] M. J. C. Gordon and T. M. Melham, *Introduction to HOL: A theorem proving environment for higher order logics*, Cambridge, 1993.

[15] R. Hennicker and M. Wirsing, *Proof systems for structured algebraic specifications: An overview*, Fundamentals of Computation Theory, LNCS, vol. 1279, Springer, 1997, pp. 19–37.

[16] B. Klin, P. Hoffman, A. Tarlecki, L. Schröder, and T. Mossakowski, *Checking amalgamability conditions for* CASL *architectural specifications*, Mathematical Foundations of Computer Science, LNCS, Springer, 2001, to appear.

[17] T. F. Melham, *A package for inductive relation definitions in* HOL, International Workshop on the HOL Theorem Proving System and its Applications, IEEE Computer Society Press, 1992, pp. 350–357.

[18] T. Mossakowski, S. Autexier, and D. Hutter, *Extending development graphs with hiding*, Fundamental Aspects of Software Engineering, LNCS, vol. 2029, Springer, 2001, pp. 269–283.

[19] W. Reif, G. Schellhorn, and A. Thums, *Flaw detection in formal specifications*, International Joint Conference on Automated Reasoning, LNCS, vol. 2083, Springer, 2001, pp. 642–657.

[20] Markus Roggenbach, Till Mossakowski, and Lutz Schröder, *Basic datatypes in* CASL, Note L-12 in [7], current version 0.7 available at http://www.informatik.uni-bremen.de/cofi/CASL/lib/basic, March 2001.

[21] Markus Roggenbach and Lutz Schröder, *Towards trustworthy specifications II: Testing by proof*, work in progress.

[22] L. Schröder, T. Mossakowski, and A. Tarlecki, *Amalgamation in* CASL *via enriched signatures*, International Colloquium on Automata, Languages and Programming, LNCS, vol. 2076, Springer, 2001, pp. 993–1004.

[23] J. R. Shoenfield, *Mathematical logic*, Addison-Wesley, 1967.

The Common Framework Initiative
for Algebraic Specification
and Development of Software: Recent Progress

Donald Sannella

Laboratory for Foundations of Computer Science
University of Edinburgh, UK

Abstract. The Common Framework Initiative (CoFI) is an open international collaboration which aims to provide a common framework for algebraic specification and development of software. The central element of the Common Framework is a specification language called CASL for formal specification of functional requirements and modular software design which subsumes many previous algebraic specification languages. This paper is a brief summary of progress on CoFI during the period 1998–2001, when CoFI received funding from the European Commission as a Working Group under the Esprit programme.

1 Introduction

Algebraic specification is one of the most extensively-developed approaches in the formal methods area. The most fundamental assumption underlying algebraic specification is that programs are modelled as *many-sorted algebras* consisting of a collection of sets of data values together with functions over those sets. This level of abstraction is commensurate with the view that the correctness of the input/output behaviour of a program takes precedence over all its other properties. Another common element is that specifications of programs consist mainly of logical *axioms*, usually in a logical system in which equality has a prominent role, describing the properties that the functions are required to satisfy. A wide variety of different approaches to algebraic specification take these two principles as their starting point.

The past 25 years has seen a great deal of research on the theory and practice of algebraic specification. Overviews of this material include [5,16,23,54,88,89,95]. Developments on the foundational side have been balanced by work on applications, but despite a number of success stories, industrial adoption has so far been limited. The proliferation of *algebraic specification languages* is seen as a significant obstacle to the dissemination and use of these techniques. Despite extensive past collaboration between the main research groups involved and a high degree of agreement concerning the basic concepts, the field has given the appearance of being extremely fragmented, with no *de facto* standard specification language, let alone an international standard. Moreover, although many tools supporting the use of algebraic techniques have been developed in the academic

M. Cerioli and G. Reggio (Eds.): WADT/CoFI 2001, LNCS 2267, pp. 328–343, 2002.

community, none of them has gained wide acceptance, at least partly because of their isolated usability: each tool uses a different specification language.

Since late 1995, work has been underway in an attempt to remedy this situation. The *Common Framework Initiative* (abbreviated CoFI) is an open international collaboration which aims to provide a common framework for algebraic specification and development of software. The Common Framework is intended to be attractive to researchers in the field as a common basis for their work, and to ultimately become attractive for use in industry. The central element of the Common Framework is a specification language called CASL (the Common Algebraic Specification Language), intended for formal specification of functional requirements and modular software design and subsuming many previous specification languages. Development of prototyping and verification tools for CASL leads to them being interoperable, i.e. capable of being used in combination rather than in isolation.

CoFI began as an unfunded collaboration, but modest funding was obtained from the European Commission under the Esprit programme during the period October 1998 – April 2001 to support CoFI meetings and CoFI-related travel within Europe (project 29432, CoFI WG). Before this point, most effort in CoFI had concentrated on the design of CASL, which was complete in almost all respects in 1998. Building on this progress, the goals of the CoFI Working Group were: to coordinate the completion of and disseminate the Common Framework; to demonstrate its practical applicability in industrial contexts; and to establish the infrastructure needed for future European collaborative research in algebraic techniques. Activity is organized under six Task Groups, as follows:

- Language Design (coordinator: Bernd Krieg-Brückner until April 2001, since then Peter Mosses)
- Semantics (coordinator: Andrzej Tarlecki)
- Methodology (coordinator: Michel Bidoit)
- Tools (coordinator: Hélène Kirchner until April 2001, since then Bernd Krieg-Brückner and Till Mossakowski)
- Reactive Systems (coordinator: Egidio Astesiano, joined by Heinrich Hussmann from October 1999)
- External Relations (coordinator: Peter Mosses)

Overall coordination of CoFI was by Peter Mosses until August 1998, since then by Don Sannella.

This paper is a summary of progress on CoFI since 1998, with pointers to publications and other material produced by CoFI members during this period. Previous summaries of CoFI are [66,71,86]. Presentations of CASL are [2,17,31,68]. The CASL tutorial [17] is especially recommended for newcomers, with more details, rationale and examples available from [2].

2   Language Design

In October 1998, the Language Design of CASL version 1.0 was published [29]. Subsequent work on polishing the design concerned syntactic extensions for literals, syntactic annotations for parsing and precedence of (mixfix) operator and predicate symbols, semantic annotations for proof obligations arising from various kinds of conservative extensions [81], details of concrete syntax affecting the form and position of comments and annotations, and details concerning the semantics of architectural specifications. A modification to version 1.0 incorporating many of these minor adjustments was released in July 1999 [30] and further adjustments were incorporated in CASL version 1.0.1 [31], released in March 2001. No further revisions are planned. The design of CASL v1.0.1 has been formally approved by IFIP WG1.3, see below under External Relations. The Semantics, Tools, and Methodology Task Groups, as well as the IFIP WG1.3 reviewers, who also made detailed comments on a previous version of CASL, all provided essential feedback regarding language design proposals. The rationale for aspects of the language design has been presented in [2,18,24,25].

Another major effort has been the development of a library of Basic Datatypes for CASL which has also produced further methodological insight. Apart from a specification of natural, integer and rational numbers and their standard algebraic properties, the usual types such as sets, collections and lists have been defined. Moreover, a first attempt has been made at a first-order specification of real numbers, which appears to be the first of its kind. These types have also been related to approximate numeric types as used in computers. The library of basic datatypes has been revised several times following feedback from the members of the Language Design and Methodology Task Groups. The final result is documented in [82,84].

Various sublanguages of CASL — total, many-sorted, equational — have been defined, often corresponding closely to embeddings of the specification languages of other frameworks into CASL [55,56,59,63]. The relation of CASL to other specification languages such as ACT ONE, ASF, HEP-theories, LSL, OBJ3 and (functional) CafeOBJ has been clarified [63,67,69]. The logic underlying CASL has been translated to first-order logic (or second-order logic, when sort generation constraints are considered). This allows the re-use of first-order and higher-order theorem provers for CASL [63].

A procedure for the stepwise approval of sublanguages and language extensions has been established. The proposals for sublanguages and a higher-order extension [41] have received preliminary approval such that the definition of a proper semantics is now the next step. Work on parametrised and polymorphic (higher-order) types is ongoing [90] and there is a proposal for an extension of CASL with a mechanism for late binding [1].

Work on object-oriented and reactive system extensions to CASL has been carried out by the Reactive Systems Task Group, see below.

3   Semantics

The activities of the Semantics Task Group have centered around the development of a complete, formal mathematical semantics for CASL [32]. This development was accompanied by some work on the borderline between semantic and design issues for CASL, devoted to the careful study of some concepts introduced in the formalism: partiality has been presented in detail in [25], architectural specifications in [18], and the details of the category of signatures in [57,93]. The early version of the semantics allowed us to identify some problems with the language design, which led to CASL version 1.0.1, as described above. A serious revision of the entire semantics is about to be completed [11]. Perhaps the most visible changes have involved the semantics of CASL architectural specifications, as presented in [49,91,92], due to problems pointed out for instance in [42,48].

One key issue in the development of the final version of the semantics was to make it fully institution-independent, as studied in [61] for structured specifications and extended to architectural specifications in [92]. See [65] for a study of the impact of institution independence on development of tools for CASL.

With the semantics essentially ready, work has started on proof systems and proof support for CASL. This includes a study of foundations for translation of CASL into some well-supported logical frameworks [20] and verification of architectural specifications [43]; some key aspects of the logical system underlying CASL have been thoroughly investigated as well [19] cf. [36,37]. An important stream of work addressed the issues of a careful, semantics-based comparison of CASL with other specification formalisms [8,20,59,63].

We have also discussed some possible extensions of CASL, notably by higher-order operations, where a complete formal proposal is under development on the basis of [41], and behavioural equivalence of higher-order models (represented by pre-logical and lax logical relations) was investigated [44,45,72]. Further work has taken place on the borderline between semantics and methodology [10].

4   Methodology

The aim of the Methodology Task Group is to enrich the formalism designed under the Common Framework Initiative with ideas on the methodology of system specification and development that we would like to support, encourage and propagate. It is expected that these methodological views will influence the use of the CASL language and help its dissemination, and they have already influenced the overall design of the CASL language.

The activities of the Methodology Task Group have been organized around the following issues:

User manual and tutorials. A CASL User Manual is currently under preparation. It is intended to reuse the corresponding material for the preparation of a CASL Electronic Tutorial that will be made available on the Web. The CASL User Manual will provide guidelines for writing specifications of individual system modules and for using the various features of the CASL language. Preliminary draft versions have been used for CASL tutorials in Berlin during ETAPS 2000 and in Genova during ETAPS 2001 [17]. Another, independent, CASL tutorial has been presented at a NASA formal methods workshop in June 2000 [79]. Related to this are studies on how to write consistent specifications [60,83], on the proper use of CASL features [26], and on the addition of annotations to CASL specifications [81]. Further studies discuss the use of CASL during the software development specification and design phases [14,27].

Case studies. A Web site with case studies illustrating the use and the benefits of CASL in various projects has been set up [9] to which CoFI participants have contributed. Extensive work has been done on basic datatypes for CASL in collaboration with the Language Design Task Group, see above for details. In addition, there have been studies on the use of CASL for computational and geometric modeling [40,50,51,52,53] and multimedia presentations [13]. CASL has also been proposed for use as a general meta-notation in semantic descriptions [70].

Formal software development based on algebraic specifications. Architectural specifications, one of the most novel concepts of CASL, have been further studied [18]. Several studies have investigated refinement in various frameworks [44,45,87]. Further work related to CASL and the software development process are [7,15,38,39].

5   Tools

CASL offers a flexible syntax including *mixfix* notation, which requires advanced parsing technology. Several CASL tools described in [22] have been built using the algebraic specification formalism ASF+SDF and the ASF+SDF Meta-Environment. CASL supports user-defined syntax which is non-trivial to process, and ASF+SDF offers a powerful parsing technology called Generalized LR. Its interactive development environment facilitates rapid prototyping complemented by early detection and correction of errors. A number of core technologies developed for the ASF+SDF Meta-Environment have been reused in the context of CASL.

Interoperability of CASL and existing tools is a major goal of the Tools Task Group. The first step has been to propose an exchange format that can be accepted as input and produced as output by every tool. The starting idea was to adopt basically abstract syntax trees with annotations providing specific information to communicate with various tools (parsers, rewrite engines, proof tools, etc.). An instantiation of a generic format developed for the representation of ASF+SDF specifications and terms provides a CASL-specific exchange format. In [21], the abstract data type of Annotated Terms (*ATerms*) is defined and their design, implementation and application are discussed. A comprehensive procedural interface enables creation and manipulation of ATerms in C or Java. The ATerm implementation is based on maximal subterm sharing and automatic garbage collection. A binary exchange format for the concise representation of ATerms (with sharing preserved) allows the fast exchange of ATerms between applications. Work is also in progress to provide XML as an external exchange format, with translations back and forth between XML and ATerms. Based on these low-level formats, high-level formats such as CasFix [22] (for abstract syntax trees of CASL specifications), CasEnv (for global environments containing signature information etc.) and FCasEnv (a flattened version of CasEnv, for use with tools that do not support structured specifications) have been developed. Formats for storing proofs and developments will follow.

One main achievement has been the integration of several tools in the CASL Tool Set CATS [62]. This combines a parser, a static checker, a LATEX pretty printer, facilities for printing signatures of specifications and structure graphs of CASL specifications, with links to various verification and development systems. To experiment with CASL specifications, the CATS system provides different user interfaces: a Web-based interface, and a compact stand-alone version (with both a command-line and a window interface). A repository with successfully and unsuccessfully parsed specifications is under development.

Existing rewrite engines provide a good basis for prototyping (parts of) CASL specifications. The problem of executing some CASL specifications using the rewrite engine provided by the ELAN system, which implements rewriting in a very efficient way, is addressed in [47]. The class of CASL specifications that can be executed are those having equational axioms (including possibly associative-commutative operators), that are oriented as conditional rewrite rules. The equality predicate is used to express the congruence on terms whilst the equivalence connective allows defining the congruence on expressions built over predicates. Subsorting and partiality features are not considered for now, but basic and structured CASL specifications are supported. The mapping from CASL to ELAN is performed by translating the CASL abstract syntax into the abstract syntax developed for ELAN. The current implementation needs the CASL Tool Set to parse a CASL specification and to generate the "flattened" FCasEnv format. By using the translation tool and then the ELAN compiler, an executable program is produced which computes normal forms with respect to a given CASL specification.

The standalone version of CATS also contains an encoding into several other logics. The encoding transforms a CASL specification into second-order logic step by step. First, partiality is encoded via error elements inhabiting a supersort; second, subsorting is encoded via injections; and third, sort generation constraints are expressed via second-order induction axioms. It is possible to stop after the first or second step if one wants to use a tool supporting subsorting or sort generation constraints directly. For details, see [63], where alternative encodings are also described. In this way, CATS allows CASL to interface with a large number of first- and higher-order theorem provers.

The HOL-CASL system, being built on top of CATS, uses the encoding of CASL into second-order logic to connect CASL to the Isabelle theorem prover and the generic graphical user interface IsaWin. This approach to encoding CASL in proof systems such as Isabelle allows verification and program transformation [63].

Various verification tools have already been developed for algebraic specifications, and can be reused for specific subsets of CASL: equational, conditional, full first-order logic with total functions, total functions with subsorts, partial functions, etc. The INKA system provides an integrated specification and theorem proving environment for a sub-language of CASL that excludes partial functions (with the encoding provided by CATS, it will also be useable with full CASL). CATS has been connected to the development graph management component of the INKA theorem proving system. Structured CASL specifications in the CasEnv format are translated to development graphs [7,64]. The development graph supports the management of theories and proof obligations that arise from CASL specifications in a theorem prover-independent way. Moreover, it provides an efficient way of managing change, allowing re-use of those parts of proofs that are not affected by the change of a specification.

All tools developed in the Tools Task Group are made available to the community, after validation by the group. A Web page for tools describing on-going work, giving access to available tools, and giving guidelines on how to propose a new tool, is available at http://www.tzi.de/cofi/Tools.

6 Reactive Systems

The aim of the Reactive Systems Task Group is to develop an extension of the Common Framework to deal with reactive, concurrent and parallel systems; object-oriented techniques for dealing with reactiveness have also been considered. The specification framework deals with all phases from requirement to design, including the intermediate steps.

From the beginning, the goal has been to have an extension which is: based on state-of-the art techniques; compatible and integrated with the CASL language proposal; mathematically rigorous; able to deal with a wide range of significant systems; sufficiently friendly for practical use by a wide community; and guided and complemented by considerations concerning methodology and tools.

The technical work of the Task Group has proceeded in two tracks:

Track 1: Autonomous Extensions of CASL towards Reactive Systems. These extensions are individual contributions aligned to the Common Framework. The following proposals have been developed within the Task Group or in close connection to it.

CASL-Charts: A formalism integrating CASL and state charts has been developed for giving discrete models of the functional and dynamic behaviour of reactive systems [77,78].

State-based CASL extension: An extension of CASL to deal with internal states has been developed [12].

Integration of CASL with process specification languages: Proposals have been made to bring CASL together with most of the known approaches to formal specification of communicating processes: CASL/CCS [85]; CASL/CSP; CASL/Petri Nets.

CASL-LTL/Design: A method for the design specification of concurrent systems has been developed, based on a formalism of (structured) conditional specifications defining processes in terms of labelled transition systems [73].

CoFI-LTL/Requirements: This is essentially a many-sorted first-order temporal logic [3], cf. [34].

JTN (Java Targeted Notation): A visual formal notation was developed for the specification of reactive, parallel, concurrent systems. It is essentially a restricted subcase of CASLLTL/Design, which is automatically translatable into Java [33].

Track 2: Coordinated Effort. In a joint effort of the members of the Task Group, work centered around the OMG standard language UML, the Unified Modelling Language. The motivation for this decision was that UML is an industry standards for specification and design of complex systems, including reactive systems, so that a link to UML will definitely enhance the practical accessibility of CASL, and also that UML already contains a number of graphical formalisms dealing with concurrency and reactiveness. It was observed that the UML standard has reached a level of maturity which is close to formal methods, but is missing a solid semantic foundation [46,94].

The basic idea of the joint work is to adopt CASL as a language for annotating UML static and dynamic diagrams, enhancing and possibly replacing the Object Constraint Language of UML. In order to achieve an integration between CASL and UML, a systematic translation between UML diagrams and CASL specifications was defined and documented. Class diagrams, representing static aspects of the system, are translated into standard CASL. Statechart diagrams are dealt with by a translation into CASLLTL [3,34,74]. First attempts were also made to translate sequence diagrams into CASLLTL. Finally, a multiview approach was worked out for the semantics of UML, integrating static and dynamic aspects. The core idea of this approach is to use CASL as a metalanguage which helps to express the semantics of various diagrams in a uniform and mathematically rigorous language [6,75]. This way, an important contribution also to the better formal underpinning of the semantics of UML diagrams was produced [76], cf. [4].

7   External Relations

The main tasks concerning external relations were establishing the relationship between CASL and previous frameworks, developing a tutorial online presentation of CASL, and liaison with IFIP Working Group 1.3.

Relationship between CASL and previous frameworks. At the level of specification-in-the-small, ACT ONE, ASF+SDF, (functional) CafeOBJ, LSL and OBJ3 [59,63,69] have been examined. The outcome was that apart from the rather pathological case of algebras with empty carriers, all of these languages can be translated to sublanguages of CASL. Moreover, the translation is always straightforward, with the exception of OBJ3's retracts (their translation to CASL had previously been studied [59]). For some languages, the corresponding sublanguage of CASL is indicated in [55,63]. At the level of specification-in-the-large, a first informal look had previously been taken at LSL, ACT TWO, ASF+SDF, and OBJ3 [58]. Although some structuring constructs of these languages cannot be translated literally to CASL, it is always possible to find a circumscription in CASL. CoFI WG has not had sufficient resources for defining embeddings of other existing languages into CASL. It is hoped that a forthcoming informal sketch of the relationship between basic specifications in CASL and various other languages may stimulate further detailed investigations, leading to the provision of translators from other languages into CASL.

Developing a tutorial online presentation of CASL. It was decided to combine the CASL Tutorial and the CASL User's Manual into a single publication presented via different media as described above under Methodology. The CASL Tutorial will be based on the same explanations and examples as given in the User's Manual, supplemented by exercises (with interactivity provided by means of some of the available CASL tools). The examples and main points from the Tutorial/User's Manual have already been used in presentations of CASL at ETAPS in 2000 and 2001 [17].

Liaison with IFIP WG1.3. Following a review of a previous version of CASL, IFIP WG1.3 was asked to review the final design of CASL version 1.0.1 in May 2000, which was presented to the IFIP WG1.3 meeting at Stanford in June 2000. The report of the IFIP referees [35] was made available to CoFI and presented at the IFIP WG1.3 meeting in Genova in March 2001. On the basis of the report of the reviewers and the subsequent response of the CASL designers, IFIP WG1.3 decided to formally approve the design of CASL version 1.0.1. The IFIP WG1.3 reviewers made some useful recommendations concerning the enhancement of the documents describing CASL, especially concerning the need for a rough indication of the relationship between CASL and existing languages at the level of basic specifications.

8   Dissemination and Take-Up

One of the principal goals of CoFI WG has been wide dissemination and awareness of the Common Framework. The main target audience at this stage has been fellow researchers in the use of Formal Methods, with some awareness among relevant industrial groups as a secondary goal. Activities include:

Publication of the scientific results in a variety of forms: in conferences and journals, via the CoFI web pages [28], and on CD ROMs distributed with the FM'99 and LFM 2000 proceedings, with the former including some prototype CASL tools and the latter including an annotated list of CoFI-related URLs [80]. Publication of a book on CASL is planned.

Tutorials and presentations on CASL and CoFI at a variety of venues and to a range of audiences. The tutorial material [17,79] has been used in teaching at a number of universities.

Workshops organized jointly with other relevant workshops (WADT in 1999 and 2001) and/or as satellites of major conferences (FM in 1999, ETAPS in 1999, 2000 and 2001). This boosts the visibility of the workshops, providing opportunities for fellow researchers to learn about (and perhaps join) CoFI.

It is rather early to expect significant industrial take-up. But a concrete start in this direction is represented by recent initiatives at Zühlke Engineering AG in Zürich where LSL has been used in the past. An internal study comparing LSL and CASL led to the decision to switch from LSL to CASL for future projects, and industrial training material on formal methods based on CASL is in preparation.

CASL will play a central role in the German multi-site "MultiMedia Instruction in Safe and Secure Systems" project, funded by the BMBF, which will develop a range of educational materials concerning formal specifications, covering an M.Sc. curriculum in Safe and Secure Systems. Logic and algebraic specification (as embodied in CASL) will be the major foundation and CASL will be the specification language for all issues involving data.

It is worth noting that the connection with CoFI and CASL is highlighted by Springer in its marketing material for [5], suggesting that some degree of "market penetration" has already been achieved. Several presentations at WADT/CoFI 2001 reported projects where CASL had been adopted for practical use. Together with some presentations concerning CASL itself, this gave the impression that CASL is already recognized as a well-established *de facto* standard language for algebraic specification.

9   Invitation

CoFI has accomplished a great deal since its inception, but more remains to be done. It is an open collaboration, and new participants are welcome to join at any time. Anybody who wishes to contribute is warmly invited to visit the CoFI web site at http://www.brics.dk/Projects/CoFI/ where all CoFI documents are freely available. Announcements of general interest to CoFI participants are broadcast on the low-volume mailing list cofi-list@brics.dk and each Task Group has its own mailing list; see the CoFI web site for subscription instructions. All of these mailing lists are moderated.

Acknowledgements

Thanks to the participants of CoFI for all their unselfish contributions; without them there would be no CoFI activity to describe. Special thanks to the coordinators of the various CoFI Task Groups, listed in the introduction. Extra special thanks to Peter Mosses, who got CoFI off the ground on behalf of IFIP WG1.3 and COMPASS WG and acted as overall coordinator until mid-1998. The work described here was supported by the ESPRIT-funded CoFI Working Group (project 29432, CoFI WG).

References

1. D. Ancona, M. Cerioli and E. Zucca. Extending CASL by late binding. *Recent Trends in Algebraic Development Techniques: Selected Papers from WADT'99*, Bonas. Springer LNCS 1827, 53–72 (2000).

2. E. Astesiano, M. Bidoit, H. Kirchner, B. Krieg-Brückner, P. Mosses, D. Sannella and A. Tarlecki. CASL: The common algebraic specification language. *Theoretical Computer Science* (2002), to appear.

3. E. Astesiano, M. Broy and G. Reggio. Algebraic specification of concurrent systems. In E. Astesiano, H.-J. Kreowski and B. Krieg-Brückner (eds.), *Algebraic Foundations of Systems Specifications*, 467–520. Springer (1999).

4. E. Astesiano, M. Cerioli and G. Reggio. Plugging data constructs into paradigm-specific languages: towards an application to UML. *Proceedings of AMAST 2000*, Iowa City. Springer LNCS 1816, 273–292 (2000).

5. E. Astesiano, H.-J. Kreowski and B. Krieg-Brückner (eds.). *Algebraic Foundations of Systems Specification.* Springer (1999).

6. E. Astesiano and G. Reggio. UML as heterogeneous multiview notation: Strategies for a formal foundation. *Proc. of OOPSLA'98 Workshop "Formalizing UML. Why? How?"* (1998).

7. S. Autexier, D. Hutter, H. Mantel and A. Schairer. Towards an evolutionary formal software-development using CASL. *Recent Trends in Algebraic Development Techniques: Selected Papers from WADT'99*, Bonas. Springer LNCS 1827, 73–88 (2000).

8. H. Baumeister. Relating abstract datatypes and Z-schemata. *Recent Trends in Algebraic Development Techniques: Selected Papers from WADT'99*, Bonas. Springer LNCS 1827, 366–382 (2000).

9. H. Baumeister. CASL case studies. `http://www.informatik.uni-muenchen.de/~baumeist/CoFI/case.html` (2001).

10. H. Baumeister and D. Bert. Algebraic specification in CASL. *Software Specification Methods: An Overview Using a Case Study*, FACIT: Formal Approaches to Computing and Information Technology. Springer (2000).

11. H. Baumeister, M. Cerioli, A. Haxthausen, T. Mossakowski, P. Mosses, D. Sannella and A. Tarlecki. CASL: The common algebraic specification language. semantics. Version 1.0, to be completed (2001).

12. H. Baumeister and A. Zamulin. State-based extension of CASL. *Proc. IFM 2000* (2000).

13. D. Bert and S. Lo Presti. Algebraic specification of operator-based multimedia scenarios. *Recent Trends in Algebraic Development Techniques: Selected Papers from WADT'99*, Bonas. Springer LNCS 1827, 382–399 (2000).

14. M. Bidoit. From requirements to software in CASL. Invited talk given at WADT'99, Bonas (1999).

15. M. Bidoit, R. Hennicker, F. Tort and M. Wirsing. Correct realizations of interface constraints with OCL. *Proc. UML'99*. Springer LNCS 1723 (1999).

16. M. Bidoit, H.-J. Kreowski, P. Lescanne, F. Orejas and D. Sannella (eds.). *Algebraic System Specification and Development: A Survey and Annotated Bibliography.* Springer LNCS 501 (1991).

17. M. Bidoit and P. Mosses. A gentle introduction to CASL. Tutorial, WADT/CoFI Workshop at the 4th European Joint Conferences on Theory and Practice of Software (ETAPS 2001), Genova (2001). Available from `http://www.lsv.ens-cachan.fr/~bidoit/CASL/`.

18. M. Bidoit, D. Sannella and A. Tarlecki. Architectural specifications in CASL. *Formal Aspects of Computing* (2002), to appear. A preliminary version appeared under the same title in *Proc. 7th Intl. Conference on Algebraic Methodology and Software Technology, AMAST '98*, Manaus. Springer LNCS 1548, 341–357 (1998).

19. T. Borzyszkowski. Generalized interpolation in CASL. *Information Processing Letters* 76:19–24 (2000).

20. T. Borzyszkowski. Higher-order logic and theorem proving for structured specifications. *Recent Trends in Algebraic Development Techniques: Selected Papers from WADT'99*, Bonas. Springer LNCS 1827, 401–418 (2000).

21. M. van den Brand, H. de Jong, P. Klint and P. Olivier. Efficient annotated terms. *Software, Practice & Experience* 30:259–291 (2000).

22. M. van den Brand and J. Scheerder. Development of parsing tools for CASL using generic language technology. *Recent Trends in Algebraic Development Techniques: Selected Papers from WADT'99*, Bonas. Springer LNCS 1827, 89–105 (2000).

23. M. Cerioli, M. Gogolla, H. Kirchner, B. Krieg-Brückner, Z. Qian and M. Wolf (eds.). *Algebraic System Specification and Development: Survey and Annotated Bibliography.* 2nd edition. Monographs of the Bremen Institute of Safe Systems 3. Shaker (1998).

24. M. Cerioli, A. Haxthausen, B. Krieg-Brückner and T. Mossakowski. Permissive subsorted partial logic in CASL. *Proceedings of AMAST'97*, Sydney. Springer LNCS 1349, 91–107 (1997).

25. M. Cerioli, T. Mossakowski and H. Reichel. From total equational to partial first order logic. In E. Astesiano, H.-J. Kreowski and B. Krieg-Brückner (eds.), *Algebraic Foundations of Systems Specifications*, 31–104. Springer (1999).

26. M. Cerioli and G. Reggio. Basic CASL at work: a compass for the labyrinth of partiality, subsorting and predicates. Presented at WADT'99, Bonas (1999).

27. C. Choppy and G. Reggio. Using CASL to specify the requirements and the design: A problem specific approach. *Recent Trends in Algebraic Development Techniques: Selected Papers from WADT'99*, Bonas. Springer LNCS 1827, 104–123 (2000).

28. CoFI. The Common Framework Initiative for algebraic specification and development, electronic archives. Notes and Documents accessible from `http://www.brics.dk/Projects/CoFI/`.

29. CoFI Language Design Task Group. CASL – The CoFI Algebraic Specification Language – Summary, version 1.0. `Documents/CASL/Summary-v1.0`, in [28] (1998).

30. CoFI Language Design Task Group. CASL – The CoFI Algebraic Specification Language – Summary, version 1.0 (revised). `Documents/CASL/v1.0/Summary`, in [28] (1999).

31. CoFI Language Design Task Group. CASL – The CoFI Algebraic Specification Language – Summary, version 1.0.1. `Documents/CASL/v1.0.1/Summary`, in [28] (2001).

32. CoFI Semantics Task Group. CASL – The CoFI Algebraic Specification Language – Semantics. `Documents/CASL/Semantics` (version 0.96), in [28] (1999).

33. E. Coscia and G. Reggio. JTN: A Java-targeted graphic formal notation for reactive and concurrent systems. *Fundamental Approaches to Software Engineering (FASE'99), European Joint Conferences on Theory and Practice of Software*, Amsterdam. Springer LNCS 1577, 77–97 (1999).

34. G. Costa and G. Reggio. Specification of abstract dynamic datatypes: A temporal logic approach. *Theoretical Computer Science* 173(2):513–554 (1997).

35. H. Ehrig, J. Meseguer, U. Montanari, F. Orejas, F. Parisi-Presicce and M. Wirsing. Recommendations on the revised design of CASL. Internal IFIP WG1.3 document (2001).

36. R. Gumb. Model sets in a nonconstructive logic of partial terms with definite descriptions. *Automated Reasoning with Analytic Tableaux and Related Methods, TABLEAUX 2000.* Springer LNAI 1847, 268–278 (2000).

37. R. Gumb. An extended joint consistency theorem for a nonconstructive logic of partial terms with definite descriptions. *Studia Logica.* To appear.

38. M. Haveraaen. Domain specific languages and software architectures – a challenge for CASL. Presented at CoFI meeting, Amsterdam (1999).

39. M. Haveraaen. A 2-tiered software process model for utilizing CASL. Technical Report 208, Department of Informatics, University of Bergen (2000).

40. M. Haveraaen, H.A. Friis and T.A. Johansen. Formal software engineering for computational modeling. *Nordic Journal of Computing* 6:241–270 (1999).

41. A. Haxthausen, B. Krieg-Brückner and T. Mossakowski. Subsorted partial higher-order logic as an extension of CASL. CoFI Note L-10, in [28] (1998).

42. P. Hoffman. Semantics of architectural specifications (in Polish). Master's thesis, Warsaw University (2000).

43. P. Hoffman. Verifying architectural specifications. *Recent Trends in Algebraic Development Techniques: Selected Papers from WADT 2001,* Genova. Springer LNCS, to appear (this volume).

44. F. Honsell, J. Longley, D. Sannella and A. Tarlecki. Constructive data refinement in typed lambda calculus. *Proc. 3rd Intl. Conf. on Foundations of Software Science and Computation Structures (FOSSACS 2000), European Joint Conferences on Theory and Practice of Software,* Berlin. Springer LNCS 1784, 149–164 (2000).

45. F. Honsell and D. Sannella. Pre-logical relations. *Information and Computation* (2002), to appear. A preliminary version appeared under the same title in *Proc. Computer Science Logic, CSL'99,* Madrid. Springer LNCS 1683, 546–561 (1999).

46. H. Hussmann, M. Cerioli, G. Reggio and F. Tort. Abstract data types and UML models. Technical Report DISI-TR-99-15, Università di Genova (1999).

47. H. Kirchner and C. Ringeissen. Executing CASL equational specifications with the ELAN rewrite engine. CoFI Note T-9 (revised version), in [28] (2000).

48. B. Klin. An implementation of static semantics for architectural specifications in CASL (in Polish). Master's thesis, Warsaw University (2000).

49. B. Klin, P. Hoffman, A. Tarlecki, L. Schröder and T. Mossakowski. Checking amalgamability conditions for CASL architectural specifications. *Proc. Intl. Symp. on Mathematical Foundations of Computer Science, MFCS 2001.* Springer LNCS 2136, 451–463 (2001).

50. F. Ledoux, A. Arnould, P. Le Gall and Y. Bertrand. A high-level operation in 3D modeling: a CASL case study. Report LaMI 52, Université d'Evry-Val d'Essonne, Evry (2000).

51. F. Ledoux, A. Arnould, P. Le Gall and Y. Bertrand. Geometric modeling with CASL. *Recent Trends in Algebraic Development Techniques: Selected Papers from WADT 2001,* Genova. Springer LNCS, to appear (this volume).

52. F. Ledoux, J.-M. Mota, A. Arnould, C. Dubois, P. Le Gall and Y. Bertrand. Formal specification for a mathematics-based application domain: geometric modeling. Report LaMI 51, Université d'Evry-Val d'Essonne, Evry (2000).

53. F. Ledoux, J.-M. Mota, A. Arnould, C. Dubois, P. Le Gall and Y. Bertrand. Spécifications formelles du chanfreinage. In *Approches Formelles dans l'Assistance au Développement de Logiciels (AFADL), Nancy* (2001).

54. J. Loeckx, H.-D. Ehrich and M. Wolf. *Specification of Abstract Data Types.* Wiley (1996).

55. T. Mossakowski. Sublanguages of CASL. CoFI Note L-7, in [28] (1997).

56. T. Mossakowski. Two "functional programming" sublanguages of CASL. CoFI Note L-9, in [28] (1998).
57. T. Mossakowski. Colimits of order-sorted specifications. *Recent Trends in Algebraic Development Techniques: Selected Papers from WADT'97*, Tarquinia. Springer LNCS 1376, 316–332 (1998).
58. T. Mossakowski. Translating other specification languages into CASL. Presented at WADT'98, Lisbon (1998). Handwritten notes are available from the author.
59. T. Mossakowski. Translating OBJ3 to CASL: The institution level. *Recent Trends in Algebraic Development Techniques: Selected Papers from WADT'98*, Lisbon. Springer LNCS 1589, 198–214 (1999).
60. T. Mossakowski. How to write consistent CASL design specifications. CoFI Note M-8, in [28] (2000).
61. T. Mossakowski. Specification in an arbitrary institution with symbols. *Recent Trends in Algebraic Development Techniques: Selected Papers from WADT'99*, Bonas. Springer LNCS 1827, 252–270 (2000).
62. T. Mossakowski. CASL: From semantics to tools. *Tools and Algorithms for the Construction and Analysis of Systems (TACAS 2000), European Joint Conferences on Theory and Practice of Software*, Berlin. Springer LNCS 1785, 93–108 (2000).
63. T. Mossakowski. Relating CASL with other specification languages: the institution level. *Theoretical Computer Science* (2002), to appear.
64. T. Mossakowski, S. Autexier and D. Hutter. Extending development graphs with hiding. *Fundamental Approaches to Software Engineering (FASE 2001), European Joint Conferences on Theory and Practice of Software*, Genova. Springer LNCS 2029, 269–283 (2001).
65. T. Mossakowski and B. Klin. Institution-independent static analysis for CASL. *Recent Trends in Algebraic Development Techniques: Selected Papers from WADT 2001*, Genova. Springer LNCS, to appear (this volume).
66. P. Mosses. CoFI: the common framework initiative for algebraic specification and development. *Proc. 7th Intl. Joint Conf. on Theory and Practice of Software Development*, Lille. Springer LNCS 1214, 115–137 (1997).
67. P. Mosses. CASL for ASF+SDF users. *ASF+SDF '97, Proc. 2nd Intl. Workshop on the Theory and Practice of Algebraic Specifications*, volume ASFSDF-97 of *Electronic Workshops in Computing*. British Computer Society, http://www.ewic.org.uk/ewic/workshop/list.cfm (1997).
68. P. Mosses. CASL: a guided tour of its design. *Recent Trends in Algebraic Development Techniques: Selected Papers from WADT'98*, Lisbon. Springer LNCS 1589, 216–240 (1999).
69. P. Mosses. CASL for CafeOBJ users. Chapter 6 of K. Futatsugi, A. T. Nakagawa and T. Tamai (eds.), *CAFE: An Industrial-Strength Algebraic Formal Method*, 121–144. Elsevier (2000).
70. P. Mosses. CASL and Action Semantics. In *AS 2000*, number NS-00-6 in Notes Series, 62–78, BRICS, Univ. of Aarhus (2000).
71. P. Mosses. CoFI: The Common Framework Initiative for algebraic specification and development. In G. Păun, G. Rozenberg and A. Salomaa (eds.), *Current Trends in Theoretical Computer Science: Entering the 21st Century*, 153–163. World Scientific (2001). An earlier version appeared in *Bull. EATCS* 59:127–132 (1996).
72. G. Plotkin, J. Power, D. Sannella and R. Tennent. Lax logical relations. *Proc. 27th Intl. Colloq. on Automata, Languages and Programming*, Geneva. Springer 1853, 85–102 (2000).

73. G. Reggio, E. Astesiano and C. Choppy. CASLLTL: A CASL extension for dynamic reactive systems – summary. Technical Report DISI-TR-99-34, Università di Genova (1999).

74. G. Reggio, E. Astesiano, C. Choppy and H. Hussmann. Analysing UML active classes and associated state machines – a lightweight approach. *Fundamental Approaches to Software Engineering (FASE 2000), European Joint Conferences on Theory and Practice of Software*, Berlin. Springer LNCS 1783, 127–146 (2000).

75. G. Reggio, M. Cerioli and E. Astesiano. An algebraic semantics of UML supporting its multiview approach. *Proc. 2nd AMAST Workshop on Algebraic Methods in Language Processing (AMILP 2000)*, number 16 in Twente Workshop on Language Processing, Enschede (2000).

76. G. Reggio, M. Cerioli and E. Astesiano. Towards a rigorous semantics of UML supporting its multiview approach. *Fundamental Approaches to Software Engineering (FASE 2001), European Joint Conferences on Theory and Practice of Software*, Genova. Springer LNCS 2029, 171–186 (2001).

77. G. Reggio and L. Repetto. CASLCHART: Syntax and semantics. Technical Report DISI-TR-00-1, Università di Genova (2000).

78. G. Reggio and L. Repetto. CASLCHART: A combination of statecharts and of the algebraic specification language CASL. *Proceedings of AMAST 2000*, Iowa City. Springer LNCS 1816, 243–257 (2000).

79. M. Roggenbach. CASL tutorial at LFM 2000 (2000). Slides available at http://www.tzi.de/~roba/tutorial.ps.

80. M. Roggenbach. An annotated list of URLs related with CoFI and CASL. On a CD ROM distributed to participants of the Fifth NASA Langley Formal Methods Workshop (2000). Also available from http://www.informatik.uni-bremen.de/~roba/URL.txt.

81. M. Roggenbach and T. Mossakowski. Proposal of some annotations and literal syntax in CASL. CoFI Note L-11, in [28] (1999).

82. M. Roggenbach and T. Mossakowski. Basic datatypes in CASL. CoFI Note L-12, version 0.4.1, in [28] (2000).

83. M. Roggenbach and L. Schröder. Towards trustworthy specifications I: Consistency checks. *Recent Trends in Algebraic Development Techniques: Selected Papers from WADT 2001*, Genova. Springer LNCS, to appear (this volume).

84. M. Roggenbach, L. Schröder and T. Mossakowski. Specifying real numbers in CASL. *Recent Trends in Algebraic Development Techniques: Selected Papers from WADT'99*, Bonas. Springer LNCS 1827, 146–161 (2000).

85. G. Salaün, M. Allemand and C. Attiogbé. Formal combination of the CCS process algebra with the CASL algebraic specification language. Presented at WADT/CoFI 2001, Genova (2001).

86. D. Sannella. The Common Framework Initiative for algebraic specification and development of software. *Proc. 3rd Intl. Conf. on Perspectives of System Informatics (PSI'99)*, Novosibirsk. Springer LNCS 1755, 1–9 (2000).

87. D. Sannella. Algebraic specification and program development by stepwise refinement. *Proc. 9th Intl. Workshop on Logic-based Program Synthesis and Transformation, LOPSTR'99*, Venice. Springer LNCS 1817, 1–9 (2000).

88. D. Sannella and A. Tarlecki. Essential concepts of algebraic specification and program development. *Formal Aspects of Computing* 9:229–269 (1997).

89. D. Sannella and A. Tarlecki. *Foundations of Algebraic Specifications and Formal Program Development*. Cambridge Univ. Press, to appear.

90. L. Schröder and T. Mossakowski. HasCASL: Towards integrated specification and development of Haskell programs. Submitted for publication (2001).

91. L. Schröder, T. Mossakowski and A. Tarlecki. Amalgamation in CASL via enriched signatures. *Proc. 28th Intl. Colloq. on Automata, Languages and Programming.* Springer 2076, 993–1004 (2001).

92. L. Schröder, T. Mossakowski, A. Tarlecki, P. Hoffman and B. Klin. Semantics of architectural specifications in CASL. *Fundamental Approaches to Software Engineering (FASE 2001), European Joint Conferences on Theory and Practice of Software,* Genova. Springer LNCS 2029, 253–268 (2001).

93. E. Wagner. On the category of CASL signatures. Presented at WADT'99, Bonas (1999).

94. R. Wieringa, E. Astesiano, G. Reggio, A. Le Guennec, H. Hussmann, K. van den Berg and P. van den Broek. Is it feasible to construct a semantics for all of UML? Dynamic behaviour and concurrency. *ECOOP Workshop Reader: UML Semantics FAQ* (1999).

95. M. Wirsing. Algebraic specification. *Handbook of Theoretical Computer Science* (J. van Leeuwen, ed.). North-Holland (1990).

Author Index

Lecture Notes in Computer Science

For information about Vols. 1–2185
please contact your bookseller or Springer-Verlag

Vol. 2226: K.P. Jantke, A. Shinohara (Eds.), Discovery Science. Proceedings, 2001. XII, 494 pages. 2001. (Subseries LNAI).

Vol. 2227: S. Boztaş, I.E. Shparlinski (Eds.), Applied Algebra, Algebraic Algorithms and Error-Correcting Codes. Proceedings, 2001. XII, 398 pages. 2001.

Vol. 2228: B. Monien, V.K. Prasanna, S. Vajapeyam (Eds.), High Performance Computing – HiPC 2001. Proceedings, 2001. XVIII, 438 pages. 2001.

Vol. 2229: S. Qing, T. Okamoto, J. Zhou (Eds.), Information and Communications Security. Proceedings, 2001. XIV, 504 pages. 2001.

Vol. 2230: T. Katila, I.E. Magnin, P. Clarysse, J. Montagnat, J. Nenonen (Eds.), Functional Imaging and Modeling of the Heart. Proceedings, 2001. XI, 158 pages. 2001.

Vol. 2231: A. Pasetti, Software Frameworks and Embedded Control Systems. XIV, 293 pages. 2002.

Vol. 2232: L. Fiege, G. Mühl, U. Wilhelm (Eds.), Electronic Commerce. Proceedings, 2001. X, 233 pages. 2001.

Vol. 2233: J. Crowcroft, M. Hofmann (Eds.), Networked Group Communication. Proceedings, 2001. X, 205 pages. 2001.

Vol. 2234: L. Pacholski, P. Ružička (Eds.), SOFSEM 2001: Theory and Practice of Informatics. Proceedings, 2001. XI, 347 pages. 2001.

Vol. 2235: C.S. Calude, G. Păun, G. Rozenberg, A. Salomaa (Eds.), Multiset Processing. VIII, 359 pages. 2001.

Vol. 2236: K. Drira, A. Martelli, T. Villemur (Eds.), Cooperative Environments for Distributed Systems Engineering. IX, 281 pages. 2001.

Vol. 2237: P. Codognet (Ed.), Logic Programming. Proceedings, 2001. XI, 365 pages. 2001.

Vol. 2239: T. Walsh (Ed.), Principles and Practice of Constraint Programming – CP 2001. Proceedings, 2001. XIV, 788 pages. 2001.

Vol. 2240: G.P. Picco (Ed.), Mobile Agents. Proceedings, 2001. XIII, 277 pages. 2001.

Vol. 2241: M. Jünger, D. Naddef (Eds.), Computational Combinatorial Optimization. IX, 305 pages. 2001.

Vol. 2242: C.A. Lee (Ed.), Grid Computing – GRID 2001. Proceedings, 2001. XII, 185 pages. 2001.

Vol. 2243: G. Bertrand, A. Imiya, R. Klette (Eds.), Digital and Image Geometry. VII, 455 pages. 2001.

Vol. 2244: D. Bjørner, M. Broy, A.V. Zamulin (Eds.), Perspectives of System Informatics. Proceedings, 2001. XIII, 548 pages. 2001.

Vol. 2245: R. Hariharan, M. Mukund, V. Vinay (Eds.), FST TCS 2001: Foundations of Software Technology and Theoretical Computer Science. Proceedings, 2001. XI, 347 pages. 2001.

Vol. 2246: R. Falcone, M. Singh, Y.-H. Tan (Eds.), Trust in Cyber-societies. VIII, 195 pages. 2001. (Subseries LNAI).

Vol. 2247: C. P. Rangan, C. Ding (Eds.), Progress in Cryptology – INDOCRYPT 2001. Proceedings, 2001. XIII, 351 pages. 2001.

Vol. 2248: C. Boyd (Ed.), Advances in Cryptology – ASIACRYPT 2001. Proceedings, 2001. XI, 603 pages. 2001.

Vol. 2249: K. Nagi, Transactional Agents. XVI, 205 pages. 2001.

Vol. 2250: R. Nieuwenhuis, A. Voronkov (Eds.), Logic for Programming, Artificial Intelligence, and Reasoning. Proceedings, 2001. XV, 738 pages. 2001. (Subseries LNAI).

Vol. 2251: Y.Y. Tang, V. Wickerhauser, P.C. Yuen, C.Li (Eds.), Wavelet Analysis and Its Applications. Proceedings, 2001. XIII, 450 pages. 2001.

Vol. 2252: J. Liu, P.C. Yuen, C. Li, J. Ng, T. Ishida (Eds.), Active Media Technology. Proceedings, 2001. XII, 402 pages. 2001.

Vol. 2253: T. Terano, T. Nishida, A. Namatame, S. Tsumoto, Y. Ohsawa, T. Washio (Eds.), New Frontiers in Artificial Intelligence. Proceedings, 2001. XXVII, 553 pages. 2001. (Subseries LNAI).

Vol. 2254: M.R. Little, L. Nigay (Eds.), Engineering for Human-Computer Interaction. Proceedings, 2001. XI, 359 pages. 2001.

Vol. 2255: J. Dean, A. Gravel (Eds.), COTS-Based Software Systems. Proceedings, 2002. XIV, 257 pages. 2002.

Vol. 2256: M. Stumptner, D. Corbett, M. Brooks (Eds.), AI 2001: Advances in Artificial Intelligence. Proceedings, 2001. XII, 666 pages. 2001. (Subseries LNAI).

Vol. 2257: S. Krishnamurthi, C.R. Ramakrishnan (Eds.), Practical Aspects of Declarative Languages. Proceedings, 2002. VIII, 351 pages. 2002.

Vol. 2258: P. Brazdil, A. Jorge (Eds.), Progress in Artificial Intelligence. Proceedings, 2001. XII, 418 pages. 2001. (Subseries LNAI).

Vol. 2259: S. Vaudenay, A.M. Youssef (Eds.), Selected Areas in Cryptography. Proceedings, 2001. XI, 359 pages. 2001.

Vol. 2260: B. Honary (Ed.), Cryptography and Coding. Proceedings, 2001. IX, 416 pages. 2001.

Vol. 2262: P. Müller, Modular Specification and Verification of Object-Oriented Programs. XIV, 292 pages. 2002.

Vol. 2264: K. Steinhöfel (Ed.), Stochastic Algorithms: Foundations and Applications. Proceedings, 2001. VIII, 203 pages. 2001.

Vol. 2267: M. Cerioli, G. Reggio (Eds.), Recent Trends in Algebraic Development Techniques. Proceedings, 2001. X, 345 pages. 2001.

Vol. 2272: D. Bert, J.P. Bowen, M.C. Henson, K. Robinson (Eds.), ZB 2002: Formal Specification and Development in Z and B. Proceedings, 2002. XII, 535 pages. 2002.

Vol. 2273: A.R. Coden, E.W. Brown, S. Srinivasan (Eds.), Information Retrieval Techniques for Speech Applications. XI, 109 pages. 2002.

Vol. 2274: D. Naccache, P. Paillier (Eds.), Public Key Cryptography. Proceedings, 2002. XI, 385 pages. 2002.

Vol. 2275: N.R. Pal, M. Sugeno (Eds.), Advances in Soft Computing – AFSS 2002. Proceedings, 2002. XVI, 536 pages. 2002. (Subseries LNAI).